Pitman Research Notes in Mathematics Series

Submission of proposals for consideration
Suggestions for publication, in the form of outlines and representative samples, are invited by the Editorial Board for assessment. Intending authors should approach one of the main editors or another member of the Editorial Board, citing the relevant AMS subject classifications. Alternatively, outlines may be sent directly to the publisher's offices. Refereeing is by members of the board and other mathematical authorities in the topic concerned, throughout the world.

Preparation of accepted manuscripts
On acceptance of a proposal, the publisher will supply full instructions for the preparation of manuscripts in a form suitable for direct photo-lithographic reproduction. Specially printed grid sheets can be provided and a contribution is offered by the publisher towards the cost of typing. Word processor output, subject to the publisher's approval, is also acceptable.

Illustrations should be prepared by the authors, ready for direct reproduction without further improvement. The use of hand-drawn symbols should be avoided wherever possible, in order to maintain maximum clarity of the text.

The publisher will be pleased to give any guidance necessary during the preparation of a typescript, and will be happy to answer any queries.

Important note
In order to avoid later retyping, intending authors are strongly urged not to begin final preparation of a typescript before receiving the publisher's guidelines. In this way it is hoped to preserve the uniform appearance of the series.

Longman Group Ltd
Longman House
Burnt Mill
Harlow, Essex, CM20 2JE
UK
(Telephone (0) 1279 426721)

Titles in this series. A full list is available from the publisher on request.

100 Optimal control of variational inequalities
V Barbu
101 Partial differential equations and dynamical systems
W E Fitzgibbon III
102 Approximation of Hilbert space operators Volume II
C Apostol, L A Fialkow, D A Herrero and D Voiculescu
103 Nondiscrete induction and iterative processes
V Ptak and F-A Potra
104 Analytic functions – growth aspects
O P Juneja and G P Kapoor
105 Theory of Tikhonov regularization for Fredholm equations of the first kind
C W Groetsch
106 Nonlinear partial differential equations and free boundaries. Volume I
J I Díaz
107 Tight and taut immersions of manifolds
T E Cecil and P J Ryan
108 A layering method for viscous, incompressible L_p flows occupying R^n
A Douglis and E B Fabes
109 Nonlinear partial differential equations and their applications: Collège de France Seminar. Volume VI
H Brezis and J L Lions
110 Finite generalized quadrangles
S E Payne and J A Thas
111 Advances in nonlinear waves. Volume II
L Debnath
112 Topics in several complex variables
E Ramírez de Arellano and D Sundararaman
113 Differential equations, flow invariance and applications
N H Pavel
114 Geometrical combinatorics
F C Holroyd and R J Wilson
115 Generators of strongly continuous semigroups
J A van Casteren
116 Growth of algebras and Gelfand–Kirillov dimension
G R Krause and T H Lenagan
117 Theory of bases and cones
P K Kamthan and M Gupta
118 Linear groups and permutations
A R Camina and E A Whelan
119 General Wiener–Hopf factorization methods
F-O Speck
120 Free boundary problems: applications and theory. Volume III
A Bossavit, A Damlamian and M Fremond
121 Free boundary problems: applications and theory. Volume IV
A Bossavit, A Damlamian and M Fremond
122 Nonlinear partial differential equations and their applications: Collège de France Seminar. Volume VII
H Brezis and J L Lions
123 Geometric methods in operator algebras
H Araki and E G Effros
124 Infinite dimensional analysis–stochastic processes
S Albeverio
125 Ennio de Giorgi Colloquium
P Krée
126 Almost-periodic functions in abstract spaces
S Zaidman
127 Nonlinear variational problems
A Marino, L Modica, S Spagnolo and M Degliovanni
128 Second-order systems of partial differential equations in the plane
L K Hua, W Lin and C-Q Wu
129 Asymptotics of high-order ordinary differential equations
R B Paris and A D Wood
130 Stochastic differential equations
R Wu
131 Differential geometry
L A Cordero
132 Nonlinear differential equations
J K Hale and P Martinez-Amores
133 Approximation theory and applications
S P Singh
134 Near-rings and their links with groups
J D P Meldrum
135 Estimating eigenvalues with *a posteriori/a priori* inequalities
J R Kuttler and V G Sigillito
136 Regular semigroups as extensions
F J Pastijn and M Petrich
137 Representations of rank one Lie groups
D H Collingwood
138 Fractional calculus
G F Roach and A C McBride
139 Hamilton's principle in continuum mechanics
A Bedford
140 Numerical analysis
D F Griffiths and G A Watson
141 Semigroups, theory and applications. Volume I
H Brezis, M G Crandall and F Kappel
142 Distribution theorems of L-functions
D Joyner
143 Recent developments in structured continua
D De Kee and P Kaloni
144 Functional analysis and two-point differential operators
J Locker
145 Numerical methods for partial differential equations
S I Hariharan and T H Moulden
146 Completely bounded maps and dilations
V I Paulsen
147 Harmonic analysis on the Heisenberg nilpotent Lie group
W Schempp
148 Contributions to modern calculus of variations
L Cesari
149 Nonlinear parabolic equations: qualitative properties of solutions
L Boccardo and A Tesei
150 From local times to global geometry, control and physics
K D Elworthy

151 A stochastic maximum principle for optimal control of diffusions
U G Haussmann
152 Semigroups, theory and applications. Volume II
H Brezis, M G Crandall and F Kappel
153 A general theory of integration in function spaces
P Muldowney
154 Oakland Conference on partial differential equations and applied mathematics
L R Bragg and J W Dettman
155 Contributions to nonlinear partial differential equations. Volume II
J I Díaz and P L Lions
156 Semigroups of linear operators: an introduction
A C McBride
157 Ordinary and partial differential equations
B D Sleeman and R J Jarvis
158 Hyperbolic equations
F Colombini and M K V Murthy
159 Linear topologies on a ring: an overview
J S Golan
160 Dynamical systems and bifurcation theory
M I Camacho, M J Pacifico and F Takens
161 Branched coverings and algebraic functions
M Namba
162 Perturbation bounds for matrix eigenvalues
R Bhatia
163 Defect minimization in operator equations: theory and applications
R Reemtsen
164 Multidimensional Brownian excursions and potential theory
K Burdzy
165 Viscosity solutions and optimal control
R J Elliott
166 Nonlinear partial differential equations and their applications: Collège de France Seminar. Volume VIII
H Brezis and J L Lions
167 Theory and applications of inverse problems
H Haario
168 Energy stability and convection
G P Galdi and B Straughan
169 Additive groups of rings. Volume II
S Feigelstock
170 Numerical analysis 1987
D F Griffiths and G A Watson
171 Surveys of some recent results in operator theory. Volume I
J B Conway and B B Morrel
172 Amenable Banach algebras
J-P Pier
173 Pseudo-orbits of contact forms
A Bahri
174 Poisson algebras and Poisson manifolds
K H Bhaskara and K Viswanath
175 Maximum principles and eigenvalue problems in partial differential equations
P W Schaefer
176 Mathematical analysis of nonlinear, dynamic processes
K U Grusa
177 Cordes' two-parameter spectral representation theory
D F McGhee and R H Picard
178 Equivariant K-theory for proper actions
N C Phillips
179 Elliptic operators, topology and asymptotic methods
J Roe
180 Nonlinear evolution equations
J K Engelbrecht, V E Fridman and E N Pelinovski
181 Nonlinear partial differential equations and their applications: Collège de France Seminar. Volume IX
H Brezis and J L Lions
182 Critical points at infinity in some variational problems
A Bahri
183 Recent developments in hyperbolic equations
L Cattabriga, F Colombini, M K V Murthy and S Spagnolo
184 Optimization and identification of systems governed by evolution equations on Banach space
N U Ahmed
185 Free boundary problems: theory and applications. Volume I
K H Hoffmann and J Sprekels
186 Free boundary problems: theory and applications. Volume II
K H Hoffmann and J Sprekels
187 An introduction to intersection homology theory
F Kirwan
188 Derivatives, nuclei and dimensions on the frame of torsion theories
J S Golan and H Simmons
189 Theory of reproducing kernels and its applications
S Saitoh
190 Volterra integrodifferential equations in Banach spaces and applications
G Da Prato and M Iannelli
191 Nest algebras
K R Davidson
192 Surveys of some recent results in operator theory. Volume II
J B Conway and B B Morrel
193 Nonlinear variational problems. Volume II
A Marino and M K V Murthy
194 Stochastic processes with multidimensional parameter
M E Dozzi
195 Prestressed bodies
D Iesan
196 Hilbert space approach to some classical transforms
R H Picard
197 Stochastic calculus in application
J R Norris
198 Radical theory
B J Gardner
199 The C^*-algebras of a class of solvable Lie groups
X Wang
200 Stochastic analysis, path integration and dynamics
K D Elworthy and J C Zambrini

201 Riemannian geometry and holonomy groups
S Salamon
202 Strong asymptotics for extremal errors and polynomials associated with Erdős type weights
D S Lubinsky
203 Optimal control of diffusion processes
V S Borkar
204 Rings, modules and radicals
B J Gardner
205 Two-parameter eigenvalue problems in ordinary differential equations
M Faierman
206 Distributions and analytic functions
R D Carmichael and D Mitrovic
207 Semicontinuity, relaxation and integral representation in the calculus of variations
G Buttazzo
208 Recent advances in nonlinear elliptic and parabolic problems
P Bénilan, M Chipot, L Evans and M Pierre
209 Model completions, ring representations and the topology of the Pierce sheaf
A Carson
210 Retarded dynamical systems
G Stepan
211 Function spaces, differential operators and nonlinear analysis
L Paivarinta
212 Analytic function theory of one complex variable
C C Yang, Y Komatu and K Niino
213 Elements of stability of visco-elastic fluids
J Dunwoody
214 Jordan decomposition of generalized vector measures
K D Schmidt
215 A mathematical analysis of bending of plates with transverse shear deformation
C Constanda
216 Ordinary and partial differential equations. Volume II
B D Sleeman and R J Jarvis
217 Hilbert modules over function algebras
R G Douglas and V I Paulsen
218 Graph colourings
R Wilson and R Nelson
219 Hardy-type inequalities
A Kufner and B Opic
220 Nonlinear partial differential equations and their applications: Collège de France Seminar. Volume X
H Brezis and J L Lions
221 Workshop on dynamical systems
E Shiels and Z Coelho
222 Geometry and analysis in nonlinear dynamics
H W Broer and F Takens
223 Fluid dynamical aspects of combustion theory
M Onofri and A Tesei
224 Approximation of Hilbert space operators. Volume I. 2nd edition
D Herrero
225 Operator theory: proceedings of the 1988 GPOTS–Wabash conference
J B Conway and B B Morrel
226 Local cohomology and localization
J L Bueso Montero, B Torrecillas Jover and A Verschoren
227 Nonlinear waves and dissipative effects
D Fusco and A Jeffrey
228 Numerical analysis 1989
D F Griffiths and G A Watson
229 Recent developments in structured continua. Volume II
D De Kee and P Kaloni
230 Boolean methods in interpolation and approximation
F J Delvos and W Schempp
231 Further advances in twistor theory. Volume I
L J Mason and L P Hughston
232 Further advances in twistor theory. Volume II
L J Mason, L P Hughston and P Z Kobak
233 Geometry in the neighborhood of invariant manifolds of maps and flows and linearization
U Kirchgraber and K Palmer
234 Quantales and their applications
K I Rosenthal
235 Integral equations and inverse problems
V Petkov and R Lazarov
236 Pseudo-differential operators
S R Simanca
237 A functional analytic approach to statistical experiments
I M Bomze
238 Quantum mechanics, algebras and distributions
D Dubin and M Hennings
239 Hamilton flows and evolution semigroups
J Gzyl
240 Topics in controlled Markov chains
V S Borkar
241 Invariant manifold theory for hydrodynamic transition
S Sritharan
242 Lectures on the spectrum of $L^2(\Gamma\backslash G)$
F L Williams
243 Progress in variational methods in Hamiltonian systems and elliptic equations
M Girardi, M Matzeu and F Pacella
244 Optimization and nonlinear analysis
A Ioffe, M Marcus and S Reich
245 Inverse problems and imaging
G F Roach
246 Semigroup theory with applications to systems and control
N U Ahmed
247 Periodic-parabolic boundary value problems and positivity
P Hess
248 Distributions and pseudo-differential operators
S Zaidman
249 Progress in partial differential equations: the Metz surveys
M Chipot and J Saint Jean Paulin
250 Differential equations and control theory
V Barbu

251 Stability of stochastic differential equations with respect to semimartingales
X Mao
252 Fixed point theory and applications
J Baillon and M Théra
253 Nonlinear hyperbolic equations and field theory
M K V Murthy and S Spagnolo
254 Ordinary and partial differential equations. Volume III
B D Sleeman and R J Jarvis
255 Harmonic maps into homogeneous spaces
M Black
256 Boundary value and initial value problems in complex analysis: studies in complex analysis and its applications to PDEs 1
R Kühnau and W Tutschke
257 Geometric function theory and applications of complex analysis in mechanics: studies in complex analysis and its applications to PDEs 2
R Kühnau and W Tutschke
258 The development of statistics: recent contributions from China
X R Chen, K T Fang and C C Yang
259 Multiplication of distributions and applications to partial differential equations
M Oberguggenberger
260 Numerical analysis 1991
D F Griffiths and G A Watson
261 Schur's algorithm and several applications
M Bakonyi and T Constantinescu
262 Partial differential equations with complex analysis
H Begehr and A Jeffrey
263 Partial differential equations with real analysis
H Begehr and A Jeffrey
264 Solvability and bifurcations of nonlinear equations
P Drábek
265 Orientational averaging in mechanics of solids
A Lagzdins, V Tamuzs, G Teters and A Kregers
266 Progress in partial differential equations: elliptic and parabolic problems
C Bandle, J Bemelmans, M Chipot, M Grüter and J Saint Jean Paulin
267 Progress in partial differential equations: calculus of variations, applications
C Bandle, J Bemelmans, M Chipot, M Grüter and J Saint Jean Paulin
268 Stochastic partial differential equations and applications
G Da Prato and L Tubaro
269 Partial differential equations and related subjects
M Miranda
270 Operator algebras and topology
W B Arveson, A S Mishchenko, M Putinar, M A Rieffel and S Stratila
271 Operator algebras and operator theory
W B Arveson, A S Mishchenko, M Putinar, M A Rieffel and S Stratila
272 Ordinary and delay differential equations
J Wiener and J K Hale
273 Partial differential equations
J Wiener and J K Hale
274 Mathematical topics in fluid mechanics
J F Rodrigues and A Sequeira
275 Green functions for second order parabolic integro-differential problems
M G Garroni and J F Menaldi
276 Riemann waves and their applications
M W Kalinowski
277 Banach C(K)-modules and operators preserving disjointness
Y A Abramovich, E L Arenson and A K Kitover
278 Limit algebras: an introduction to subalgebras of C*-algebras
S C Power
279 Abstract evolution equations, periodic problems and applications
D Daners and P Koch Medina
280 Emerging applications in free boundary problems
J Chadam and H Rasmussen
281 Free boundary problems involving solids
J Chadam and H Rasmussen
282 Free boundary problems in fluid flow with applications
J Chadam and H Rasmussen
283 Asymptotic problems in probability theory: stochastic models and diffusions on fractals
K D Elworthy and N Ikeda
284 Asymptotic problems in probability theory: Wiener functionals and asymptotics
K D Elworthy and N Ikeda
285 Dynamical systems
R Bamon, R Labarca, J Lewowicz and J Palis
286 Models of hysteresis
A Visintin
287 Moments in probability and approximation theory
G A Anastassiou
288 Mathematical aspects of penetrative convection
B Straughan
289 Ordinary and partial differential equations. Volume IV
B D Sleeman and R J Jarvis
290 *K*-theory for real *C**-algebras
H Schröder
291 Recent developments in theoretical fluid mechanics
G P Galdi and J Necas
292 Propagation of a curved shock and nonlinear ray theory
P Prasad
293 Non-classical elastic solids
M Ciarletta and D Ieşan
294 Multigrid methods
J Bramble
295 Entropy and partial differential equations
W A Day
296 Progress in partial differential equations: the Metz surveys 2
M Chipot
297 Nonstandard methods in the calculus of variations
C Tuckey
298 Barrelledness, Baire-like- and (LF)-spaces
M Kunzinger
299 Nonlinear partial differential equations and their applications. Collège de France Seminar. Volume XI
H Brezis and J L Lions
300 Introduction to operator theory
T Yoshino

301 Generalized fractional calculus and applications
V Kiryakova
302 Nonlinear partial differential equations and their applications. Collège de France Seminar Volume XII
H Brezis and J L Lions
303 Numerical analysis 1993
D F Griffiths and G A Watson
304 Topics in abstract differential equations
S Zaidman
305 Complex analysis and its applications
C C Yang, G C Wen, K Y Li and Y M Chiang
306 Computational methods for fluid-structure interaction
J M Crolet and R Ohayon
307 Random geometrically graph directed self-similar multifractals
L Olsen
308 Progress in theoretical and computational fluid mechanics
G P Galdi, J Málek and J Necas
309 Variational methods in Lorentzian geometry
A Masiello
310 Stochastic analysis on infinite dimensional spaces
H Kunita and H-H Kuo
311 Representations of Lie groups and quantum groups
V Baldoni and M Picardello
312 Common zeros of polynomials in several variables and higher dimensional quadrature
Y Xu
313 Extending modules
N V Dung, D van Huynh, P F Smith and R Wisbauer
314 Progress in partial differential equations: the Metz surveys 3
M Chipot, J Saint Jean Paulin and I Shafrir
315 Refined large deviation limit theorems
V Vinogradov
316 Topological vector spaces, algebras and related areas
A Lau and I Tweddle
317 Integral methods in science and engineering
C Constanda
318 A method for computing unsteady flows in porous media
R Raghavan and E Ozkan
319 Asymptotic theories for plates and shells
R P Gilbert and K Hackl
320 Nonlinear variational problems and partial differential equations
A Marino and M K V Murthy
321 Topics in abstract differential equations II
S Zaidman
322 Diffraction by wedges
B Budaev
323 Free boundary problems: theory and applications
J I Diaz, M A Herrero, A Liñan and J L Vazquez
324 Recent developments in evolution equations
A C McBride and G F Roach
325 Elliptic and parabolic problems: Pont-à-Mousson 1994
C Bandle, J Bemelmans, M Chipot, J Saint Jean Paulin and I Shafrir
326 Calculus of variations, applications and computations: Pont-à-Mousson 1994
C Bandle, J Bemelmans, M Chipot, J Saint Jean Paulin and I Shafrir
327 Conjugate gradient type methods for ill-posed problems
M Hanke
328 A survey of preconditioned iterative methods
A M Bruaset
329 A generalized Taylor's formula for functions of several variables and certain of its applications
J-A Riestra
330 Semigroups of operators and spectral theory
S Kantorovitz
331 Boundary-field equation methods for a class of nonlinear problems
G N Gatica and G C Hsiao
332 Metrizable barrelled spaces
J C Ferrando, M López Pellicer and L M Sánchez Ruiz
333 Real and complex singularities
W L Marar

W L Marar (Editor)
ICMSC-USP, Brazil

Real and complex singularities

LONGMAN

Copublished in the United States with
John Wiley & Sons Inc., New York.

Longman Group Limited
Longman House, Burnt Mill, Harlow
Essex CM20 2JE, England
and Associated Companies throughout the world.

Copublished in the United States with
John Wiley & Sons Inc., 605 Third Avenue, New York, NY 10158

First published 1995

AMS Subject Classifications: (Main) 32S, 58C, 14C
(Subsidiary) 57Q, 35B

ISSN 0269-3674

ISBN 0 582 27780 9

British Library Cataloguing in Publication Data

A catalogue record for this book is
available from the British Library

Library of Congress Cataloging-in-Publication Data

Real and complex singularities / W.L. Marar (editor).
p. cm. -- (Pitman research notes in mathematics series ; ???)
Papers from the 3rd International Workshop on Real and Complex Singularities held Aug. 1-5, 1994, in São Carlos, Brazil.
1. Singularities (Mathematics)--Congresses. I. Marar, W. L. II. International Workshop on Real and Complex Singularities (3rd : 1994 : São Carlos, São Paulo, Brazil) III. Series.
QA614.58.R43 1995
514'.74--dc20 95-9093
CIP

Printed and bound in Great Britain
by Biddles Ltd, Guildford and King's Lynn

Contents

Preface

The 3rd International Workshop on Real and Complex Singularities took place at the Instituto de Ciências Matemáticas de São Carlos of the Universidade de São Paulo from the 1st to the 5th of August, 1994.

This volume contains 13 papers presented at the workshop. They constitute a fine collection of geometrical aspects of Singularity Theory and its applications.

Thanks are due to Luis Fávaro and Maria Ruas for their help in organizing the meeting, to Aron Simis, Janey Daccach, Milton da Costa Lopes Filho and the staff at Longman for their help in the preparation of this volume, and to the scientific committee: Bill Bruce, Jim Damon, Terry Gaffney, David Mond, Andrew duPlessis, Jorge Sotomayor and Terry Wall.

Thanks are also due to Fapesp, CNPq, SBM and USP for providing the essential funds to cover all the participants' local expenses.

I am indebted to the late Maria Helena Derigi (Lene) whose organizational skills and devotion to her work provided the meeting with the best support one could expect.

Ton Marar

São Carlos, July 1995

List of contributors

J. Basto Gonçalves
Grupo de Matemática Aplicada
R. das Taipas 135
4000 Porto
Portugal

Jean-Paul Brasselet
SiGma CNRS
CIRM Luminy Case 916
F-13288 Marseille Cedex 9
France

Shirley Bromberg
Universidad Autonoma Metropolitana
Iztapalapa
Departamento de Matematicas
Apartado Postal 55-534
C.P. 09340 Mexico (DF)
Mexico

J. W. Bruce
University of Liverpool
Department of Pure Mathematics
P O Box 147
Liverpool L69 3BX
England

J. Scott Carter
University of South Alabama
Department of Mathematics
Mobile, Alabama 36688
USA

Terence Gaffney
Northeastern University
Department of Mathematics
Boston, Ma 02115
USA

Peter Giblin
University of Liverpool
Department of Pure Mathematics
P O Box 147
Liverpool L69 3BX
England

Shiuichi Izumiya
Hokkaido University
Faculty of Science
Department of Mathematics
Sapporo 060
Japan

Tomohiro Kawakami
Department of Liberal Arts
Osaka Prefectural College of Technology
Neyagawa Osaka 572
Japan

Santiago López de Medrano
Universidad Nacional Autonoma
Instituto de Matematicas
Ciudad Universitaria
C.P. 04510 Mexico (DF)
Mexico

David Mond
University of Warwick
Mathematics Institute
Coventry CV4 7AL
England

Juan José Nuño Ballesteros
Universitat de Valencia
Dep. de Geometria i Topologia
Campus de Burjasot
46100 Burjasot (Valencia)
Spain

Diana Ratcliffe
University of Warwick
Mathematics Institute
Coventry CV4 7AL
England

Maria Aparecida Soares Ruas
Universidade de São Paulo
Instituto de Ciências Matemáticas
Caixa Postal 668
13560-970 São Carlos
Brazil

Osamu Saeki
Hiroshima University
Faculty of Science
Department of Mathematics
Higashi - Hiroshima 724
Japan

Marcelo José Saia
Universidade Estadual Paulista
IGCE
Rua 10 2527
13506-496 Rio Claro (SP)
Brazil

Masahiro Saito
University of Texas
Department of Mathematics
Austin, Texas 78712
USA

Jorge Sotomayor
Universidade de São Paulo
Instituto de Matemática e Estatística
Departamento de Matemática Aplicada
Caixa Postal 66281 - Ag. Cidade
05368-970 São Paulo (SP)
Brazil

Farid Tari
University of Liverpool
Department of Pure Mathematics
P O Box 147
Liverpool L69 3BX
England

C.T.C. Wall
University of Liverpool
Department of Pure Mathematics
P O Box 147
Liverpool L69 3BX
England

C.T.C.Wall

Real rational quartic curves

Introduction

Our object is to give a complete listing of the real types of irreducible rational quartic curves in the projective plane, to discuss the geometrical properties of each type, and to illustrate both the partition of the space of curves and the individual types by pictures.

We define two quartic curves to be equisingular if we can join them by a (connected, 1-parameter) family of curves, each having the same list of singularities. Our first objective is to give a precise listing of the equisingularity types of rational quartics. Similarly we define strict equisingularity by requiring also that the duals of the curves in the family each have the same list of singularities. According to the general results of [5] this is equivalent to saying that the number of hyperflexes in the family is constant. We concentrate on the case when this number is 0, but our results are detailed enough to cover the other cases.

The geometric properties we have in mind include discussion of reality conditions for bitangents, flexes and eventually hyperflexes and flecnodes.

We begin by recalling from [5] the description of normal forms for equations, and hence the geometry of rational quartics in the complex case, but only using such normalisations of coordinates as will also be valid over $\mathbb{R}$: more precisely, here the normalisations of coordinates are valid over any field of characteristic $\neq 2$.

We will go through each case defining a precise normal form over $\mathbb{R}$ (this is necessary for the discussions of connectedness), giving the equisingularity classification, describing the topological type of each type of curve, and finally listing the singularities of the dual - this will demand a further refinement of the equisingularity classification.

Each section (r) deals with a group of from five to a dozen cases; at the end we give a table (Table r) summarising the main properties of each case, and figures (Figure r) illustrating the curves and - where relevant - the auxiliary conics, and in several cases also a figure (Figure rm) illustrating the parameter spaces.

Results in the complex case

We first fix notation. We take coordinates (x, y, z) in the complex projective plane; X stands for the point $(1,0,0)$; similarly for Y and Z. The equation

$$F(x,y,z) \equiv ax^2 + by^2 + cz^2 + 2fyz + 2gzx + 2hxy = 0,$$

with discriminant

$$\Delta \equiv abc + 2fgh - af^2 - bg^2 - ch^2 \neq 0,$$

defines a conic S. Following [5] we group rational quartics Γ into four types. For the first three, a quadratic transformation defined by a net of conics maps Γ into the (nonsingular) conic S.

Case α

Set $\alpha(x,y,z) = (yz, zx, xy)$. If Γ is a quartic with double points at X, Y and Z, $\alpha(\Gamma)$ is a conic S passing through none of the points X, Y and Z, so $a, b, c \neq 0$.

Here, our calculations will distinguish subcases according as a,b and c all have the same sign or not. It is also necessary to treat the cases when the types of the 3 singular points are $3A_1$, A_2+2A_1, $2A_2+A_1$ and $3A_2$ individually, though the first is by far the most complicated.

Case α^*

Quartics with 3 double points which are not all defined over the ground field require separate discussion. The only case arising over $\mathbb{R}$ is when two of them are complex conjugate. We then take them as $(1, \pm i, 0)$, and call them I and J; and take the third point as Z $(0,0,1)$. As quadratic transformation with these as base points we take $\alpha^*(x,y,z) = (xz, yz, x^2+y^2)$. Then $\alpha^*(\Gamma)$ is again a real conic S.

Case β

Now take $\beta(x,y,z) = (xz, yz, y^2)$. A quartic Γ with a singularity of type A_1 or A_2 at Z and one of type A_3 or A_4 at X, with tangent XY, is transformed by β to a conic S (with a, c nonzero).

Case γ

A quartic Γ with a singular point of type A_5 or A_6 at X, with tangent XY has equation of the form

$$a(y^2 - pxz)^2 + 2hyz(y^2 - pxz) + 2gz^2(y^2 - pxz) + by^2z^2 + 2fyz^3 + cz^4 = 0,$$

with $a, p \neq 0$. Setting $px = x'$ reduces p to 1; we then have $\Gamma = \gamma(S)$, where $\gamma(x,y,z) = (y^2 - xz, yz, z^2)$, and $a \neq 0$.

Case δ

If Γ has a triple point, we take the equation of Γ in the form $x\phi(y,z) + \psi(y,z) = 0$, where ϕ may be taken to be one of $y^3 \pm yz^2$, y^2z and y^3, and ψ may be further normalised.

The above normal forms for α, β, γ and α^* have all been chosen to be involutory, so that we have (e.g.) $\Gamma = \alpha(S)$ together with $S = \alpha(\Gamma)$.

To determine the singularities of the dual we again begin by recalling the results of [5]. Write $\Gamma^\vee$ for the dual of Γ; n, $n^\vee$ for their degrees: $n^\vee$ is also known as the *class* of Γ. Then $\Gamma^\vee$ has the same list of singularities as Γ , except that (i) the singularities of type A_1 and A_2 and triple points of Γ do not contribute to singularities of $\Gamma^\vee$, (ii) $\Gamma^\vee$ may have singularities of types A_1, A_2 and E_6 not arising from singularities of Γ.

If $\Gamma^\vee$ has no E_6 points, the numbers of A_1 and A_2 singularities on $\Gamma^\vee$ may now be determined from Plücker's formulae for the class and genus. Each E_6 point counts for this purpose as $2A_2 + A_1$. We thus have a complete determination of singularities of $\Gamma^\vee$ once we know the number of its E_6 points (hyperflexes of Γ).

In fact we may determine the bitangents explicitly as follows. Let τ denote any of α, α^*, β, γ. If L is bitangent to Γ, $\tau(L)$ is a conic having double contact with S and belonging to the net defining τ. Hence there exists a linear form λ such that $F = \lambda^2$ is the equation of this conic. To determine λ we use the 3 conditions defining the net. For example, in case α, the coefficients of x^2, y^2 and z^2 in $F - \lambda^2$ all vanish; moreover this has nonzero discriminant. We will say that the bitangent $L = \alpha(F - \lambda^2)$ *corresponds to* the linear form λ.

The following table is taken from [5] but gives only the genus 0 cases. The first entry is $n^\vee$ and an entry $X \to Y$ (N) means that if Γ has singularities of types X and no hyperflexes, then $\Gamma^\vee$ has singularities of types Y; and N is the maximum possible number of hyperflexes of Γ (for each hyperflex of Γ we must replace $2A_2 + A_1$ by E_6 in the list of singularities of $\Gamma^\vee$).

	Class 6		
$3A_1$	$\to$	$6A_2 + 4A_1$	(3)
D_4	$\to$	$6A_2 + 4A_1$	(2)
$A_3 + A_1$	$\to$	$A_3 + 6A_2 + 2A_1$	(2)
A_5	$\to$	$A_5 + 6A_2 + A_1$	(1)
	Class 5		
$A_2 + 2A_1, D_5$	$\to$	$4A_2 + 2A_1$	(1)
$A_3 + A_2$	$\to$	$A_3 + 4A_2$	(0)
$A_4 + A_1$	$\to$	$A_4 + 3A_2 + A_1$	(1)
A_6	$\to$	$A_6 + 3A_2$	(0)
	Class 4		
$2A_2 + A_1, E_6$	$\to$	$2A_2 + A_1$	(1)
$A_4 + A_2$	$\to$	$A_4 + A_2$	(0)
	Class 3		
$3A_2$	$\to$	A_1	(0)

A simple node (A_1 singularity) of Γ is called a flecnode if one of the tangent lines to Γ at the point is an inflexional tangent to the corresponding branch; it is a biflecnode if this condition holds for each tangent line. It is simple to check for the existence of flecnodes, since the nodal tangents at A_1 singularities of Γ are at once found from our normal forms, and we just need to check for the fourth intersection with Γ.

Methods of calculation in the real case

The division into cases, and basic method of handling each case are as before, save that - as noted above - we need to consider a variant form α^* of α.

We recall that the singularities A_{2k-1} now have to be distinguished as A_{2k-1}^-, with 2 real branches, and A_{2k-1}^+, with 0 - i.e. giving an isolated point. Similarly we have types D_4^-, with 3 (smooth, transverse) real branches, and D_4^+, with just 1.

As Γ has at most one singularity other than simple nodes and cusps, such a singularity necessarily occurs at a real point. For duality, we know that singularities A_n $(n \geq 3)$ for Γ and $\Gamma^\vee$ correspond, so have the same numbers of real branches, so have the same type as each other over $\mathbb{R}$. The remaining singularities of $\Gamma^\vee$ have type A_1, A_2 or possibly E_6.

Not all of these need occur at real points. Our notation for k pairs of complex conjugate singular points of type A_1 will be $(2kA_1)$ (and similarly for A_2).

We define the type of a real bitangent to be -1 if the two points of contact are real and distinct; 0 if they coincide, so we have a hyperflex (alternatively called a 'point of undulation'); and +1 if the points of contact are complex conjugate. This notation is chosen so that type -1 yields an A_1^- singularity (or *crunode*: distinct real tangents) of $\Gamma^\vee$, type $+1$ an A_1^+ (or *acnode*: an isolated point of the real locus), and type 0 an E_6.

We have already seen how to find the bitangents in each case. For a bitangent corresponding to the linear form λ, use the equation of λ to substitute for one of the variables in F. This gives a binary quadratic with discriminant Φ, say: there are then 2 real points of contact if and only if $\Phi > 0$, and in general the type of the bitangent is $-sgn(\Phi)$.

It remains to determine how many of the A_2 points of $\Gamma^\vee$ (flexes of Γ) are real. Rather than resort to calculation here, we may obtain the answers directly from Klein's equation. We recall from [6] the general form of this.

For each real singular point P of Γ write $m'(P) = m(P) - r(P)$, where $m(P)$ is the multiplicity of Γ at P, $r(P)$ the number of real branches. Then if $n'(\Gamma) = n - \sum_P m'(P)$, we have $n'(\Gamma) = n'(\Gamma^\vee)$.

In the present situation, this formula may be simplified. For as singularities $A_n (n \geq 3)$ are the same for both Γ and $\Gamma^\vee$, the corresponding terms cancel. The remaining real singular points of Γ contribute as follows:

Singularity S	A_1^-	A_1^+	A_2	D_4^-	D_4^+	D_5	E_6
$m'(S)$	0	2	1	0	2	1	2

Thus if (as in Klein's case) Γ has no triple point, then twice the number of A_1^+ (or E_6) on $\Gamma^\vee$ plus the number of A_2 is equal to the same for Γ, added to $n^\vee - n$.

We also observe that if Γ is varied within a stratum so that the type of a bitangent changes from $-$ through 0 to $+$, the corresponding singularities on $\Gamma^\vee$ change from $2A_2 + A_1^-$ through E_6 to $(2A_2) + A_1^+$: a 'dovetail' perestroika.

1 Case α: three nodes

Recall that $\alpha(x, y, z) = (yz, zx, xy)$. We have a quartic Γ with double points at X, Y and Z, and $\alpha(\Gamma)$ is a conic S passing through none of the points X, Y and Z.

We have two subcases according as a,b and c all have the same sign (replacing F by $-F$ if necessary, we may take this as $+$) or not.

$a, b, c > 0$

Adjust the coordinates x, y and z by scalar factors, to ensure $a = b = c = 1$. We then have $\Delta \equiv 1 + 2fgh - (f^2 + g^2 + h^2) \neq 0$. The singular point X of Γ has type A_1 unless $f = \pm 1$, when it has type A_2. More precisely, X is an acnode (type A_1^+) if $|f| < 1$; X is a crunode (type A_1^-) if $|f| > 1$. Corresponding remarks apply to Y, Z.

Since there is symmetry between x, y and z, we may suppose $|h| \leq |g| \leq |f|$; since changing the sign of x, y or z changes the sign of two of f, g and h we may suppose $0 \leq g \leq f$. We now have a unique normal form for each curve.

Removing the planes $|f| = 1$ (and similarly for g, h) partitions our moduli space into five regions:

(a)	(b)	(c)	(d)	(e)
$\lvert h\rvert \leq g \leq f < 1$	$\lvert h\rvert \leq g < 1 < f$	$\lvert h\rvert < 1 < g \leq f$	$1 < -h \leq g \leq f$	$1 < h \leq g \leq f$

Moreover, Δ is nonzero. Now $\Delta = (1 - f^2)(1 - h^2) - (g - fh)^2$ is always < 0 in cases (b), (c) and (d); it can take either sign in (a) and (e). The above 5 cases thus yield 7, and we separate (a) into (α1), with $\Delta > 0$ and (α2), with $\Delta < 0$; rename (b), (c) and (d) as (α3), (α4) and (α5) respectively, and divide (e) into cases (α6), with $\Delta < 0$, and (α7), with $\Delta > 0$.

Connectivity of the regions (a) ... (e) is clear. Now consider the surface $\Delta = 0$ in (a). We can solve the equation as $h = fg \pm \sqrt{(1 - f^2)(1 - g^2)}$: call these values V_+, V_-. We have $-1 < V_- < g < V_+$ throughout the region $0 < g \leq f < 1$. The strata (α1) ($V_- < h \leq g$) and (α2) ($-1 < h < V_-$) are thus connected: they are fibred over a triangle with connected fibres.

This will be the pattern of our proof of connectivity in all cases. For case (e) above, $\Delta = 0$ gives $h = fg \pm \sqrt{(f^2 - 1)(g^2 - 1)}$: call these values V_+, V_-, so that (α7) is given by $V_- < h < V_+$. Clearly $V_+ \geq g$; also $V_- \geq 1$, with equality only when

$f = g$. Further, $V_- < g$ if and only if $f < 2g^2 - 1$. Thus the projection of $(\alpha 7)$ on the (f, g)-plane is the set $1 < g < f < 2g^2 - 1$, which is connected; again we have connected fibres; and $(\alpha 6)$ too (where $h < V_-$) is connected.

We begin study of the topology of Γ by considering the signature of the quadratic form F giving S. This is determined by the signs of $a(= 1)$, $ab - h^2(= 1 - h^2)$ and Δ, or by symmetry with h replaced by f or g. Thus if $\Delta < 0$ (cases $(\alpha 2) - (\alpha 6)$), the signature is $(+, +, -)$; if $\Delta > 0$ it is $(+, +, +)$ in case $(\alpha 1)$; it is $(+, -, -)$ in case $(\alpha 7)$. In particular, for $(\alpha 1)$ S has no real points; the only real points of Γ are the acnodes.

Otherwise, the interior of S is given by $F > 0$ for $(\alpha 7)$, $F < 0$ for the rest. Thus it contains none of X, Y, Z in $(\alpha 2)$ - $(\alpha 6)$, but all of them for $(\alpha 7)$. We can thus determine how S is placed with respect to the triangle (Figure 1). To confirm the picture for $(\alpha 6)$, note that the points where S meets $x = 0$ are the roots of $y^2 + 2fyz + z^2 = 0$, so at them both, y/z has the opposite sign to f; similarly for the other sides.

From these we can infer the topology of the curve Γ. For Γ is the image of S by a map with identifies the two intersection points of S with each side ($x = 0$ etc.) of the triangle (and if there is no real intersection, adjoins an acnode). In particular the cases $(\alpha 5)$, $(\alpha 6)$ and $(\alpha 7)$ are distinguished by these patterns of identification.

We now begin our study of the dual curves: most of the work is an analysis of the bitangents.

We seek linear forms $\lambda \equiv px + qy + rz$ such that $F - \lambda^2$ vanishes at X, Y and Z, thus $p^2 = q^2 = r^2 = 1$. We normalise $p = 1$ (but leave p in some formulae for symmetry): we have 4 real bitangents $L_{q,r}$. Applying α we see that these have equations

$$(f - qr)x + (g - rp)y + (h - pq)z = 0.$$

Calculating the points of contact of $L_{q,r}$ leads us to the square root of

$$\Phi_{q,r} = f^2 + g^2 + h^2 - 3 + 2(f - gh)qr + 2(g - hf)rp + 2(h - fg)pq;$$

if $\Phi_{q,r} > 0$, the points of contact are real, so $L_{q,r}$ has type $-$; in general, the type of $L_{q,r}$ is $-sgn(\Phi_{q,r})$.

Observe that $\Delta + \Phi_{q,r} = -2(1 - qrf)(1 - rpg)(1 - pqh)$, and that solving $\Phi_{q,r} = 0$ for h leads to

$$0 = (pqh - rpg - qrf + 1)^2 + (-4 - 4pqfg + 4qrf + 4rpg);$$

$$pqh - 1 = (rpg - 1) + (qrf - 1) \pm 2\sqrt{(rpg - 1)(qrf - 1)};$$

and so eventually to

$$0 = \pm\sqrt{(qrf - 1)} \pm \sqrt{(rpg - 1)} \pm \sqrt{(pqh - 1)};$$

in particular,

(C) $1-qrf$, $1-rpg$, $1-pqh$ must all have the same sign.

Denote this sign by ϵ. The same calculation shows that if these expressions do not all have the same sign, then $\Phi_{q,r}>0$, so $L_{q,r}$ has type $-$.

Cases ($\alpha 1$) and ($\alpha 6$) are decided by the remark that $(\Delta+\Phi_{q,r})$ has sign $-\epsilon$, so that if $sgn(\Delta)=\epsilon$ we must have $sgn(\Phi_{q,r})=-\epsilon$; but in general we must calculate.

We next check for cases (a)...(e) above which values of (q,r) satisfy condition (C). We find for (a) all 4 values; for (b), $(1,-1)$ and $(-1,1)$; for (c) and (d) just $(1,-1)$ and for (e) just $(1,1)$; with $\epsilon=+1$ in all but the last of these cases.

For (a) these hypersurfaces are given by the vanishing of

$$\begin{array}{lcl}(\Phi_{++})\ \phi_1 & = & \sqrt{((1-f)}+\sqrt{(1-g)}-\sqrt{(1-h)},\\ (\Phi_{-+})\ \phi_2 & = & -\sqrt{((1+f)}+\sqrt{(1-g)}+\sqrt{(1+h)},\\ (\Phi_{--})\ \phi_3 & = & \sqrt{((1-f)}-\sqrt{(1+g)}+\sqrt{(1+h)},\\ (\Phi_{+-})\ \phi_4 & = & -\sqrt{((1+f)}+\sqrt{(1+g)}+\sqrt{(1-h)},\end{array}$$

where the sign outside the square root is, in each case, adjusted to be $-$ for the largest of the 3 terms and $+$ for the other two. If we change the sign of one or both of the latter, ϕ becomes negative. Since the product of these 4 expressions is Φ, we have $sgn\ \phi=-sgn\ \Phi$.

We have $\phi_4>0$, $\phi_2\geq\phi_3$ (with equality when $f^2=g^2$) and $\phi_3\geq\phi_1$ (with equality when $g^2=h^2$). For $i=1,2,3$, ϕ_i increases with h: $\phi_i<0$ at $h=-1$ and >0 at $h=g$. Thus ϕ_i vanishes at a unique value $h=H_i$, with $-1<H_2\leq H_3\leq H_1<g$; further, $H_1<V_-<g$.

We thus have connected regions $-1<h<H_2$, $H_2<h<H_3$, $H_3<h<H_1$ and $H_1<h<V_-$ in ($\alpha 2$) where the bitangents have respective types $(---+)$, $(--++)$, $(-+++)$, $(++++)$; all four have type $+$ (as they clearly must) for ($\alpha 1$).

For the transition values $h=H_i$, one of the types is 0. We have two types $=0$ ($H_2=H_3=0$) if $f=g$ and $\phi_3=0$ (which meets the plane $f=g$ in an ellipse intersecting the cone $f\geq|h|$ defining (a) in an arc) or ($H_3=H_1=0$) if $\phi_3=0$ and $h=-g$ (we have $\phi_3>0$ when $h=g$); here $\phi_3=0$ meets the plane $g+h=0$ in an ellipse intersecting the cone $f\geq g$ in an arc.

Both conditions hold if $g=3/5$, and we have 3 hyperflexes. We see as before that each of these transition strata is also connected; and that the strata may be uniquely labelled by the list of types of the bitangents.

For case ($\alpha 3$) the relevant functions are

$$\phi_5=\sqrt{(1+g)}+\sqrt{(1-h)}-\sqrt{(1+f)},$$

$$\phi_6=\sqrt{(1-g)}+\sqrt{(1+h)}-\sqrt{(1+f)};$$

we have $\phi_5\geq\phi_6$, with equality when $g^2=h^2$. Both decrease to $-\infty$ as f increases to ∞. Since $\sqrt{(1+g)}+\sqrt{(1-h)}>0$, and the condition $\sqrt{(1-g)}+\sqrt{(1+h)}-\sqrt{(1+g)}>0$

defines the interior of an ellipse, which thus intersects $\{(g,h)||h| < g < 1\}$ in a connected set, the regions $\phi_6 > 0$, $\phi_5 > 0 > \phi_6$ and $0 > \phi_5$ are connected. The corresponding bitangent types are $(++--)$, $(+---)$ and $(----)$; the transition regions with a single hyperflex are also connected, and we can have 2 hyperflexes: the situation is similar to the above.

For cases ($\alpha 4$) and ($\alpha 5$), we have ϕ_5 as above. As f increases from g to ∞, ϕ_5 decreases from $\surd(1-h) > 0$ to $-\infty$, so we have connected regions $\phi_5 > 0$ (with types $(---+)$) and $\phi_5 < 0$ (with types $(----)$) and a connected transition stratum.

Finally for type (e) we have

$$\phi = \surd(1+g) + \surd(1+h) - \surd(1+f).$$

As in the previous two cases, the sign of ϕ gives the desired classification; however, we must also take note of Δ. But by a remark above, for ($\alpha 6$) as $sgn(\Delta + \Phi) = -\epsilon = 1$ and $\Delta < 0$, we have $\Phi > 0$ (and $\phi < 0$).

Our discussion of the dual is summarised in Table 1.

We conclude this subsection with a discussion of the possibility of flecnodes: here it is convenient to relax the condition $|h| \leq g \leq f$. The condition for a flecnode at X is $0 = g^2 - 2fgh + h^2$; and for a biflecnode is $0 = g = h$. If X is a biflecnode and $|f| > 1$, we have case ($\alpha 3$) and X is a crunode; if $|f| < 1$, we have case ($\alpha 1$) and an acnode; in particular if $f = g = h = 0$ we have 3 such.

If X and Y are both flecnodes, it follows that $f^2 = g^2$, and we may write $f = g = (1+t^2)/2t$, $h = (1+t^2)/2$, with $t > 0$; so $\Delta < 0$. If $t < 1$ this gives case ($\alpha 4$); if $t > 1$ it belongs to ($\alpha 6$).

We abstain from a complete analysis of the cases with a single flecnode, which would be somewhat lengthy.

a, b, c not all of same sign

Here we may normalise to $a = b = 1$, $c = -1$. Interchanging x and y if necessary, let $|f| \geq |g|$; adjust the signs of x, y and z to have $h \geq 0$, $f \geq 0$.

S always intersects XZ and YZ, so the points X, Y on Γ have type A_1^-. The point Z has type A_1^- if $|h| > 1$; A_1^+ if $|h| < 1$; we have $h^2 \neq 1$ (else Z has type A_2). If $h^2 < 1$, then the discriminant $\Delta = -1 + 2fgh - f^2 - g^2 + h^2 < 0$. We thus have cases ($\alpha 8$), where $0 \leq h < 1$; if $h > 1$ we have ($\alpha 9$) if $\Delta < 0$ and ($\alpha 10$) if $\Delta > 0$.

Now $\Delta = 0$ leads to $h = \pm\surd(1+f^2)(1+g^2) - fg$. Taking the minus sign here gives a negative root; the positive sign gives a number $V \geq 1$ (with equality if $f = g$). The cases ($\alpha 8$) ($0 \leq h < 1$), ($\alpha 9$) ($1 < h < V$) and ($\alpha 10$) ($V < h$) thus each give connected strata.

The quadratic form F has signature $(++-)$ for ($\alpha 8$) and ($\alpha 9$); $(+--)$ for ($\alpha 10$). The interior domain of S is given by $F < 0$ in the first two cases and then contains the vertex Z; it is given by $F > 0$ for $(\alpha 10)$ and contains the vertices X and Y. These cases are sketched in Figure 1.

In these cases there are no real bitangents.

X is a flecnode if $g^2 - 2fgh - h^2 = 0$. This implies $\Delta = -1 - f^2 < 0$, so we have case $(\alpha 8)$ if $h < 1$, in particular if X is a biflecnode (so $h = 0$); and case $(\alpha 9)$ if $h > 1$. Both cases may occur with both X and Y being flecnodes.

Z is a flecnode if $f^2 - 2fgh + g^2 = 0$. This implies $\Delta = -1 + h^2$, so we have case $(\alpha 8)$ if $h < 1$ and case $(\alpha 9)$ if $h > 1$: both cases also allow Z a biflecnode ($f = g = 0$). If X and Z are both flecnodes, however, we must have $f = g = h = 0$ and 3 biflecnodes and case $(\alpha 8)$.

Our discussion of the duals of cases $(\alpha 1) - (\alpha 10)$ is summarised in Table 1.

Table 1

Name	c	(f,g,h)	Δ	*Singularities*	*Bitangents*	*Flexes*
$(\alpha 1)$	1	$\|h\| \le g \le f < 1$	> 0	$3A_1^+$	$+ + ++$	0
$(\alpha 2)$	1	$\|h\| \le g \le f < 1$	< 0	$3A_1^+$	$+ * **$	0
$(\alpha 3)$	1	$\|h\| \le g < 1 < f$	< 0	$2A_1^+ + A_1^-$	$* * --$	2
$(\alpha 4)$	1	$\|h\| < 1 < g \le f$	< 0	$A_1^+ + 2A_1^-$	$* - --$	2
$(\alpha 5)$	1	$1 < -h \le g \le f$	< 0	$3A_1^-$	$* - --$	0
$(\alpha 6)$	1	$1 < h \le g \le f$	< 0	$3A_1^-$	$- - --$	2
$(\alpha 7)$	1	$1 < h \le g \le f$	> 0	$3A_1^-$	$* - --$	0
$(\alpha 8)$	-1	$0 \le h < 1$	< 0	$A_1^+ + 2A_1^-$	*None*	4
$(\alpha 9)$	-1	$h > 1$	< 0	$3A_1^-$	*None*	2
$(\alpha 10)$	-1	$h > 1$	> 0	$3A_1^-$	*None*	2

In Table 1, we have $a = b = 1$ throughout; if $c = 1$ then $|h| \le g \le f$ and if $c = -1$ then $h \ge 0,\ f \ge |g|$. The conditions in the columns labelled c, (f, g, h), Δ characterise the different cases.

The number of real flexes on Γ, or equivalently of A_2 points (real cusps) on the dual, is inferred from Klein's equation. The list of strata for the refined stratification is not made explicit, but is inferred from the following convention which we adopt in subsequent tables also.

In the column headed 'Bitangents', we give a sequence of k, say, symbols $+, *, -$ (appearing in this order). This means that each of the curves in the stratum has k real bitangents. For each symbol $*$ we may substitute $+$, 0 or $-$: we do this in all possible ways such that in the resulting sequence any $+$ symbols precede any 0 symbols which in turn precede any $-$ symbols. For each such sequence, the set of normal forms of curves in the family whose bitangent types coincide with the sequence is non-empty and connected. The number of flexes given is the number of real flexes on Γ in the case when each $*$ is replaced by a $+$: this should be increased by 1 (hyperflex) for each replacement by 0 and by 2 for each by a $-$.

Our pictures of the curves Γ and the auxiliary conics S in these cases are presented in Figure 1. We have included a picture of the triangle $xyz = 0$ as part of the picture: this has the advantage of pinpointing the acnodes, and also clarifying the relation between the topology of Γ and the position of S with respect to the triangle.

To convert the projective curves to affine form, I have aimed to convert the conic $y^2 = xz$ to a circle, and so have used the substitution

$$x = \sqrt{3}u + v - \tfrac{2}{3}, \; y = -2v - \tfrac{2}{3}, \; z = -\sqrt{3}u + v - \tfrac{2}{3},$$

with $\sqrt{3}$ approximated by $\frac{7}{4}$ for speed of calculation. Then define F as usual, $G := \alpha(F)$, $TF := xyzF$, $TG := xyzG$ and apply the above substitution to obtain curves in the (u, v)-plane. In principle we can take values of the parameters (a, b, c, d, e, f) given by the normalisations above. However to obtain nice pictures - and an overriding consideration has been to draw all real points of Γ in the chosen region of the plane where possible - it has been found essential to change signs in several cases. I have also on occasion permuted x and y for the sake of appearance. Finally it has not always been possible to choose the parameters so as to give simultaneously good pictures for both S and Γ. I have taken care that in such cases, at least the topological pictures of the two correspond.

These pictures, and the others in this paper have been drawn using the Silicon Graphics machines (Iris and Indigo) obtained by our research group using the SERC grant GR/H/59855.

I now discuss how the general shape of Γ can be inferred from the position of S with respect to the base triangle. First observe that the 3 lines $xyz = 0$ divide the projective plane into 3 triangular regions. The transformation α preserves the interior of each of these regions; blows up the 3 vertices; and collapses each edge to the opposite vertex. Thus if S does not meet $x = 0$, X is an acnode of Γ. If it meets it in two points both lying in the region $yz > 0$, then Γ has a crunode at X, with both tangents meeting the triangular region where x, y and z are all positive; similarly in other cases.

All intersections of Γ with $xyz = 0$ lie at the 3 vertices, so are described as above. These are joined by arcs which are the images by α of the arcs into which S is divided by these lines (and which thus lie in the same triangular regions). As no further crossings are permitted, there is a (topologically) unique way to do this.

A similar analysis will apply for the remaining (α) cases, save that here tangencies of S with sides of the triangle give cusps of Γ, with tangent lines inferred as above; again we can see into which region each cusp points, and then join up the pieces.

Figure 1

(α2) (α3) (α4)

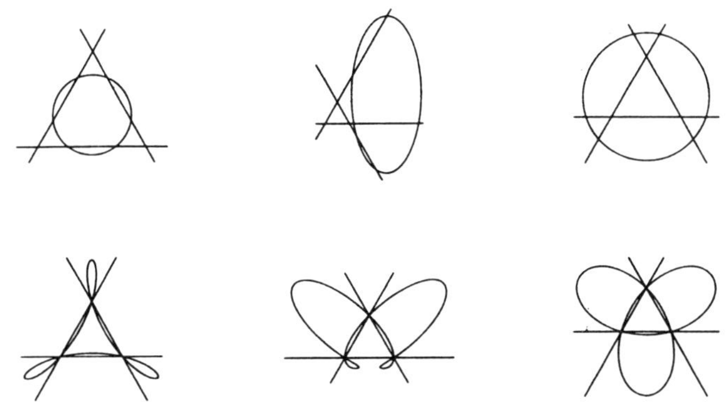

(α5) (α6) (α7)

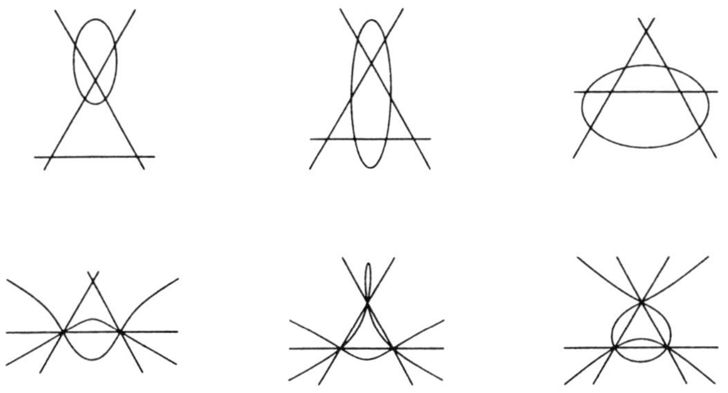

(α8) (α9) (α10)

2 Case α; at least one cusp

$A_2 + 2A_1$; $a, b, c > 0$

We may take $a = b = c = 1$, $h = -1$; we have $|f| \neq 1$, $|g| \neq 1$, and $\Delta \equiv -(f+g)^2 \neq 0$. Changing the sign of z will change those of f, g; we may thus suppose $f + g > 0$. We also have symmetry in f and g, so may suppose $|f| \geq |g|$, hence $f \geq |g|$. Cases are thus (α11) $|g| \leq f < 1$; (α12) $|g| < 1 < f$; (α13) $1 < -g \leq f$, and (α14) $1 < g \leq f$.

Clearly each of these regions is connected.

In each case, F has signature $(+,+,-)$; *Int S* is given by $F < 0$ and contains none of X, Y, Z; S meets YZ (XZ) if $|f| > 1$ ($|g| > 1$). These conics are sketched in Figure 2.

For these cases we have two real bitangents $(f-r)x + (g-r)y - 2z = 0$ obtained as above but with $p = q = 1$, $r = \pm 1$; $\Phi_r = (f-g)^2 + 4r(f+g) - 4$, and the type is $-sgn(\Phi_r)$. The parabola $\Phi_1 = 0$ separates the region defining (α11); the parabola $\Phi_{-1} = 0$ separates both regions (α12) and (α13). This is shown on Figure 2m, where we also indicate the types of the two bitangents corresponding to each region. As usual, these correspond to curves Γ with a hyperflex; it is not possible here to have 2 hyperflexes. The connectedness of the various parts of the moduli space is clear from this figure.

The quartic Γ has a flecnode at Y if $f^2 + 2fg + 1 = 0$ and at X if $g^2 + 2fg + 1 = 0$. These curves also are marked on Figure 2m, and the data summarised in Table 2.

$A_2 + 2A_1$; a, b, c not all of the same sign

We may take $a = b = 1$, $c = -1$, and $h = -1$. This already prohibits S from touching XZ or YZ. We have $\Delta = -(f+g)^2$. As above we may take $f \geq |g| \geq 0$. This is a connected region and we have just one case, (α15) say, with singularities $A_2 + 2A_1^-$.

F has signature $(++-)$; the interior region $F < 0$ of S contains the point Z. There are no real bitangents, hence 2 real flexes. There is a flecnode at X if $g^2 + 2fg = 1$ and one at Y if $f^2 + 2fg = 1$; if $f = g = 3^{-1/2}$ we have both. These curves are shown in Figure 2m.

$2A_2 + A_1$

Here we necessarily have a, b, c of the same sign, and may take $a = b = c = 1$ and $g = h = -1$. Then $f \neq \pm 1$; $\Delta = -(f-1)^2 < 0$. We thus have three cases (α16) $f < -1$, (α17) $-1 < f < 1$, (α18) $f > 1$; each is clearly connected.

S touches XY, XZ; it meets YZ if $|f| > 1$, i.e in cases (α16), (α18). F has signature $(++-)$; the interior $F < 0$ of S contains none of X, Y, Z. These cases (and also (α15), (α19)) are illustrated in Figure 2.

There is a unique bitangent $(f-1)x - 2y - 2z = 0$, which corresponds to $x + y + z$. It has type $-sgn(f^2 + 6f - 7)$, which is $-$ in case (α18), $+$ in case (α17), while (α16) is subdivided to $f < -7$ (with $-$), $-7 < f < -1$ (with $+$). No flecnodes are possible in this case.

$3A_2$

We may take $a = b = c = 1$ and $f = g = h = -1$: there is a unique case (α19), say. The unique bitangent $x + y + z = 0$ corresponds to $x + y + z$ and has type $+$. There are no flexes.

Table 2

Name	c	(f,g)	*Singularities*	*Bitangents*	*Flexes*
$(\alpha 11)$	1	$\lvert g\rvert \leq f < 1$	$A_2 + 2A_1^+$	$+*$	2
$(\alpha 12)$	1	$\lvert g\rvert < 1 < f$	$A_2 + A_1^+ + A_1^-$	$*-$	2
$(\alpha 13)$	1	$1 < -g \leq f$	$A_2 + 2A_1^-$	$--$	2
$(\alpha 14)$	1	$1 < g \leq f$	$A_2 + 2A_1^-$	$*-$	0
$(\alpha 15)$	-1	$\lvert g\rvert \leq f$	$A_2 + 2A_1^-$	*None*	2
$(\alpha 16)$	1	$g = -1,\ f < -1$	$2A_2 + A_1^-$	$*$	0
$(\alpha 17)$	1	$g = -1,\ -1 < f < 1$	$2A_2 + A_1^+$	$+$	2
$(\alpha 18)$	1	$g = -1,\ f > 1$	$2A_2 + A_1^-$	$-$	2
$(\alpha 19)$	1	$f = g = -1$	$3A_2$	$+$	0

In Table 2 we have $a = b = 1$, $h = -1$ throughout; the columns headed c, (f,g) give the conditions characterising the cases.

The (2-dimensional) moduli spaces for $(\alpha 11)$ - $(\alpha 15)$ are illustrated in Figure 2m. In this, and in later figures of moduli spaces, the solid curves separate the different cases. The dashed curves correspond to hyperflexes, and separate the diagram into regions where the types of the bitangents take the constant values marked in the figure. The dotted curves correspond to flecnodes.

Pictures of the conics S and curves Γ for cases $(\alpha 11)$ – $(\alpha 19)$ are presented in Figure 2, to which the same remarks as for Figure 1 apply.

Figure 2m

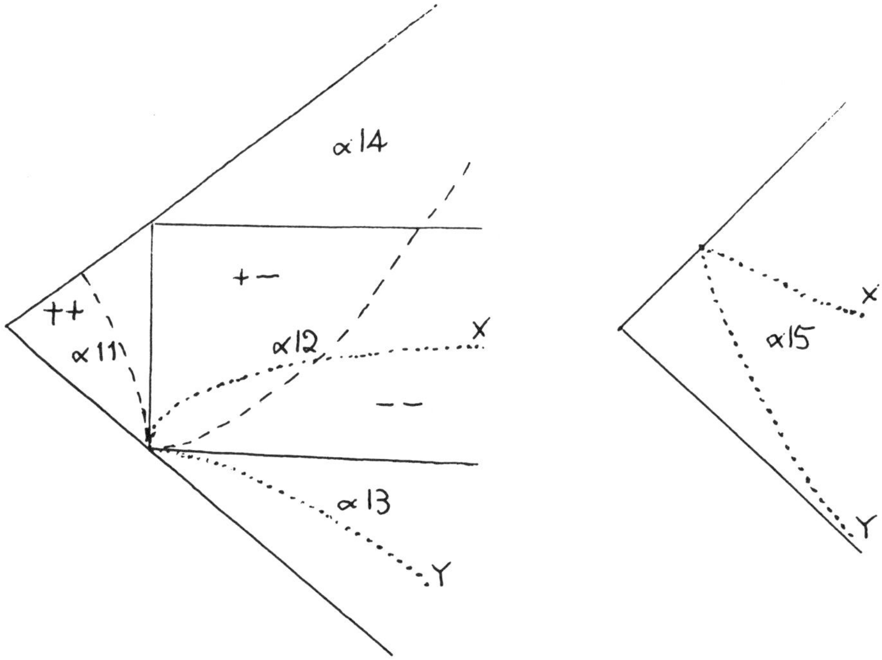

Figure 2

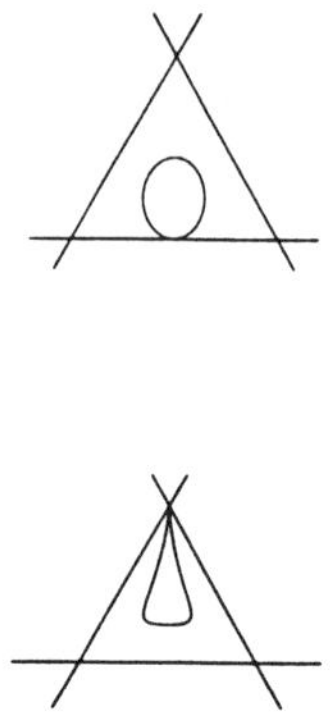

(α11)

(α12)

(α13)

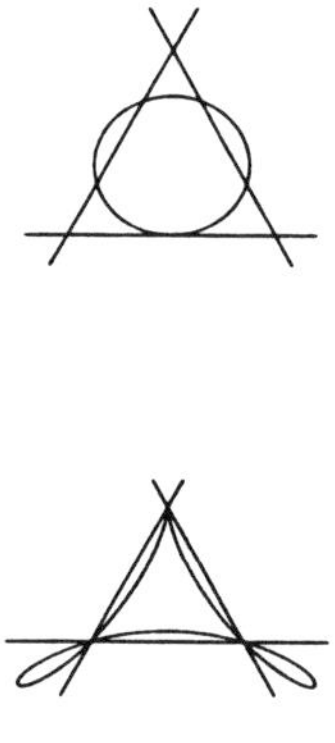

(α14)

(α15)

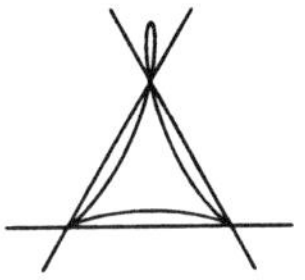

(α16)

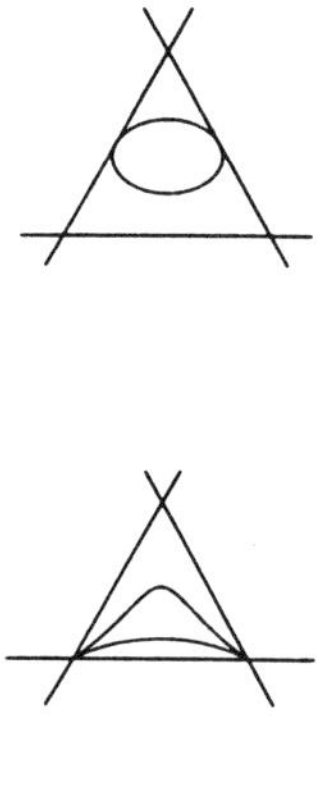

(α17)

(α18)

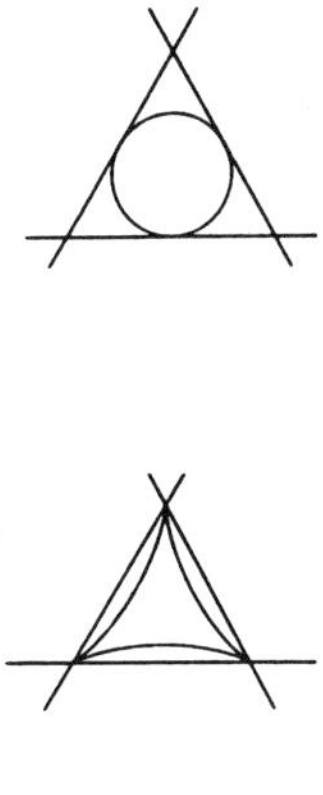

(α19)

3 Case α^*

Here we have the quadratic transformation $\alpha^*(x,y,z) = (xz, yz, x^2+y^2)$ with base points $I, J = (1, \pm i, 0)$, and Z. Although at first sight this may seem the least interesting case (since the singular points are not all visible) it is natural to consider the affine plane $z = 1$, with $(1, \pm i, 0)$ being the 'circular points at infinity' I and J. We then recognise α^* as inversion in the unit circle with centre the origin

$$(x,y) \mapsto \left(\frac{x}{x^2+y^2}, \frac{y}{x^2+y^2}\right).$$

The quartic curves Γ obtained by inversion from conics S are bicircular quartics, and were a major type of interest in e.g. [1].

In general, a quartic Γ is termed bicircular if it has singular points at I and J. When we have a further double point Z, Γ is obtained by inverting a conic S with centre of inversion at Z, and Z has type A_1^-, A_2 or A_1^+ according as S is a hyperbola, parabola or ellipse. (If S is a line pair, Γ consists of two circles through Z, which may explain the term 'bicircular'.) In general the singularities of Γ at I and J are nodes, and the quartic is termed a limaçon if Z is a node and a cardioid if it is a cusp (though the name 'cardioid' is often reserved for the case when I and J also are cusps). The case with 3 biflecnodes is known as the 'leminiscate of Bernoulli'. The points I and J are cusps if S touches IZ and JZ, i.e. if Z is a focus of S.

In terms of the equation, it is convenient first to rotate coordinates to make the coordinate axes parallel to the axes of S, so that $h = 0$. As S does not pass through Z, $c \neq 0$; as it does not contain I and J, $a \neq b$. Multiplying by suitable constants, and interchanging x and y if necessary, we may suppose $c = a - b = 1$. This fixes coordinates up to sign changes of x and y, so we further normalise $f, g \geq 0$.

We have $0 \neq \Delta = b(b+1) - (b+1)f^2 - bg^2$. S touches IJ if $ab = 0$ (making Z a cusp on Γ), and touches IZ and JZ if $(f+ig)^2 = -1$, so $f = 0$ and $g = 1$.

$A_1 + (2A_1)$

We have $b \neq 0, -1$, $(f,g) \neq (0,1)$ (defining a subset which does not separate the parameter space), and $0 \neq \Delta$. We have 5 cases: if $b < -1$ (hence $\Delta > 0$), (α^*1); if $-1 < b < 0$ then if $\Delta > 0$, (α^*2) but if $\Delta < 0$, (α^*3); while for $b > 0$ if $\Delta > 0$ we have (α^*4) and if $\Delta < 0$, (α^*5).

Each of these sets of conditions gives a connected stratum: this is immediate for (α^*1). For the others, observe that for each fixed b, $\Delta = 0$ defines a conic with centre the origin in the (f,g)-plane, which divides the quadrant $f, g \geq 0$ into two connected pieces, which fit to give connected regions in 3-space.

S meets IJ in real points - or equivalently, the affine S is a hyperbola - in cases (α^*2), (α^*3). In case (α^*4), S has no real points; in cases (α^*1), (α^*5), S is an ellipse.

The signature of the quadratic form F is $(+--)$ for (α^*1), (α^*2); here the interior $F>0$ of S contains Z. The signature is $(+++)$ for (α^*4), and $(++-)$ for (α^*3) and (α^*5): here the interior $F<0$ of S does not contain Z.

Bitangents correspond to linear forms $px+qy+rz$ such that $F-(px+qy+rz)^2$ is a linear combination of $xz,\ yz,\ x^2+y^2$. We may take $r=1$; then $pq=0$ and $p^2-q^2=a-b=1$. Thus $q=0$ and $p=\pm 1$, giving 2 real bitangents $2(g-p)x+2fy+bz=0$. The bitangent given by $px+z$ has type $-sgn\{f^2-b(b+2-2pg)\}$.

Thus (b,f,g)-space is now partitioned by

$$\begin{aligned} f^2 &= V_1 \equiv b-bg^2/(b+1) \quad (\text{giving } \Delta=0),\\ f^2 &= V_2 \equiv b(b+2+2g),\\ f^2 &= V_3 \equiv b(b+2-2g). \end{aligned}$$

Observe that $\frac{(b+1)}{b}(V_2-V_1)=(b+1+g)^2\geq 0$, and similarly for V_3-V_1. In particular, where equality occurs, $V_1=-b^2<0$.

For (α^*1) $b<-1$, $V_1\leq V_2<V_3$. As $V_1<0$, the cases are as follows (at the end of each line we indicate the types of the bitangents):

$$\begin{array}{rll} f^2<V_2, & \text{projecting to} \quad 0\leq g<-1-\frac{b}{2}, & ++\\ V_2<f^2<V_3, & \text{projecting to} \quad g>1+\frac{b}{2}, & +-\\ f^2>V_3, & \text{projecting to} \quad b<-1; & -- \end{array}$$

and we see that each of these, and the surfaces separating them, are connected; we have 2 hyperflexes if $g=0,\ f^2=b(b+2)$.

If $-1<b<0$, then $V_2<V_3\leq V_1$ and $V_2<0$; we have (α^*2) if $\Delta>0$, i.e. if $f^2<V_1$ and (α^*3) if $f^2>V_1$. Cases are

$$\begin{array}{lrll} (\alpha^*2) & V_2<f^2<V_3, & \text{projecting to} \quad g>1+\frac{b}{2}, & +-\\ (\alpha^*2) & V_3<f^2<V_1, & \text{projecting to} \quad g>\sqrt{(1+b)}, & --\\ (\alpha^*3) & f^2>V_1, & \text{projecting to} \quad -1<b<0; & -- \end{array}$$

again all are connected.

If $0<b$, then $V_1\leq V_3<V_2$; we have (α^*4) if $\Delta>0$, i.e. if $f^2<V_1$, and (α^*5) if $f^2>V_1$. Cases are

$$\begin{array}{lrll} (\alpha^*4) & f^2<V_1, & \text{projecting to} \quad g<\sqrt{(1+b)}, & ++\\ (\alpha^*5) & V_1<f^2<V_3, & \text{projecting to} \quad g<1+\frac{b}{2}, & ++\\ (\alpha^*5) & V_3<f^2<V_2, & \text{projecting to} \quad b>0, & +-\\ (\alpha^*5) & f^2>V_2, & \text{projecting to} \quad b>0; & -- \end{array}$$

and again connectedness follows. In this case also we can have 2 hyperflexes. These regions in the (b,g)-plane are illustrated in Figure 3m. The results are summarised in Table 3.

Taking reality considerations into account, we see that the only possibilities for flecnodes are:

- flecnode at Z, which occurs if $af^2 + bg^2 = 0$;
- flecnodes at I and J, which occurs if either $f = 0$, $g^2 = \frac{(2b+1)^2}{4b+1}$ or $g = 0$, $f^2 = \frac{(2b+1)^2}{4b+3}$;
- biflecnode at Z, which occurs if $f = g = 0$;
- 3 biflecnodes, which occurs only if $a = \frac{1}{2}$, $b = -\frac{1}{2}$, $f = g = 0$.

We may check these conditions either by transforming the conditions obtained for Case (α) or by calculating afresh.

Now if a and b have the same sign, $af^2 + bg^2$ can only vanish if f and g do: so in cases (α^*1) and (α^*4) (acnodes) if Z is a flecnode, it is a biflecnode; otherwise if Z is a flecnode, we have case (α^*3). The case $f = 0$, $g^2 = \frac{(2b+1)^2}{4b+1}$ arises if $-\frac{1}{4} < b < 0$ (Case (α^*2)) and if $0 < b$ (Case (α^*4)). The case $g = 0$, $f^2 = \frac{(2b+1)^2}{4b+3}$ arises with either $-\frac{3}{4} < b < 0$ (Case (α^*3)) or $0 < b$ (Case (α^*5)). The final case with 3 biflecnodes belongs (as just noted) to case (α^*3).

$A_2 + (2A_1)$

We retain our normalisations, but now $ab = 0$, so we have 2 cases. If (case (α^*7)) $a = 0$, $\Delta = -bg^2 = g^2 > 0$; if (case (α^*6)) $b = 0$, $\Delta = -af^2 = -f^2 < 0$. Clearly each is connected.

The conic S contains Z in its interior in case (α^*7) but not in case (α^*6).

Bitangents correspond to $px + z$ with $p^2 = 1$. However in case (α^*6) this leads to no real bitangent. For (α^*7), the bitangent $2(g-p)x + 2fy - z = 0$ has type $sgn\{-1 + 2gp - f^2\}$. The parabola $2g = 1 + f^2$ has $(0,1)$ as focus and $g = 0$ as directrix, and divides the half-plane $g > 0$ into two connected regions: in the interior $f^2 < 2g - 1$, the bitangents have types $--$; in $f^2 > 2g - 1$, they have types $-+$.

It follows from the above that I and J cannot be flecnodes for (α^*7); for (α^*6), only if $g = 0$, $f^2 = \frac{1}{3}$.

$A_1 + (2A_2)$

We now have the previously excluded case $(f, g) = (0, 1)$. It follows that $\Delta = b^2 > 0$. Since neither b nor a may vanish, we have cases (α^*8) $(b < -1)$, (α^*9) $(-1 < b < 0)$, and (α^*10) $(b > 0)$. Clearly each of these is connected.

There is just one bitangent $4x + bz = 0$, corresponding to $x - z$; it has type $-sgn\{b(b+4)\}$, so is $-$ for (α^*10), $+$ for (α^*9) while (α^*8) is divided into $b > -4$ with $+$ and $b < -4$, with $-$. There is no flecnode in this case.

$A_2 + (2A_2)$

We have a unique case $S = z^2 - y^2 + 2xz$, which we name (α^*11). The picture of S is the same as for (α^*7). The unique bitangent $4x = z$ corresponds to $x - z$ and has type $-$. There are no flexes.

We will not illustrate the conics S in this case, since the only relevant geometry is the relation of S to the point Z and to the line IJ at infinity. The conic S has no real points in cases $(\alpha^*4), (\alpha^*10)$; is an ellipse in cases $(\alpha^*1), (\alpha^*5), (\alpha^*8)$; is a parabola in cases $(\alpha^*6), (\alpha^*7), (\alpha^*11)$; and is a hyperbola in cases $(\alpha^*2), (\alpha^*3), (\alpha^*9)$. In the cases when S is non-empty, Z is in the interior region in all cases except $(\alpha^*3), (\alpha^*5), (\alpha^*6)$.

For the quartic curve Γ we draw the natural affine picture with $z = 1$. We omit, however, the cases when there is only one real point. Moreover, since when I and J are cusps, a small change in f or g will change what happens at I and J but the qualitative picture of S and Γ remains the same, in Figure 3 one picture suffices for both, and we choose the one with cusps at I and J.

We again conclude the section with a summary table.

Table 3

Name	b	(f, g)	*Singularities*	*Bitangents*	*Flexes*
(α^*1)	$b < -1$	$\Delta > 0$	$A_1^+ + (2A_1)$	$**$	0
(α^*2)	$-1 < b < 0$	$\Delta > 0$	$A_1^- + (2A_1)$	$*-$	0
(α^*3)	$-1 < b < 0$	$\Delta < 0$	$A_1^- + (2A_1)$	$--$	2
(α^*4)	$0 < b$	$\Delta > 0$	$A_1^+ + (2A_1)$	$++$	0
(α^*5)	$0 < b$	$\Delta < 0$	$A_1^+ + (2A_1)$	$**$	0
(α^*6)	$b = 0$	$f > 0$	$A_2 + (2A_1)$	None	2
(α^*7)	$b = -1$	$g > 0$	$A_2 + (2A_1)$	$*-$	0
(α^*8)	$b < -1$	$f = 0,\ g = 1$	$A_1^+ + (2A_2)$	$*$	0
(α^*9)	$-1 < b < 0$	$f = 0,\ g = 1$	$A_1^- + (2A_2)$	$-$	0
(α^*10)	$b > 0$	$f = 0,\ g = 1$	$A_1^+ + (2A_2)$	$+$	0
(α^*11)	$b = -1$	$f = 0,\ g = 1$	$A_2 + (2A_2)$	$-$	0

In Table 3, $a - b = c = 1$, $h = 0$, $f, g \geq 0$ throughout; the conditions defining the cases are given in the columns headed b, (f, g).

Figure 3m

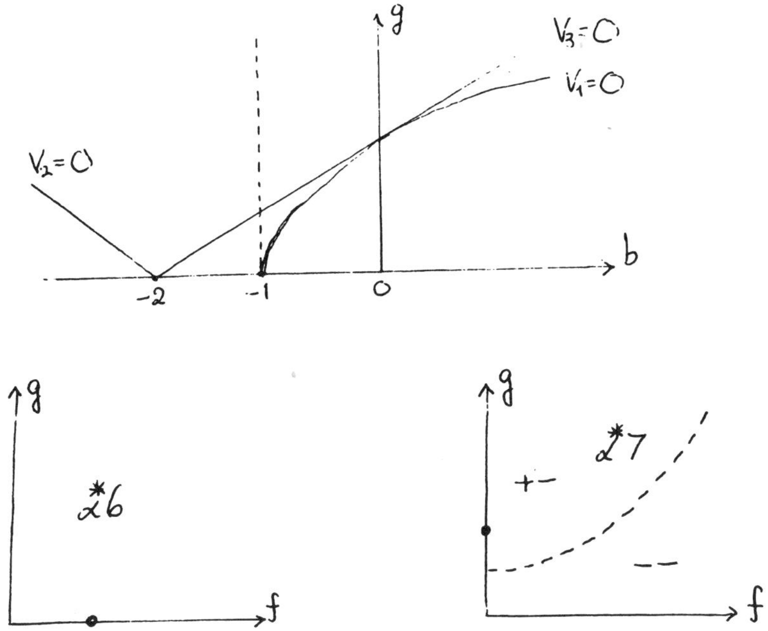

Figure 3

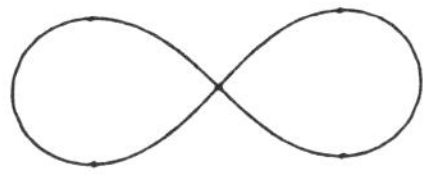

(α^*3)

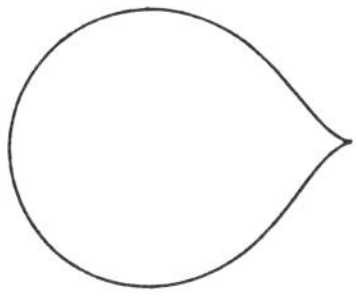

(α^*6)

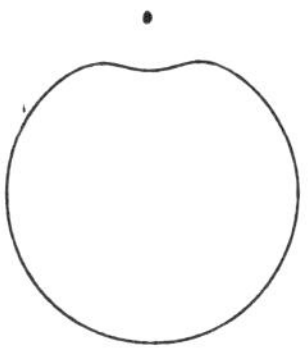

(α^*5)

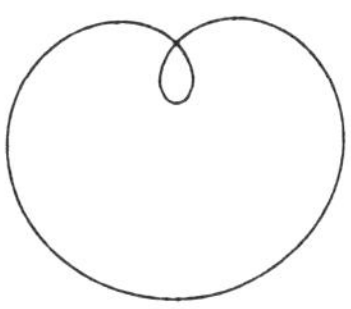

($\alpha^*2, 9$)

($\alpha^*7, 11$)

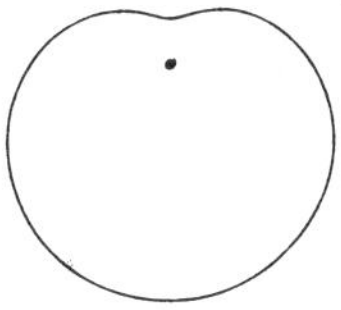

($\alpha^*1, 8$)

4 Case β : $A_3 + A_1$

We have $\beta(x,y,z) = (xz, yz, y^2)$; if Γ is a quartic with an A_1 or A_2 singularity at Z and an A_3 or A_4 singularity at X, with tangent XY, $\beta(\Gamma)$ is a conic S with $a \neq 0, b \neq 0$, and conversely. Here the singular points are necessarily real; we may normalise $a = 1$, and have $h = 0$, but must allow $c = \pm 1$. Changing the signs of x and y, if necessary, we suppose $f, g \geq 0$.

Since we have an A_1 and not an A_2, $b \neq 0$; scaling, we may suppose $b = \pm 1$. The branches at the A_1 singularity are real (we have A_1^-) if and only if $b < 0$. Also, since we have an A_3, $c \neq g^2$ (else we have an A_4); the branches at the A_3 singularity are real (we have A_3^-) if and only if $c < g^2$. We also require $0 \neq \Delta \equiv bc - bg^2 - f^2$.

For each pair of values of b, c (each $= \pm 1$) these conditions define a partition of the (positive quadrant of the) (f,g)-plane. The cases are listed below and illustrated in Figure 4m; each is visibly connected.

As before, bitangents correspond to lines $L \equiv px + qy + rz = 0$, with $F - L^2$ a linear combination of xz, yz and y^2. We may take $p = 1$; then $q = 0$ and $r^2 = c$. There is no real bitangent unless $c > 0$; if $c = 1$, we have two, $2(g-r)x + 2fy + bz = 0$ corresponding to $x + rz$, with $r = \pm 1$. The types of these bitangents are given by $-sgn(f^2 - 2b(1 - rg))$. This too is illustrated in Figure 4m; connectedness of each stratum is again immediate from the figure.

The node is a flecnode if it is a crunode ($b = -1$) and $f = g$: this occurs in cases ($\beta 5$), ($\beta 6$), ($\beta 8$). It is a biflecnode if $f = g = 0$, which occurs in cases ($\beta 1$), ($\beta 4$), ($\beta 5$), ($\beta 8$).

In Table 4, cases are defined by specifying b and c as well as by conditions on (f,g). Throughout the table, we have $a = 1$, $h = 0$, $f, g \geq 0$.

Table 4

Name	(b,c)	(f,g)	Δ	*Singularities*	*Bitang.*	*Flexes*
($\beta 1$)	$(1,1)$	$f^2 + g^2 < 1$	> 0	$A_3^+ + A_1^+$	$++$	0
($\beta 2$)	$(1,1)$	$g^2 < 1 < f^2 + g^2$	< 0	$A_3^+ + A_1^+$	$**$	0
($\beta 3$)	$(1,1)$	$g^2 > 1$	< 0	$A_3^- + A_1^+$	$*-$	2
($\beta 4$)	$(1,-1)$		< 0	$A_3^- + A_1^+$	*None*	4
($\beta 5$)	$(-1,1)$	$g^2 < 1$	< 0	$A_3^+ + A_1^-$	$--$	2
($\beta 6$)	$(-1,1)$	$1 < g^2 < 1 + f^2$	< 0	$A_3^- + A_1^-$	$--$	2
($\beta 7$)	$(-1,1)$	$g^2 > 1 + f^2$	> 0	$A_3^- + A_1^-$	$*-$	0
($\beta 8$)	$(-1,-1)$	$f^2 < 1 + g^2$	> 0	$A_3^- + A_1^-$	*None*	2
($\beta 9$)	$(-1,-1)$	$f^2 > 1 + g^2$	< 0	$A_3^- + A_1^-$	*None*	2

The conventions adopted in Figure 4 - and indeed also in Figure 5 - are the same as in Figure 1, except that the triangle is drawn with Y as the top vertex and XZ as the lower edge. The vertex Y and the edge $x = 0$ of the basic triangle no longer have special significance, but are drawn for convenience; we have an A_1 (or, eventually, A_2) at Z, whose nature is determined as before, and a higher singularity at X, with tangent $z = 0$. By considering the graded tangent cone at X, with equation $z^2 + 2gy^2z + cy^4 = 0$, we see not only that we have an A_3^+ if $c > g^2$, an A_4 if equality occurs and an A_3^- with the opposite inequality; but also that in the latter case, the two branches lie on opposite sides of the tangent if $c < 0$, on the same side if $c > 0$; this side including points with x, y, z all positive if $g < 0$ (and not otherwise). All intersections of Γ with $yz = 0$ are determined by these rules; those with $x = 0$ may be checked ad hoc. Like α, β preserves each of the four triangles determined by these 3 lines.

Figure 4m

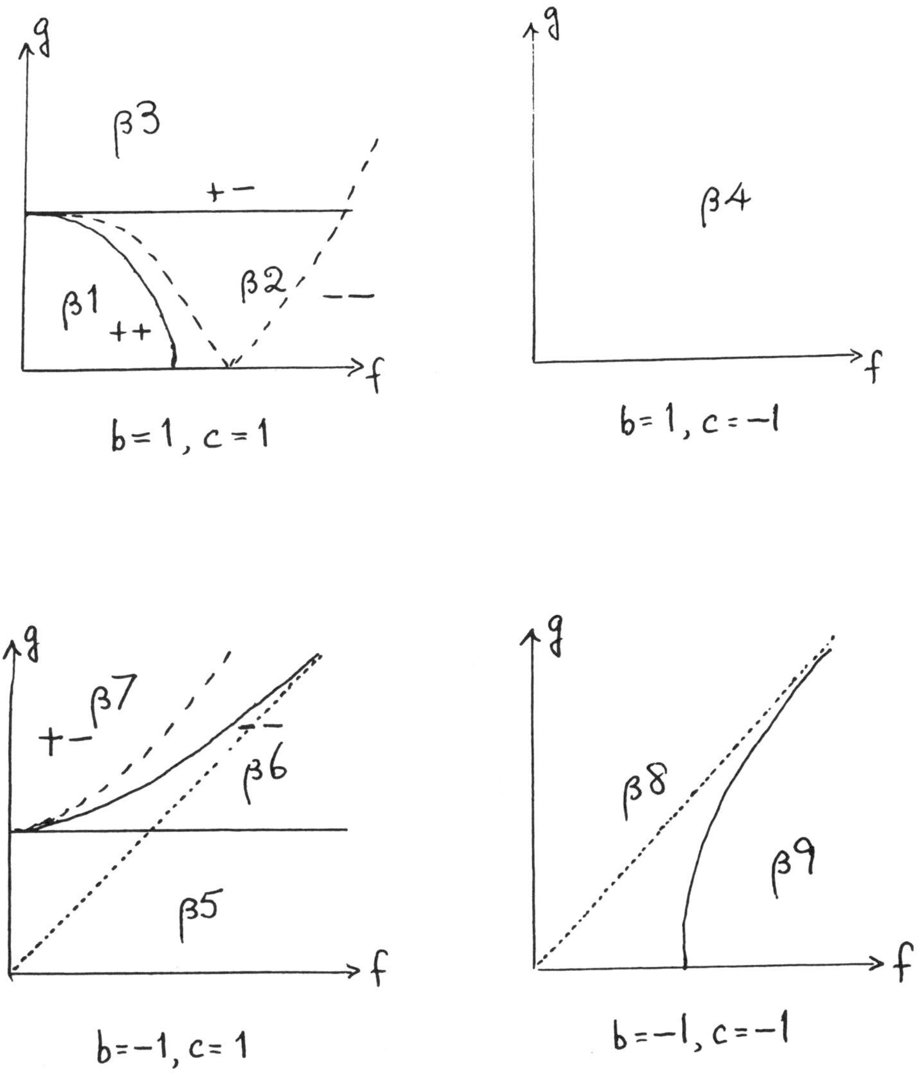

Figure 4

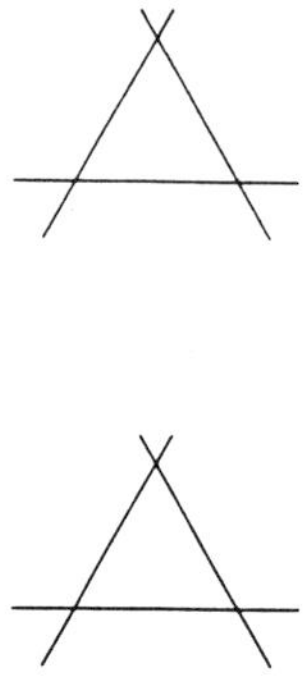

(β1)

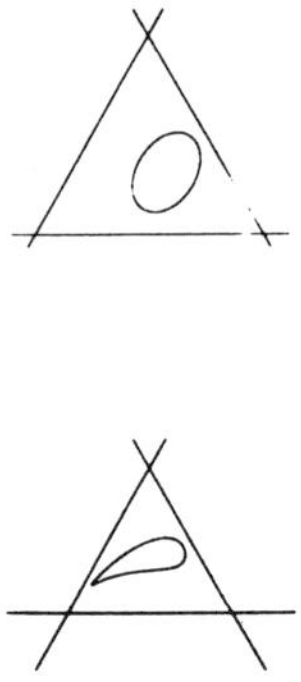

(β2)

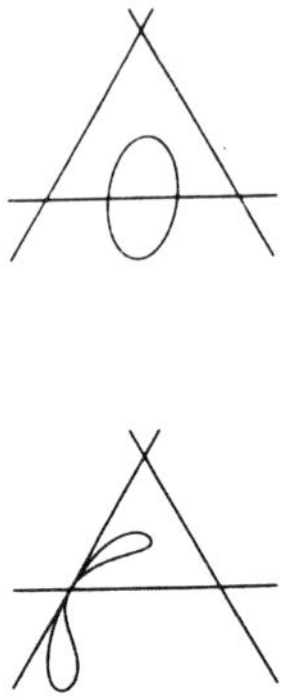

(β3)

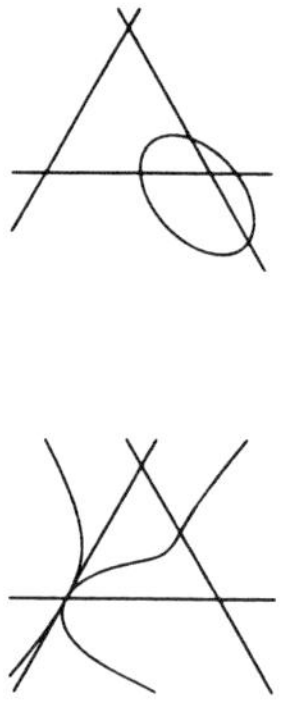

(β4)

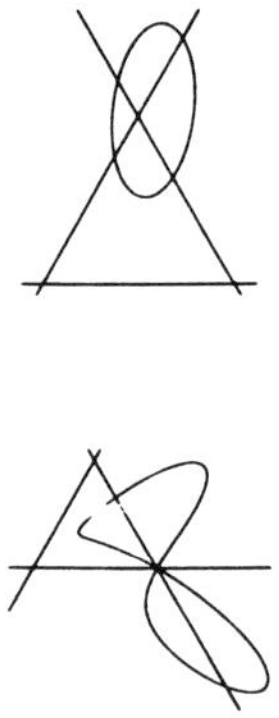

(β5)

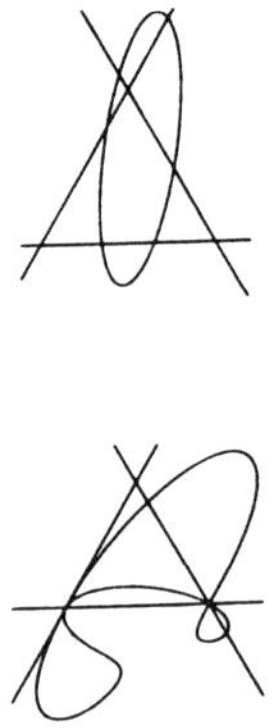

(β6)

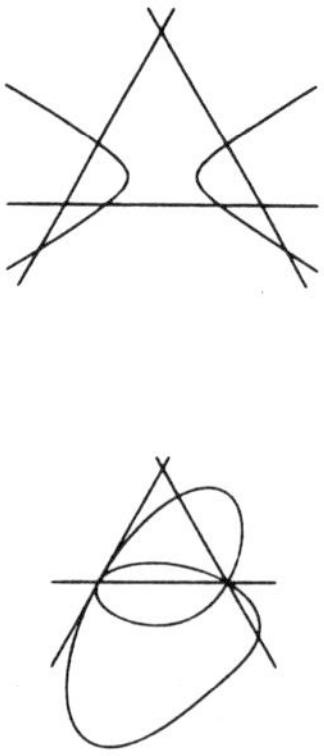

(β7)

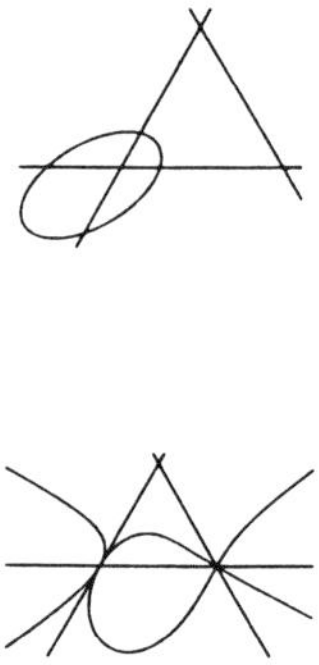

(β8)

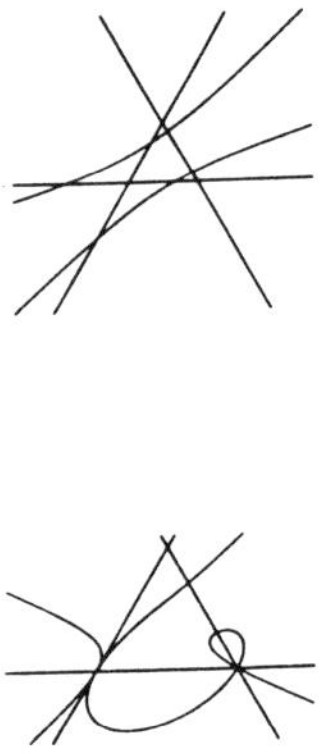

(β9)

5 Further subcases of Case β

$A_3 + A_2$

We now have $b = 0$. We may normalise $f = 1$; then $\Delta = -1$. We have $c \neq 0$, g^2, and can take $c = \pm 1$, giving cases (β10), with $c > g^2$, (β11), with $0 < c < g^2$, and (β12), with $c < 0$.

There is no bitangent in these cases. In each case Γ has 2 real flexes.

$A_4 + A_1$

Here $c = g^2 \neq 0$ and we may take $c = g = 1$, $b \neq 0$, $\Delta = -f^2$, so $f \neq 0$ and we may also take $f = 1$; we have cases (β13), with $b > 0$ and (β14), with $b < 0$.

There is one bitangent $4x + 2y + bz = 0$, corresponding to $x - z$; it has type $sgn(4b - 1)$. The node is a flecnode if $b = -1$; it cannot be a biflecnode.

$A_4 + A_2$

Here we have a unique normal form $b = 0$, $c = f = g = 1$, which we label (β15). There is no bitangent and just one flex.

In Table 5, we have $a = f = 1$, $h = 0$, $g \geq 0$ throughout.

Table 5

Name	b	(c, g)	*Singularities*	*Bitangents*	*Flexes*
(β10)	$b = 0$	$0 < g^2 < c$	$A_3^+ + A_2$	*None*	2
(β11)	$b = 0$	$0 < c < g^2$	$A_3^- + A_2$	*None*	2
(β12)	$b = 0$	$c < 0$	$A_3^- + A_2$	*None*	2
(β13)	$b > 0$	$c = g = 1$	$A_4 + A_1^+$	$*$	1
(β14)	$b < 0$	$c = g = 1$	$A_4 + A_1^-$	$-$	1
(β15)	$b = 0$	$c = g = 1$	$A_4 + A_2$	*None*	1

Figure 5

(β10) (β11) (β12)

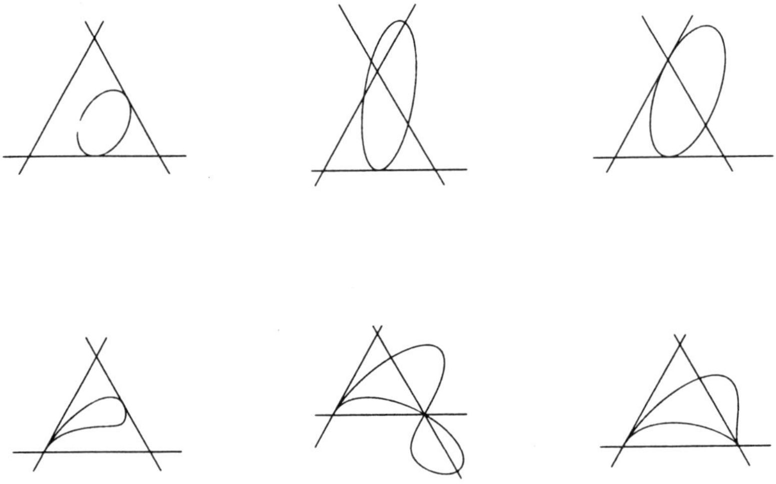

(β13) (β14) (β15)

6 Case γ

We recall that if Γ has a singular point of type A_5 or A_6, then after a change of coordinates it may be put in the form $\gamma(S)$, where $\gamma(x,y,z) = (y^2 - xz, yz, z^2)$ (so that γ^2 is the identity).

We may suppose $a = 1$. Then substitute $y = y' - \frac{1}{2}hz$ in the equation of Γ to reduce h to 0, and then $x = x' + gz$ to reduce g to 0. We still have available the substitution

$$x = x' + 2\lambda y' + \lambda^2 z'; \; y = y' + \lambda z'; \; z = z'$$

which leaves a and b unchanged and replaces f by $f + b\lambda$, c by $c + 2f\lambda + b\lambda^2$.

If the singular point has type A_5, then $b \neq 0$. We may further reduce b to ± 1 and f to 0, so $F = x^2 + by^2 + cz^2$, and the real types are determined by the signs of b and c. The branches at the singular point are real if $b < 0$, and if $b, c > 0$, then Γ has no real point apart from the singular point.

There is one bitangent $bx + (c - r^2)z = 0$ which corresponds to $2x + bz$; its type is $sgn\{b(4c + b^2)\}$.

For type A_6, we must have $b = 0$, $f \neq 0$ and can then reduce to $f = 1$, $c = 0$, so $F = x^2 + 2yz$. There is then no bitangent and just one real flex.

Table 6

Name	(b,c)	*Singularities*	*Bitangents*	*Flexes*
$(\gamma 1)$	$b, c > 0$	A_5^+	$+$	0
$(\gamma 2)$	$b > 0 > c$	A_5^+	$*$	0
$(\gamma 3)$	$c > 0 > b$	A_5^-	$-$	2
$(\gamma 4)$	$0 > b, c$	A_5^-	$*$	0
$(\gamma 5)$	$b = 0$	A_6	*None*	1.

In Table 6, $a = 1$, $g = h = 0$ throughout; we have $f \neq 0$ for $(\gamma 5)$.

The topology of Γ is determined by the position of the conic S with respect to the line $z = 0$ and the point X on it. The conic is empty for $(\gamma 1)$, is disjoint from $z = 0$ for $(\gamma 2)$, touches it for $(\gamma 5)$, and crosses it in the other two cases, with X in the interior region of the conic for $(\gamma 4)$, but not for $(\gamma 3)$.

Figure 6m

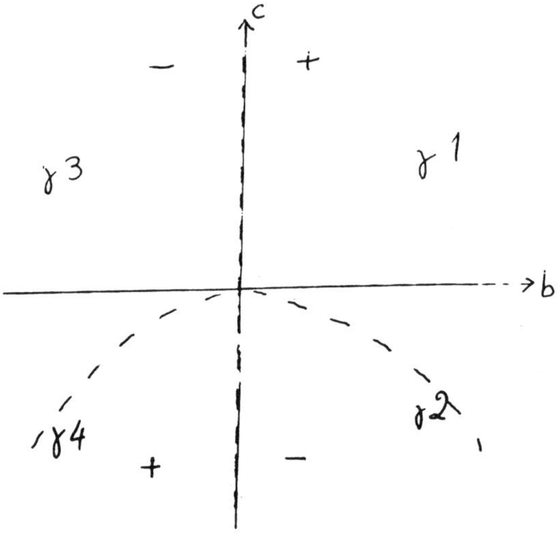

Figure 6

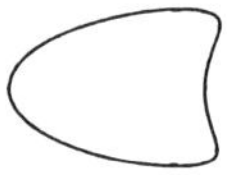

(γ2)

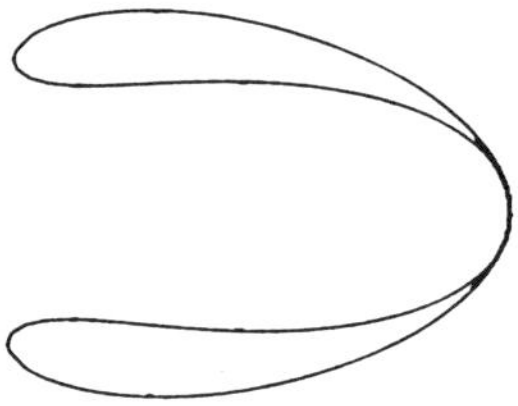

(γ3)

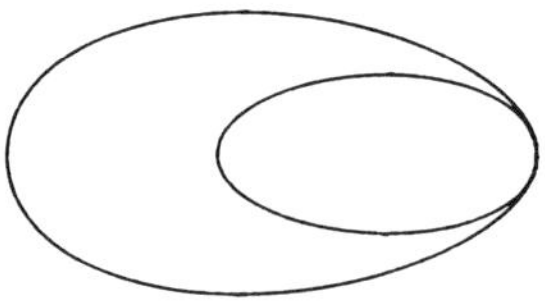

(γ4)

(γ5)

7 Case δ

D_4, 3 real tangents

We take the equation of Γ in the form $x\phi(y,z)+\psi(y,z)=0$, with $\phi=y^3-yz^2$; adding a linear combination of y and z to x we reduce to

$$x(y^3-yz^2)+ay^2z^2+byz^3+cz^4=0,$$

with c and $a\pm b+c$ nonzero. Symmetries permuting the factors of (y^3-yz^2) have the effect of permuting $16c$ and $a\pm b+c$.

There are 2 cases: ($\delta 1$), where c and $a\pm b+c$ have the same sign (say >0), and ($\delta 2$), where they do not. In case ($\delta 1$) we may normalise by $a-b+c\le a+b+c\le 16c$ and also $c=1$. Clearly the stratum is connected.

In case ($\delta 2$) we may normalise as $a\pm b+c>0>c$ and $c=-1$; rewriting this as $a>|b|+1$ shows that it too is connected.

The line $x=qy+rz$ meets Γ where $R(t)=qt^4+rt^3+(a-q)t^2+(b-r)t+c$ vanishes, so is a bitangent if this is a perfect square (up to sign). Since $R(1)=a+b+c$, $R(0)=c$, $R(-1)=a-b+c$, this happens only in case ($\delta 1$). We have 4 bitangents given by choosing square roots $2\epsilon u, 2\eta v$ of $a\pm b+c$ (where $2u, 2v$ are the positive square roots), viz. $q=(\epsilon u+\eta v-1)^2$, $r=2(\epsilon u+\eta v-1)(\epsilon u+\eta v)$. This bitangent has type $-sgnE(\epsilon,\eta)$, where $E(\epsilon,\eta)=((\epsilon u-\eta v)^2-4(\epsilon u+\eta v-1))$, so the type is $-$ unless $\epsilon=\eta=1$. The parabola $E(1,1)=0$ is shown in Figure 7m, and divides the region $0\le u\le v\le 2$ into 2 subregions.

D_4, 1 real tangent

We take the equation in the form

$$x(y^3+yz^2)+ay^2z^2+byz^3+cz^4=0,$$

with c and $-a\pm ib+c$ nonzero. This gives a single connected family ($\delta 3$). We may normalise $c=1$ and $b\le 0$.

The line $x=qy+rz$ is a bitangent if we can write

$$qt^4+rt^3+(a+q)t^2+(b+r)t+1=(u't^2+vt+1)^2,$$

so $(-a+ib+1)=(-u'+iv+1)^2$. Let $-u+iv$ be the square root of the nonzero complex number $-a+ib+1$ with $u\ge 0$ and hence (since $b\le 0$) $v\ge 0$: we can take this quadrant, less the point $(0,0)$, as the moduli space - which is connected.

The 2 square roots $\pm(-u+iv)$ now yield two real bitangents, with types given by $-sgn(v^2-4(1\pm u))$.

D_5

We may take the equation as

$$x(y^2z) + ay^4 + byz^3 + cz^4 = 0,$$

and reduce a to 1 and c to ($\delta 4$) $c = -1$ or ($\delta 5$) $c = 1$. We may also suppose $b \geq 0$.

If $c = -1$ there is no real bitangent. Otherwise we set as before $x = qy + rz$, and then need $R(t) = t^4 + bt^3 + rt^2 + qt + 1$ to be a perfect square, necessarily $(t^2 + \frac{1}{2}bt \pm 1)^2$, so have 2 bitangents. The case with a $-$ sign has type $-$, the other has type $sgn(4-b)$.

E_6

The equation of Γ may be taken as

$$xy^3 + 2cy^2z^2 + z^4 = 0,$$

(where c may further be reduced to ± 1 or 0): this case we label ($\delta 6$).

The unique bitangent is $x = c^2y$; it has type $-sgn(c)$.

Table 7

Name	*Singularities*	*Bitangents*	*Flexes*
($\delta 1$)	D_4^-	$* - --$	0
($\delta 2$)	D_4^-	*None*	2
($\delta 3$)	D_4^+	$**$	0
($\delta 4$)	D_5	*None*	2
($\delta 5$)	D_5	$*-$	0
($\delta 6$)	E_6	$*$	0

Figure 7m

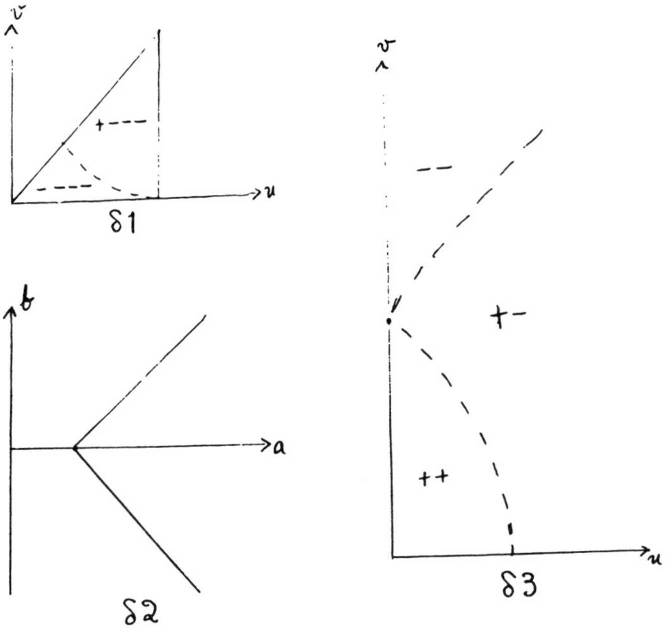

Figure 7

(δ1)

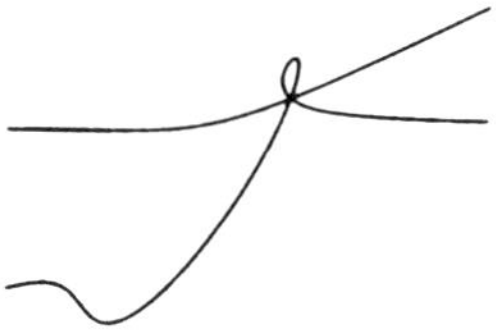

(δ2)

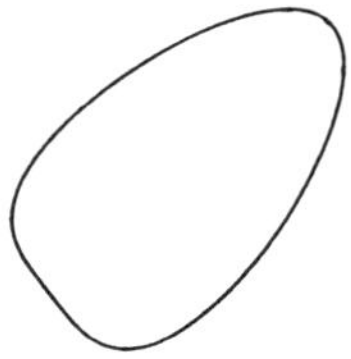

(δ3)

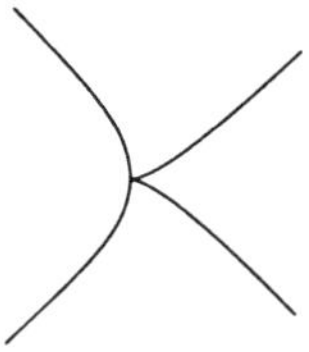

(δ4)

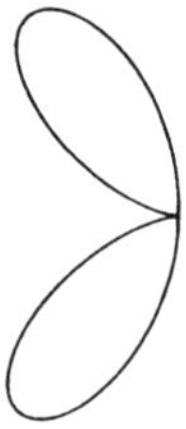

(δ5)

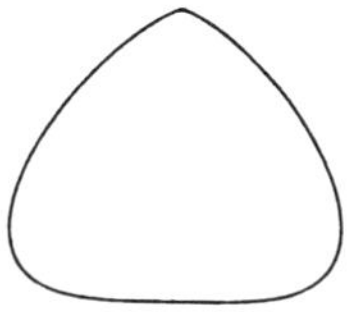

(δ6)

8 Summaries

Lists of curves with special properties

We can pick out several lists from our tables. Cases when $\Gamma^\vee$ has a real E_6 singularity occur for Γ of types

$$\begin{array}{cccccccc}
(\alpha 2) & (\alpha 3) & (\alpha 4) & (\alpha 5) & (\alpha 7) & (\alpha 11) & (\alpha 12) & (\alpha 14) \\
(\alpha 16) & (\alpha^* 1) & (\alpha^* 2) & (\alpha^* 5) & (\alpha^* 7) & (\alpha^* 8) & (\beta 2) & (\beta 3) \\
(\beta 7) & (\beta 13) & (\gamma 2) & (\gamma 4) & (\delta 1) & (\delta 3) & (\delta 5) & (\delta 6)
\end{array}$$

In each case the dual undergoes a transition from $(2A_2) + A_1^+$ via E_6 to $2A_2 + A_1^-$; the values of the moduli for which this occurs are given explicitly above (see also the figures). Only for

$$(\alpha 2),\ (\alpha 3),\ (\alpha^* 1),\ (\alpha^* 5),\ (\beta 2),\ (\delta 3)$$

is it possible to have 2 real hyperflexes, and only in the first of these cases can we have 3. In the opposite direction, in cases

$$\begin{array}{c}
(\alpha 8),\ (\alpha 9),\ (\alpha 10),\ (\alpha 15),\ (\alpha^* 6),\ (\beta 4),\ (\beta 8),\ (\beta 9) \\
(\beta 10),\ (\beta 11),\ (\beta 12),\ (\beta 15),\ (\gamma 5),\ (\delta 2),\ (\delta 4)
\end{array}$$

there are no real bitangents at all. In cases

$$(\alpha 1), (\alpha^* 4), (\alpha^* 10), (\beta 1), (\gamma 1)$$

the curve Γ has no real points apart from the singular points.

Curves with class at most 4

A more complete discussion of duality is possible when the dual curve $\Gamma^\vee$ also has degree ≤ 4. We first consider the case $n^\vee = 3$. A quartic curve Γ over $\mathbb{C}$ with 3 cusps is dual to a nodal cubic curve. In the real case, if $\Gamma^\vee$ has a crunode, it has only one real flex, so Γ has only one real critical point, but the contacts of the bitangent are dual to the tangents at the vertex of $\Gamma^\vee$, so are real. If $\Gamma^\vee$ is acnodal it has 3 real flexes so Γ has 3 real singular points, but the points of contact of the bitangent are complex conjugate.

We see from the table in the Introduction that if $n^\vee = 4$, then the singularities of Γ form one of the lists $2A_2 + A_1$, E_6 and $A_4 + A_2$. Over $\mathbb{R}$ there is a unique curve of the latter type, which is self-dual. The other cases are more interesting. We recall from [5] that over $\mathbb{C}$, we have the normal form $\alpha(S)$, with $a = b = c = 1$, $g = h = -1$, $f \neq \pm 1$, and that for each curve Γ with $f \neq -7$, the dual is isomorphic to Γ itself; if $f = -7$, the dual has an E_6 singularity. Over $\mathbb{R}$, as well as this we have the normal form $\alpha^*(S)$, with $a = b + 1$, $c = g = 1$, $f = h = 0$,. The curve in this family with parameter b is equivalent over $\mathbb{C}$ to the curve in the other with parameter $f = 2b + 1$.

From the first group we distinguish cases $f < -7$, $-7 < f < -1$, $-1 < f < 1$, $1 < f$, with labels $(\alpha 16i), (\alpha 16ii), (\alpha 17), (\alpha 18)$ respectively; and from the second cases $b < -4$, $-4 < b < -1$, $-1 < b < 0$, $0 < b$, with labels $(\alpha^* 8i), (\alpha^* 8ii), (\alpha^* 9)$ and $(\alpha^* 10)$. Of the former group, $(\alpha 16ii)$ has 0 real flexes; the others have 2; of the latter group, $(\alpha^* 8ii)$ has 2 real flexes; the rest have 0. We conclude that cases $(\alpha 16ii)$ and $(\alpha^* 8ii)$ are dual to each other; each curve in the other families is isomorphic over $\mathbb{R}$ to its own dual.

The curves in $(\alpha 16)$ with $f = -7$ and in $(\alpha^* 8)$ with $b = -4$ have duals with an E_6 singularity. These curves have type $(\delta 6)$, and there are just 3 normal forms in this type, given by $xy^3 + 2cy^2z^2 + z^4 = 0$, where $c = 0$ or ± 1. Since the type of the hyperflex is $-c$, the dual $\Gamma^\vee$ has type $(\alpha^* 8)$ if $c < 0$, $(\delta 6)$ if $c = 0$ and $(\alpha 16)$ if $c > 0$.

Curves with points at infinity

The author has expended considerable effort in choosing parameters for the figures so that the curves illustrated have no points at infinity. In the remaining cases this is not possible; we now discuss why this is so.

There are 10 such cases; in each one, the finite part of Γ as pictured can be described as follows. There are two lines l, m say, which are asymptotes; each has two ends e.g. l^+, l^-. G is made up of two parts, a smooth curve G_1 going from l^- to m^+, and another curve G_2 which goes from m^- to l^+, and intersects G_1 at some point in between. It follows that Γ contains a simple closed curve which starts at l^-, follows G_1 to an intersection point, and then follows G_2 from there till it closes up at l^+. This is homologous to l in the real projective plane, and thus has non-empty intersection with any projective line. Since no line is disjoint from Γ, any affine picture must omit points at infinity.

In the cases that arise, the curve G_2 has a loop in cases $(\beta 8), (\alpha 10), (\delta 2), (\alpha 9)$ and $(\beta 9)$, a cusp in $(\delta 4)$, $(\alpha 15)$ and $(\beta 12)$; and an acnode in $(\beta 4)$ and $(\alpha 8)$: G_1 passes through the singular point in the δ cases and touches G_2 at a smooth point in the β cases.

A further difficulty that arises specially in these cases is in seeing all the real inflections. If the asymptote l is not inflexional, then (since l is 2-sided in the projective plane) in an affine picture Γ will appear to go to infinity on one side of l and return on the other side (with curvature pointing in the opposite direction); with a flex at infinity, it will lie on the same side of the asymptote at both ends, and the two curvatures will point in the same direction. Corresponding remarks hold when (as in practice) only a finite amount of the picture is exhibited: if the two remote parts of the curve which are tending to the same asymptote have curvature vectors nearly in the same direction, then there is a flex on the remote part of the curve in between.

When this is taken into account, the correct number of flexes are visible in each of the pictures.

References

[1] A.B.Basset, An elementary treatise on cubic and quartic curves, Bell, Cambridge & London, 1901.

[2] C.T.C.Wall, Deformations of real singularities, Topology **29** (1990) 441-460.

[3] C.T.C.Wall, Is every quartic a conic of conics?, Math. Proc. Camb. Phil. Soc., **109** (1991) 419-424.

[4] C.T.C.Wall, Duality of singular plane curves, Jour. London Math. Soc., **50** (1994) 265-275.

[5] C.T.C.Wall, Geometry of quartic curves, Math. Proc. Camb. Phil. Soc., to appear.

[6] C.T.C.Wall, Duality of real projective plane curves: Klein's equation, Topology, to appear

J. Sotomayor[1]

Stability properties of the inverse mapping domain

1 Introduction.

Consider a smooth (i.e. C^∞) mapping $\mathbf{f}$ defined on $\mathbf{R}^n$, with values in $\mathbf{R}^n$. Assume that $\mathbf{f}'(\mathbf{x})$ is invertible for every $\mathbf{x}$ on a compact n-dimensional region $\mathbf{M}$, with smooth boundary $\partial\mathbf{M}$. Without loss of generality, it will be assumed that $\mathbf{0} \in \mathbf{M}, \mathbf{f}(\mathbf{0}) = \mathbf{0}$ and that $\partial\mathbf{M} = \{\delta = 0\}$, for a smooth function δ defined on $\mathbf{R}^n$, positive only on the interior of $\mathbf{M}$, with non-vanishing derivative on $\partial\mathbf{M}$.

The problem of determining the inverse mapping as well as to estimate the size of its domain of definition, for mappings such as $\mathbf{f}$, is of paramount interest in Mathematical Analysis and its applications. A landmark in this subject is the paper or Wazewski [W]. Other references concerning the historical and present day aspects and developements on this problem can be found in [Me, G-Z, S.1].

This paper is concerned with the investigation of the structural stability properties, under small deformations of $\mathbf{f}$, of the *right inverse* $\mathcal{R} = \mathcal{R}(\mathbf{f}, \mathbf{0})$ of $\mathbf{f}\mid_{\mathbf{M}}$ on its *maximal star-shaped domain of definition* $\mathbf{S} = \mathbf{S}(\mathbf{f}, \mathbf{0})$, *centered at* $\mathbf{0}$.

For the present considerations, a right inverse $\mathcal{R}$ on a domain $\mathbf{S}$ containing $\mathbf{0}$, with values on $\mathbf{R}^n$, is characterized by the following conditions:

$$\mathcal{R}(\mathbf{0}) = \mathbf{0}, \quad \mathcal{R}(\mathbf{S}) \subseteq \mathbf{M} \quad \text{and} \quad \mathbf{f} \circ \mathcal{R} = \mathbf{Id_S}$$

The *maximal star-shaped domain* $\mathbf{S}(\mathbf{f}, \mathbf{0})$ is the maximal element, under the inclusion order of domains, star shaped at $\mathbf{0}$, of right inverse mappings of $\mathbf{f}$, sending $\mathbf{0}$ to $\mathbf{0}$.

As any star shaped domain, $\mathbf{S}(\mathbf{f}, \mathbf{0})$ is characterized by its *radial function* $\rho = \rho(\mathbf{f}, v)$, for v on the unit $n-1$ dimensional sphere $\mathbf{S}^{n-1}$. That is,

$$\rho(\mathbf{f}, v) = \sup\{t > 0, [0, t]v \subset \mathbf{S}(\mathbf{f}, \mathbf{0})\}.$$

Initial results concerning analytic expressions of the functions ρ and $\mathcal{R}$ have been obtained in [S.1], in terms of the solutions of the *Modified Wazewski differential equation associated to* $\mathbf{f}$, $(M.W.E; \mathbf{f})$.

In the present context, it is convenient to write:

$$dx/dt = \delta(\mathbf{x})(\mathbf{f}'(\mathbf{x})^{-1})v, \quad \mathbf{x}(0) = \mathbf{0}, \quad \mathbf{x} \in \mathbf{M}, \quad v \in \mathbf{S}^{n-1}. \qquad (M.W.E; \mathbf{f})$$

[1]Partially supported by FAPESP, Grant 93/2482-7.

Denote by $\mathbf{x} = \mathbf{x}(t, v, \mathbf{f})$ the solution of this differential equation on $[0, \infty)$, the radial function of $\mathbf{S}(\mathbf{f}, \mathbf{0})$ can be expressed as:

$$\rho(\mathbf{f}, v) = \int_0^\infty \delta(\mathbf{x}(t, v, \mathbf{f}))\, dt.$$

This function is the right extreme of the maximal solution of the *Wazewski differential equation associated to* $\mathbf{f}$:

$$d\mathbf{x}/d\tau = (\mathbf{f}'(\mathbf{x})^{-1})v, \quad \mathbf{x}(0) = \mathbf{0}, \quad \mathbf{x} \in \mathbf{M}, \quad v \in \mathbf{S}^{n-1}, \qquad (W.E; \mathbf{f})$$

obtained by dividing the modified equation by $\delta(\mathbf{x})$. The relationship between the time variables (t and τ) of these equations is given by $d\tau/dt = \delta(\mathbf{x})$. Integration gives ρ.

The solution $\tau \to \mathbf{x}(\tau, v)$ of $(W.E; \mathbf{f})$ on its maximal interval $[0, \rho]$ provides the right inverse $\mathcal{R}(\mathbf{f}, \mathbf{0})$ at the point τv of the segment $[0, \rho]$v.

For the present study, concerned more with the geometric nature of the domain $\mathbf{S}(\mathbf{f}, \mathbf{0})$ and its dependence on $\mathbf{f}$, rather than with its analytic expression, it will be convenient to use directly the solutions of $(W.E; \mathbf{f})$.

Recall that a *smooth deformation* of a mapping $\mathbf{g}$ from a manifold $\mathbf{N}$ to a manifold $\mathbf{P}$ is a mapping $\mathbf{g}_\lambda$ from $\mathbf{N} \times \mathbf{R}$ to $\mathbf{P}$ such that for $\lambda = 0$, i.e. on $\mathbf{N} \times \{0\}$, coincides with $\mathbf{g}_0 = \mathbf{g}$.

A mapping $\mathbf{f}$ as above will be said to be $\mathbf{S}$-*stable* on $\mathbf{M}$, provided, for any smooth deformation $\mathbf{f}_\lambda$ (preserving the origin) there is a smooth deformation v_λ of the identity of $\mathbf{S}^{n-1}$ onto itself, and a smooth deformation r_λ, of the constant unity function on $\mathbf{S}^{n-1}$, such that

$$\rho(\mathbf{f}_\lambda, v) = r_\lambda(\vartheta_\lambda(v))\rho(\mathbf{f}, \vartheta_\lambda(v)), \quad \text{for} \quad v \in \mathbf{S}^{n-1} \quad \text{and small values of} \quad \lambda.$$

This means that the star-shaped domain $\mathbf{S}_\lambda = \mathbf{S}(\mathbf{f}_\lambda, \mathbf{0})$ for the right inverse $\mathcal{R}_\lambda = \mathcal{R}(\mathbf{f}_\lambda, \mathbf{0})$ of $\mathbf{f}_\lambda$ is a smooth deformation of $\mathbf{S}_0 = \mathbf{S}(\mathbf{f}, \mathbf{0})$ achieved by means of a deformation of the identity which preserves rays emanating from $\mathbf{0}$.

In the above definitions, the preservation of the origin (i.e. $\mathbf{f}_\lambda(\mathbf{0})$) is not essential since it can always be obtained by composition with a translation in the target space which send back $\mathbf{f}_\lambda(\mathbf{0})$ to $\mathbf{0}$.

By considering the solutions of $(W.E; \mathbf{f}_\lambda)$ for $v = \vartheta_\lambda$, the right inverse $\mathcal{R}(\mathbf{f}_\lambda, \mathbf{0})$ on $\mathbf{S}_\lambda$ can also be obtained as a deformation of $\mathcal{R}(\mathbf{f}, \mathbf{0})$.

Theorem 1 of this paper states necessary and sufficient conditions for a mapping $\mathbf{f}$ to be $\mathbf{S}$-stable.

Theorem 2 shows that the conditions imposed on Theorem 1 hold for most mappings $\mathbf{f}$.

These two results are stated and their proofs outlined in section 2.

2 On Stability and Genericity.

Recall from Singularity Theory [Go-Gu] that a multigerm of a mapping $\mathbf{g}$ from a manifold $\mathbf{N}$ to a manifold $\mathbf{P}$ at points $\{\mathbf{n}_1, \mathbf{n}_2, \dots, \mathbf{n}_k\}$, mapped by $\mathbf{g}$ onto $\mathbf{p}$, is *infinitesimally stable* if for any germ of vector field $\mathbf{W}$ at $\{\mathbf{n}_1, \mathbf{n}_2, \dots, \mathbf{n}_k\}$, over $\mathbf{g}$, there are germs of vector fields $\mathbf{X}$, at $\{\mathbf{n}_1, \mathbf{n}_2, \dots, \mathbf{n}_k\}$, and $\mathbf{Y}$, at $\mathbf{p}$, such that

$$D\mathbf{g} \circ \mathbf{X} + \mathbf{Y} \circ \mathbf{g} = \mathbf{W}.$$

From [Go-Gu], follows that the verification of these conditions depend only on the solvability of systems of linear equations whose coefficients depend only on the derivatives of order at most $(\dim \mathbf{P} + 1)$ of $\mathbf{g}$ at the points $\mathbf{n}_i$.

For compact manifolds, infinitesimal stability for all multigerms of a mapping $\mathbf{g}$ implies stability under deformations $\mathbf{g}_\lambda$ of $\mathbf{g}$, and conversely [Go-Gu]. Stability under deformations in this context means the following: for any given $\mathbf{g}_\lambda$ there are deformations s_λ and t_λ of the identities of $\mathbf{N}$ and $\mathbf{P}$ such that $\mathbf{g} = t_\lambda \circ \mathbf{g}_\lambda \circ s_\lambda$, for small λ.

The *v-visible points* of the $\mathbf{f}$-image of the boundary $\partial\mathbf{M}$ along the ray spanned by $v \in \mathbf{S}^{n-1}$ are those points $\mathbf{p} \in \partial\mathbf{M}$ for which $\mathbf{f}(\mathbf{p})$ is a positive multiple of v and the whole segment $[\mathbf{0}, \mathbf{f}(\mathbf{p})]$is contained in $\mathbf{S}$.

Notice that these are the points where the solution of $(W.E; \mathbf{f})$ meets the boundary of $\mathbf{M}$.

A mapping $\mathbf{f}$ as above (in section 1) is $\mathbf{S}$-*infinitesimally stable on* $\mathbf{M}$ if for any v in $\mathbf{S}^{n-1}$ and any collection $\mathbf{p}_i$, $i = 1, \dots, k$, of v-visible boundary points of the $\mathbf{f}$-image of the boundary of $\mathbf{M}$, the multigerm

$$\mathbf{f}/ \mid \mathbf{f} \mid : (\partial\mathbf{M}; \quad \mathbf{p}_i, i = 1, \dots, k) \to (\mathbf{S}^{n-1}, v)$$

is infinitesimally stable.

Theorem 1 (Stability). *A mapping* $\mathbf{f}$ *is* $\mathbf{S}$*-stable on* $\mathbf{M}$ *if and only if it is* $\mathbf{S}$*-infinitesimally stable on* $\mathbf{M}$.

The infinitesimal stability condition is closely related to that imposed by Sotomayor [S.2] and Percell [P] for the structural stability of differential equations on manifolds with boundary. See also [V].

In fact, to get a closer view of the relationship between these conditions and those of structural stability on manifolds with boundary it is more appropriate to consider the following modification of $(W.E; \mathbf{f})$

$$dx/d\zeta = (\mathbf{f}'(\mathbf{x})^{-1})\mathbf{f}(\mathbf{x}), \mathbf{x} \in \mathbf{M}, \qquad (W.Z; \mathbf{f})$$

considered by Zampieri [Z]. It has the same orbits as $(W.E; \mathbf{f})$, though with different parametrizations. It can be verified that the $\mathbf{S}$-infinitesimal stability is equivalent to the "boundary and transversality conditions" of [S.2] along all the orbits of $(W.Z; \mathbf{f})$

whose limit, as ζ tends to $-\infty$, is $\mathbf{0}$. In the terminology of [P], **S**-infinitesimal stability, corresponds to conditions "L and G", along any compact arc of these same orbits of $(W.Z;\mathbf{f})$. As such it can be expressed in terms of Jacobians of succesive derivarives of δ along the vector field $(W.Z;\mathbf{f})$.

The proof of Theorem 1 can be obtained from that of "infinitesimal stability implies stability under deformations", in Singularity Theory [Go-Gu]. In fact, the mapping ϑ_λ, required by **S**-stability, is the deformation of the identity in the target space (t_λ above, but for the deformation map $\mathbf{g}_\lambda = \mathbf{f}_\lambda/|\mathbf{f}|$, restricted to the visible part of the image of $\partial\mathbf{M}$).

Similarly, using now the deformation s_λ in the source space, the deformation r_λ of the unit function can be obtained as follows: $r_\lambda(v) = |\mathbf{f}(s_\lambda^{-1}(\mathbf{x}(\rho(\mathbf{f},v),v)|/\rho(\mathbf{f},v)$, where $\mathbf{x}$ is the solution of Wazevski Equation $(W.E;\mathbf{f})$.

Theorem 2 (Genericity). *Any mapping* $\mathbf{f}$ *as above, can be approximated on* $\mathbf{M}$, *in the* C^{n+1} *topology, by one which is* **S**-*infinitesimally stable on* $\mathbf{M}$ *and, therefore, is* **S**-*stable.*

Follows an outline of the proof. As a consequence of an appropriate setting in terms of Transversality Theory of the infinitesimal stability conditions, it follows from the proofs of Theorems 2.5 and 6.4 in [P] that the function δ can be approximated by one $\delta*$ for which these conditions hold for $\mathbf{f}$, relative to the manifold $\mathbf{M}* = \{\delta* \geq 0\}$. This manifold can be expressed as the image of $\mathbf{M}$ by an approximation $\mathbf{i}^*$ of the identity of $\mathbf{R}^n$. The mapping $\mathbf{f}* = \mathbf{f} \circ \mathbf{i}*$ approximates $\mathbf{f}$ and verifies the infinitesimal stability conditions relative to $\mathbf{M}$.

References

[Go-Gu] Golubitsky M., Guillemin V., Stable of Mappings and their Singularities, Springer Graduate Texts, 1974.

[G-Z] Gorni G., Zampieri G. Bull. Austr. Math. Soc. **49**, (1994).

[Me] Meisters G. Inverting Polynomial Maps on N-Space by Solving Differential Equations, Preprint, Lincoln, 1992.

[S.1] Sotomayor J., Inversion of Smooth Mappings, Journal of Appl. Math and Physics (ZAMP), **41**, 1990.

[S.2] Sotomayor J., Structural Stability on Manifolds with boundary, Global Analysis and its Applications, Vol III, IAEA, Vienna, 1973.

[P] Percell P., Structural Stability on Manifolds with boundary, Topology, **12**, 1973.

[V] Vishik S. M. , Vector fields near the boundary of a manifold, Vestnik Moscow Univ., Ser I, Math Mech. **27**, (1972).

[W] Wazewski T., Sur l'évaluation du domaine d'existence des fonctions implicites, Ann.Soc.Polon. Math. **20**, 1947.

[Z] Zampieri G., Finding Domains of invertibility for Smooth functions by means of Attraction Basins, Journ. Diff. Equat. **104**, (1993).

M. A. S. Ruas[1] & M. J. Saia

The polyhedron of equisingularity of germs of hypersurfaces

1 Introduction

Let $G : k^n \times k^r, 0 \to k, 0$ be a deformation of non-decreasing Newton filtration of a commode germ $g : k^n, 0 \to k, 0$, which has an algebraically isolated singularity at 0. The germ g is smooth or analytic when $k = \mathbb{R}$ and holomorphic when $k = \mathbb{C}$. In this article we consider the problem of determining when the variety $G^{-1}(0)$ is Whitney equisingular along the parameter space $0 \times k^r$ at zero.

We compare the polytopes that arise from approaches to topological triviality by Damon–Gaffney and Yoshinaga, following the algebraic approach used by Teissier to characterize the Whitney equisingularity through the integral closure of ideals. We show that both polytopes lead to the same convex subset of the Newton polyhedron of the germ g. Moreover, for any deformation G by terms in this subset, we show that $G^{-1}(0)$ is Whitney equisingular along $0 \times k^r$.

For the class of holomorphic curves in $\mathbb{C}^2$, defined by commode germs g with isolated singularity at 0, we show that these polytopes characterize precisely the Whitney equisingularity of the pair $\{G^{-1}(0), 0 \times k\}$.

For any g as above, Teissier defined in $[T_3]$, a Newton polygon (in the coordinates u, v) associated to the image by $v = \frac{\partial g}{\partial x_1}, u = x_1$ of the polar curve associated to the hyperplane $x_1 = 0$ (assumed generic). His definition also leads to a set which can be constructed from the information given by $C[x_1, \ldots, x_n] / \left\langle \frac{\partial g}{\partial x_i} \right\rangle$, and is an invariant of equisingularity. The approach we use here is different from his.

Acknowledgment: We are very grateful to Prof. Terence Gaffney for many valuable suggestions. Part of this work was done when the second author was at the University of Liverpool, with financial support of FAPESP. The second author thanks these institutions for their help.

2 Teissier's results

We denote by X_G the variety in $k^n \times k^r, 0$ defined by $G^{-1}(0)$. The pair $\{X_G, 0 \times k^r\}$ is Whitney equisingular at 0 if $0 \times k^r$ is a stratum of a Whitney stratification of X_G in a neighbourhood of 0 (see[T_2]).

For complex germs, Teissier uses the notion of integral closure of ideals to characterize the Whitney equisingularity of the pair $\{X_G, 0 \times k^r\}$.

[1]Financial support from CNPq, process #300066/88 – 0

Let I be an ideal in a ring A. An element $h \in A$ is said to be integral over I if it satisfies an integral dependence relation $h^n + a_1 h^{n-1} + \ldots + a_n = 0$ with $a_i \in I^i$. The set of such elements form an ideal $\overline{I}$ in A, called the integral closure of I.

When $A = \mathcal{O}_{X,x_0}$, the local ring of a complex analytic set, Teissier gives various notions equivalent to the above concept.

Proposition 2.1 *Let I be an ideal in $\mathcal{O}_{X,x_0}$, where X is a complex analytic space. The following statements are equivalent:*

i. $h \in \overline{I}$

ii. (Growth condition) *For each choice of generators $\{g_i\}$ of I there exists a neighbourhood U of x_0 and a constant $\mathcal{E} > 0$ such that for all $x \in U$:*

$$| h(x) | \leq \mathcal{E} \cdot sup_i \, |g_i(x)|$$

iii. (Valuative criterion) *For each analytic curve $\varphi: (\mathbb{C}, 0) \to (X, x_0)$, $h \circ \varphi$ lies in $(\varphi^*(I))\mathcal{O}_1$.*

iv. There exists a faithful $\mathcal{O}_{X,x_0}$ module L of finite type such that $h \cdot L \subset I \cdot L$.

Theorem 2.2 $[T_2$, *pg* 604$]$ *The pair $\{X_G, 0 \times k^r\}$ is equisingular if and only if* $\frac{\partial G}{\partial t} \in \overline{\left\langle x_i \frac{\partial G}{\partial x_j}{}_{i,j=1,\ldots,n} \right\rangle}$ *in $\mathbb{C}_{n+1}$.*

Remark 2.3 *The sufficient condition in Teissier's algebraic criterion for equisingularity of the pair $\{X_G, 0 \times k^r\}$ is also valid in the real analytic category.*

3 The approaches to topological triviality by Damon-Gaffney and Yoshinaga

We denote by C_n (in the appropriate category), the ring of germs $g : k^n, 0 \to k, 0$ which are either C^∞ or real analytic when $k = \mathbb{R}$ or holomorphic when $k = \mathbb{C}$.

Let $\Theta(g)$ denote the C_n module of germs of vector fields $\xi : k^n, 0 \to Tk$ such that $\pi \circ \xi = g$ for $\pi : Tk \to k$. $\Theta(Id_{k^n})$ will be denoted by Θ_n.

We say that a deformation $G : k^n \times k^r, 0 \to k, 0$ of $g : k^n, 0 \to k, 0$, is topologically trivial if there is a germ of a homeomorphism $\psi : k^n \times k^r, 0 \to k^n \times k^r, 0$ of the form $\psi(x,t) = (\overline{\psi}(x,t), t)$ such that

$$G \circ \psi(x,t) = g(x).$$

Damon and Gaffney describe in [DG] a method based on the Newton filtration of the germ g to prove that G is topologically trivial. The general form of their result is given by a filtration condition involving the terms $\frac{\partial G}{\partial t_i}$ and the deformation G. Their method consists of constructing controled vector fields based on the Newton diagram of the germ g. When the deformation G satisfies the filtration condition, the topological triviality is proven by solving the localized equation for triviality using such vector fields.

Definition 3.1 *The Newton polyhedron $\Gamma_+(g)$ of a germ $g : k^n, 0 \to k, 0$ is the convex hull in $\mathbb{R}^n_+$ of the set $\bigcup\{m + v,\ v \in \mathbb{R}^n_+ : a_m \neq 0$ in $j^\infty g(x)\}$, where $j^\infty g(x) = \sum_m a_m x^m$ is the Taylor expansion of the germ g.*

We shall denote by $\Gamma(g)$, the union of all the compact faces of $\Gamma_+(g)$. We call $\Gamma(g)$ the Newton diagram of g.

This polyhedron, as well as the others to be defined in the sequel, depend upon the coordinate system.

For each face $\Delta \subset \Gamma(g), C(\Delta)$ denotes the cone of half-rays emanating from 0 and passing throug Δ.

A germ g is commode if for each j there is an x_j^{mj} with nonzero coefficient in the Taylor expansion of g. Under this condition, the collection of all $C(\Delta)$, with $\Delta \subset \Gamma(g)$ gives a polyhedral decomposition for $\mathbb{R}^n_+$.

The Newton diagram of g induces a filtration on C_n, called the Newton filtration associated to g, defined via a piecewise-linear function $\phi\colon \mathbb{R}^n_+ \to \mathbb{R}$ satisfying the following conditions.

1. For each closed face Δ, ϕ is linear on the cone $C(\Delta)$.
2. ϕ takes positive integer values on the lattice points of $\mathbb{R}^n_+ - \{0\}$.
3. There exists a positive integer p such that $\phi_{|\Gamma(g)} = p$.

For any $q \in \mathbb{N}$ let $A_q = \{g \in C_n\colon\ supp\ g \subset \phi^{-1}(q + \mathbb{N})\}$. Then the decreasing sequence $C_n = A_0 \supset A_1 \supset \cdots$ is the Newton filtration of C_n associated to $\Gamma_+(g)$ (see [K] for more details).

For any monomial x^α, we define $fil(x^\alpha) = \phi(\alpha)$, where ϕ is the Newton filtration associated to $\Gamma_+(g)$. This can be extended to a germ f by defining $fil(f) = min\{\phi(\alpha) : a_\alpha \neq 0\}$.

All the concepts in this section extend to C_{n+r}, the ring of r-parameter families of germs in n- variables, by defining $fil(x^\alpha t^\beta) = fil(x^\alpha)$.

We denote by $A_{q,t}$ the set of all $g \in C_{n+r}$ such that $fil(g) \geq q$. We call a level A_l (resp. $A_{l,t}$) commode if for each j there is an x_j^{mj} which is a vertex of A_l, (resp.$x_j^{mj} t^\beta$ which is a vertex of $A_{l,t}$).

For each commode level (A_l), we let $ver(A_l) = \{x^\alpha\colon \alpha$ is a vertex of $\phi^{-1}(l)\}$.

Fixed a face $\Delta \subset \Gamma(g)$, for any germ of function f we denote $f_\Delta = \sum_{m \in \Delta \cap \mathbb{Z}^n_+} a_m x^m$, with $a_m \neq 0$ in $j^\infty f(x)$.

Definition 3.2 *A commode germ g is nondegenerate if for each* $\Delta \subset \Gamma(g)$

$$\left\{x \in k^n : x_1 \frac{\partial g}{\partial x_1}_{\Delta} = \ldots = x_n \frac{\partial g}{\partial x_n}_{\Delta} = 0\right\} \subset \{x_1 x_2 \ldots x_n = 0\}.$$

Definition 3.3 *A deformation G of a germ g satisfies a filtration condition if there exists an integer* $l > 0$ *so that*

$$ver(A_{lp})\frac{\partial G}{\partial t_i} \subset A_{lp,t}\left\langle x_j \frac{\partial G}{\partial x_j}\right\rangle, \quad 1 \leq i \leq r,$$

where $\phi_{|\Gamma(g)} = p$ *and* $A_{lp,t}\left\langle x_j \frac{\partial G}{\partial x_j}\right\rangle$ *denotes the* $A_{lp,t}$*-module generated by* $x_j \frac{\partial G}{\partial x_j}$, $1 \leq j \leq n$.

Theorem 3.4 ([DG], Theorem 2) *Let* $G : k^{n+r}, 0 \to k, 0$ *be a deformation of the commode germ g. Then, a sufficient condition that G is a topologically trivial deformation of g is that G satisfies the filtration condition for all* $i = 1, \ldots, r$.

To prove this result, Damon and Gaffney seek to solve the localized equations

$$-\rho\frac{\partial G}{\partial t_i} = \xi_i(g)$$

where $\rho: k^{n+r}, 0 \to \mathbb{R}_+$ is a control function, so that $\rho^{-1}(0) = \{0\} \times k^r$ and $\xi_i: k^{n+r}, 0 \to Tk^n$ is a smooth germ of vector field. Also they require that $\rho^{-1}\xi_i$ be controled by ρ along $\frac{\partial}{\partial t_i}$.

If the deformation G satisfies the filtration condition, for each $i = 1, \ldots, r$ and for each vertex x^α of A_l, we may write

$$-x^\alpha \frac{\partial G}{\partial t_i} = \sum_{j=i}^{n} \xi_{ij}^\alpha(G).$$

Then, for each $i = 1, \ldots, r$, the germ of vector field $\xi_i: k^{n+r}, 0 \to Tk^n$ defined by $\xi_i = \sum_\alpha \overline{x}^\alpha \xi_{ij}^\alpha$ ($\overline{x}^\alpha$ denotes the complex conjugate of x^α), solves the localized equations

$$-\rho\frac{\partial G}{\partial t_i} = \xi_i(g) \text{ where } \rho = \sum_\alpha |x^\alpha|^2 : \; x^\alpha \in ver(A_l).$$

The integrability of the vector field $\rho^{-1}\xi_i$ follows from the Lemmas 3.4 and 3.5 of [DG].

Notice that if a deformation G satisfies a filtration condition, then the folowing condition holds:

$$ver(A_{lp})\left(\frac{\partial G}{\partial t_i}\right)_{|t=0} \subset A_{lp}\left\langle x_j \frac{\partial g}{\partial x_j}\right\rangle, \quad 1 \leq i \leq r, \qquad (*)$$

We want to define a convex polyhedron $\Sigma_+(g) \in \mathbb{R}^n_+$, associated to the germ g and such that it is the largest convex set with the property that $\forall\ m \in \Sigma_+(g)$, x^m satisfies $(*)$.

For a fixed $m \in \mathbb{Z}^n_+$ consider the condition:
$(\mathcal{P})$ there exists a positive integer l so that

$$ver(A_{lp})x^m \subset A_{lp}\left\langle x_j \frac{\partial g}{\partial x_j}\right\rangle .$$

Definition 3.5 *The polyhedron $\Sigma_+(g)$ is the convex hull of the set*

$$\bigcup \{m + \mathbb{R}^n_+ \ :\ condition\ (\mathcal{P})\ holds\ \}$$

The next example illustrates the fact that monomial deformations cannot be replaced by polynomial deformations in the definition of $\Sigma_+(g)$. The computations can be done following the method of "local patching data" introduced in [DG]. But, one can also come back to this example after reading Proposition 4.4.

Example 3.6 *Let $g(x) = x^7 + y^7 + xy(x^2 - y^2)^2$.*

Applying Damon-Gaffney's methods, we see that the coordinates of the vertices of $\Sigma_+(g) \subset \mathbb{R}^2_+$ are $(7,0)$ and $(0,7)$.

On the other side, we can check that for small values of t, the deformation $G(x,t) = x^7 + y^7 + (1+t)xy(x^2 - y^2)^2$ satisfies the filtration condition, however, $xy(x^2-y^2)^2$ is not in $\Sigma_+(g)$.

The vertices of the polyhedron associated to all diretions θ such that the deformation $G(x,t) = g(x) + t\theta$ satisfies the filtration condition, are $(7,0)$, $(5,1)$, $(1,5)$ and $(0,7)$. The term $(5,1)$ is in this polyhedron, however the deformation given by $G(x,t) = g(x) + tx^5y$ is not topologically trivial.

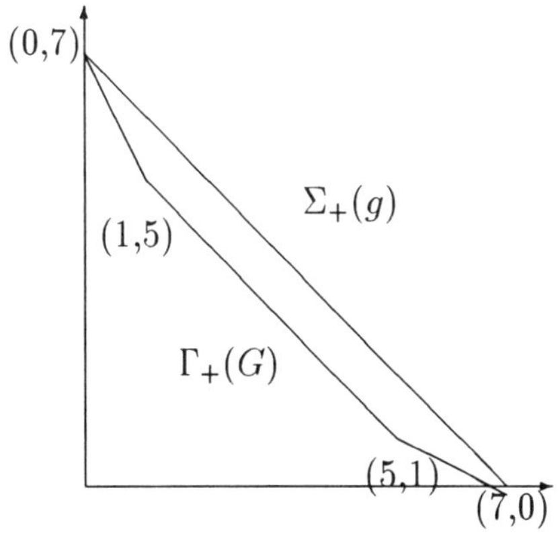

The approach of Yoshinaga [Y] to the question of topological triviality is also based on the method of controlled vector fields.

Let $| \, grad \; g|^2 \; = \; \sum_{i=1}^{n} \left| x_i \dfrac{\partial g}{\partial x_i} \right|^2$. For a fixed $m \; \in \; \mathbb{Z}_+^n$, consider the following condition:

($\mathcal{G}$) there exists a positive $\mathcal{E} \; = \; \mathcal{E}(m)$ such that $| \; grad \; g \; | \geq \; \mathcal{E}|x^m|$ in a neighbourhood of the origin in k^n.

Definition 3.7 (Y , pg 804) *The gradient polyhedron of* g, $\Lambda^+(g)$ *is the convex hull of the set*

$$\bigcup \{m + \mathbb{R}_+^n \; : \; condition \; (\mathcal{G}) \; holds\}$$

Theorem 3.8 (Y , pg. 805) (a) *If* $\Gamma_+(\theta) \; \subset \; int \; \Lambda^+(g)$, *then the family* $G(x,t) = g(x) + t\theta(x)$ *is topologically trivial for all* $t \; \in \; [0,1]$

(b) *If* $\Gamma_+(\theta) \; \subset \; \Lambda^+(g)$, *then* $G(x,t) = g(x) + t\theta(x)$ *is topologically trivial for sufficiently small values of* t.

To prove this theorem, Yoshinaga shows that the controled vector field defined by $X(x,t) = (-\overline{p(x,t)}, 1)$, if $x \neq 0$, and $X(0,t) = (0,1)$, where

$$p(x,t) = \bar{\theta} \; |Grad_x \; G|^{-2} \left(|x_1|^2 \frac{\partial G}{\partial x_1}, \ldots, |x_n|^2 \frac{\partial G}{\partial x_n}, 1 \right)$$

and $|Grad_x \; G|^2 = \sum_{i=1}^{n} \left| x_i \dfrac{\partial G}{\partial x_i} \right|^2$ is continuous and the result follows from the fact that the function $G(\varphi(u; x, t))$ is constant along each solution curve φ of the diferential equation $\left(\dfrac{dx}{du}, \dfrac{dt}{du} \right) = X(x,t)$.

Remark When the germ g is nondegenerate with respect to its Newton diagram, we find in both works similar results:

Proposition 3.9 (DG, Corollary 1) *Suppose that* $g \, : \, k^n, 0 \; \rightarrow \; k, 0$ *is a nondegenerate germ with an isolated singularity. Any deformation of* g *of non-decreasing Newton filtration* $fil(G) = fil(g)$ *is a topologically trivial deformation.*

Proposition 3.10 (Y, Theorem 1.7) $\Lambda^+(g) = \Gamma_+(g) \; \Leftrightarrow \; g$ *is nondegenerate with respect to its Newton diagram.*

4 Equivalence among the approaches to topological triviality and equisingularity

Let $G : k^n \times k, 0 \to k, 0$ denote a deformation of a commode germ $g : k^n, 0 \to k, 0$ and $X_G = G^{-1}(0)$. It follows from Teissier's equisingularity criterion (Th. 2.2) that a necessary condition for the equisingularity of the pair $\{X_G, 0 \times k\}$, is that $\frac{\partial G}{\partial t}_{|t=0} \in \left\langle \overline{x_i \frac{\partial g}{\partial x_j}_{i,j=1,\dots,n}} \right\rangle$.

As in section 3, we define:

Definition 4.1 *The polyhedron of equisingularity of a germ g, denoted by $\mathcal{E}(g)$ is the convex hull of the set*

$$\cup \left\{ m + \mathbb{R}^n_+, \text{ where } x^m \in \left\langle \overline{x_i \frac{\partial g}{\partial x_j}_{i,j=1,\dots,n}} \right\rangle \right\}.$$

Notice that Example 3.6 can also be quoted here to give an equisingular deformation $G(x,t) = g(x) + t\theta(x)$ by a direction θ such that $\Gamma_+(\theta) \not\subset \mathcal{E}(g)$.

However, when g is nondegenerate, the following result holds:

Proposition 4.2 *Let g be nondegenerate. Then, $\mathcal{E}(g)$ is the convex hull of the set*

$$\cup \left\{ \Gamma_+(\theta), \text{ where } \theta \in \left\langle \overline{x_i \frac{\partial g}{\partial x_j}_{i,j=1,\dots,n}} \right\rangle \right\}.$$

Proof: The proof is an easy consequence of Theorem 3.4 in [S]. In that paper the concept of Newton polyhedron, $\Gamma_+(I)$, of an ideal I is defined. The main result is that I is nondegenerate if and only if $\Gamma_+(I) = \Gamma_+(\bar{I})$.

We let $I = \left\langle x_i \dfrac{\partial g}{\partial x_j}_{i,j=1,\dots,n} \right\rangle$, and to obtain the result, we just observe that the ideal $\left\langle x_i \frac{\partial g}{\partial x_j} \right\rangle$ is nondegenerate whenever g is. □

A natural question is to identify the directions θ such that $\Gamma_+(\theta) \subset \mathcal{E}(g)$ and $G + t\theta$ is equisingular (of course, we mean that the pair $\{G^{-1}(0), 0 \times k\}$ is equisingular). We show that the polyhedra $\Sigma_+(g)$ and $\Lambda_+(g)$ arising respectively from Damon-Gaffney's and Yoshinaga's approaches define the same subset of $\mathcal{E}(g)$. Moreover, we also prove the equisingularity of any deformation $G(x,t) = g(x) + t\theta(x)$ by terms in this subset.

For any germ of function f, we denote by Jf, the ideal generated by $\left\{ x_i \frac{\partial f}{\partial x_i};\ i = 1, \dots, n \right\}$.

Definition 4.3 *We denote by $C(\overline{Jg})$ the convex hull of the set*

$$\bigcup \left\{ m + v, v \in \mathbb{R}^n_+ :\ x^m = x_1^{m_1} \cdot \ldots \cdot x_n^{m_n} \in \overline{Jg} \right\}.$$

Then we have:

Proposition 4.4 $C(\overline{Jg}) = \Lambda^+(g) = \Sigma^+(g)$.

Proof: A monomial $x^m \in \Lambda^+(g)$ if and only if it satisfies the condition $(\mathcal{G})$, *i.e.*, there exists a neighbourhood U of 0 such that $|\text{grad } g| \geq c|x^m|$, for some constant $c > 0$ and this condition is clearly equivalent to the condition (ii) of the Proposition 2.1. Thus, $\Lambda^+(g) = C(\overline{Jg})$.

To show the second equality, let $x^m \in \Sigma_+(g)$, then there exists a level commode A_l in C_n such that

$$ver[A_{2l}]\ x^m \subset A_{2l}\left\langle x_i \frac{\partial g}{\partial x_i}\right\rangle$$

Hence, we may write the localized equation:

$$-\rho \cdot x^m = \sum_i \xi_i(g)\ ,$$

where $\rho = \sum_\alpha |x^\alpha|^2$, x^α are all the vertices of A_l, and

$$\xi_i = \sum_j h_{i,j}\ x_j \frac{\partial}{\partial x_j}, \quad \text{with } fil\,(h_{i,j}) \geq 2l.$$

Then

$$\rho|x^m| \leq n\ \max_j |h_{i,j}|\ \max_j |x_j \frac{\partial g}{\partial x_j}|.$$

Now, since $fil\,(h_{i,j}) \geq 2l$, it follows from the Lemma 3.4 of [DG, pg 342], that there exist a constant $c > 0$ and a neighbourhood U of 0 in k^n such that $|h_{i,j}| \leq c_1 \rho$ for each $j = 1, \ldots, n,$. Hence,

$$\rho\ |x^m| \leq c\rho\ \max_\text{j} \left| x_j\ \frac{\partial g}{\partial x_j}\right|$$

Dividing both sides by ρ for $x \neq 0$, we obtain the following inequality which is also valid for $x = 0$

$$|\ x^m\ | \leq c\ \max_\text{j} \left| x_j\ \frac{\partial g}{\partial x_j}\right| \ \text{in } U\ .$$

Thus, by condition (ii) of Proposition 2.1 it follows that $x^m \in \overline{Jg}$ and $\sum^+(g) \subset C(\overline{Jg})$.

In order to prove the opposite side, let θ be a monomial in $\overline{Jg}$ with $fil\ \theta = q$. According to (ii) of the Proposition 2.1, there exists a neighbourhood U of 0 and a constant $C > 0$ such that

$$|\theta| \leq C\ sup_i \left| x_i \frac{\partial g}{\partial x_i}\right| \ for\ all\ x \in U.$$

Since g has isolated singularity at 0, the ideal Jg has finite codimension in C_n, therefore there exists an integer $l > 0$ such that the level $A_{lp} \subset Jg$. Hence for each vertex x^α of A_{lp}, we may write the equation.

$$x^\alpha = \sum_{i=1}^{n} h_{\alpha,i} x_i \frac{\partial g}{\partial x_i}$$

Thus, for each $h_{\alpha,i}$, we have

$$|h_{\alpha,i}| \; |\theta| \; \leq \; C \; sup_i \left| h_{\alpha,i} x_i \frac{\partial g}{\partial x_i} \right|$$

for all $x \in \; U$. Therefore, for each vertex x^α of A_{lp}, we may write the equation

$$x^\alpha \theta = \sum_{i=1}^{n} h_{\alpha,i} \; \theta \; x_i \frac{\partial g}{\partial x_i}.$$

with $fil\,(h_{\alpha,i}\theta) \geq lq$. Thus we conclude that all monomials of $\overline{Jg}$ satisfy the filtration condition $\mathcal{P}$ for g and $C(\overline{Jg}) \subset \Sigma_+(g)$. □

Proposition 4.5 *Let $G(x,t) = g(x) + t\theta(x)$ be a deformation of g such that $\Gamma_+(\theta) \subset C(\overline{Jg})$ (or equivalently, $\Gamma_+(\theta) \subset \Lambda_+(g)$, $\Gamma_+(\theta) \subset \Sigma_+(g)$). Then the pair $\{X_G, 0 \times k\}$ is Whitney equisingular.*

Proof: The argument is essentialy as in [Y, Lemma 2.1]. If $\Gamma_+(\theta) \subset C(\overline{J(g)})$, then $\Gamma_+(x_i \frac{\partial \theta}{\partial x_i} \subset C(\overline{J(g)})$ and this implies that

$$|t| \left| x_i \frac{\partial \theta}{\partial x_i} \right| \leq |t| \; c \; sup_i \left| x_i \frac{\partial g}{\partial x_i} \right| .$$

On the other hand

$$\left| x_i \frac{\partial G}{\partial x_i} \right| = \left| x_i \frac{\partial g}{\partial x_i} + t \cdot x_i \frac{\partial \theta}{\partial x_i} \right|$$

$$\geq \left| x_i \frac{\partial g}{\partial x_i} \right| - |t| \cdot \left| x_i \frac{\partial \theta}{\partial x_i} \right| \geq (1 - \alpha) \left| x_i \frac{\partial g}{\partial x_i} \right| ,$$

for some $0 < \alpha < 1$ and $|t| \leq \frac{\alpha}{c}$ and from this inequality it follows that $\left\langle \overline{x_i \frac{\partial g}{\partial x_i}} \right\rangle \subset \left\langle \overline{x_i \frac{\partial G}{\partial x_i}} \right\rangle$. The result now follows from the Theorem 2.2. □

Corollary 4.6 *Let $g : (\mathbb{C}^n, 0) \to (\mathbb{C}, 0)$ be a complex germ with isolated singularity at $x = 0$, and nondegenerate with respect to $\Gamma_+(g)$. Then $\Gamma_+(g) = \Gamma_+(g) \bigcap \mathcal{E}(g) = C(\overline{Jg})$.*

5 The polyhedron of equisingularity of germs of plane curves

Let $G(x_1, x_2) = g(x_1, x_2) + t\theta(x_1, x_2)$, be an equisingular deformation of a commode germ $g : (\mathbb{C}^2, 0) \to (\mathbb{C}, 0)$, defining a plane curve with isolated singularity at 0.

Proposition 5.1 *With the above hypothesis, if* $\theta(x_1, x_2) = x_1^{a_1} x_2^{a_2}$, *with* $a_i \neq 0$, $i = 1, 2$, *then* $\Gamma_+(\theta) \subset C(\overline{Jg}) \subset \Gamma_+(g)$.

Proof: Let us assume that

$$x_1^{a_1} x_2^{a_2} \notin \overline{\left\langle x_i \frac{\partial g}{\partial x_i} \right\rangle} \quad \text{and} \quad x_1^{a_1} x_2^{a_2} \in \overline{\left\langle x_i \frac{\partial g}{\partial x_j} \right\rangle}.$$

Then, according to the Theorem 2.2, item ii, there exists a curve $\phi : (\mathbb{C}, 0) \to (\mathbb{C}^2, 0)$ with

$$fil(x_1^{a_1} x_2^{a_2} \circ \phi) < fil(x_i \frac{\partial g}{\partial x_i} \circ \phi),\ i = 1, 2.$$

Hence the curve ϕ is written as $\phi(\lambda) = (c_1 \lambda^{r_1} + \phi_1,\ c_2 \lambda^{r_2} + \phi_2)$, with degree $(\phi_i) > r_i, c_i \neq 0$. Let us fix $r_1 \leq r_2$.

Since $x_1^{a_1} x_2^{a_2} \in \overline{\left\langle x_i \frac{\partial g}{\partial x_j} \right\rangle}$ and $r_1 \leq r_2$, we have:

$$(*) \qquad fil\left(x_1 \frac{\partial g}{\partial x_2} \circ \phi\right) \leq fil(x_1^{a_1} x_2^{a_2} \circ \phi) < fil\left(x_2 \frac{\partial g}{\partial x_2} \circ \phi\right)$$

Furthermore, g is commode and $c_i \neq 0$, $i = 1, 2$ hence

$$(**) \qquad fil\left(x_1 \frac{\partial g}{\partial x_2} \circ \phi\right) = fil\left(x_2 \frac{\partial g}{\partial x_2} \circ \phi\right) - (r_2 - r_1).$$

The conditions (*) and (**) imply that

$$fil\left(x_1 \frac{\partial x_1^{a_1} x_2^{a_2}}{\partial x_2} \circ \phi\right) < fil\left(x_1 \frac{\partial g}{\partial x_2} \circ \phi\right)$$

Then, we can write

$x_1 \frac{\partial g}{\partial x_2} \circ \phi = \sum_{i \geq i_0} a_i \lambda^i$, $a_{i_0} \neq 0$, $x_1 \frac{\partial x_1^{a_1} x_2^{a_2}}{\partial x_2} \circ \phi = \sum_{j \geq j_0} b_j \lambda^j$, $b_{j_0} \neq 0$ and $j_0 < i_0$.

Let $\psi : \mathbb{C}, 0 \to \mathbb{C}^2 \times \mathbb{C}, 0$ be defined by $\psi(\lambda) = (\phi(\lambda), t(\lambda))$, where $t(\lambda) = \lambda^{(i_0 - j_0)} S(\lambda). [\sum_i^q a_i \lambda^{i - i_0}]$ and $S(\lambda) = \left[\sum_j b_j\ \lambda^{j - j_0}\right]^{-1}$.

Then $fil(x_1^{a_1} x_2^{a_2} \circ \phi) = fil(x_1^{a_1} x_2^{a_2} \circ \psi)$ and $fil\left(x_1 \frac{\partial G}{\partial x_2} \circ \psi\right) = 0$, hence $x_1^{a_1} x_2^{a_2} \notin \overline{\left\langle x_i \frac{\partial G}{\partial x_j} \right\rangle}$ □

Thus, g does not have equisingular deformation by mixed terms below the Newton diagram of g.

As we can see in the example 5.2 this result does not hold in higher dimensions.

Example 5.2 *(from [A, ex.5.3]) The deformation* $G(x,y,z,t) = x^5 + y^6 + z^5 + y^3z^2 + tx^2y^2z$ *is equisingular and* $x^2y^2z \notin \Gamma_+(g)$.

The following example shows that even for complex plane curves, $\mathcal{E}(g)$ may contain $C(\overline{Jg})$ properly.

Example 5.3 $g(x_1, x_2) = x_1^a + x_1^4 x_2 + x_2^5$, $a > 5$.

The germ g is commode and $\Gamma_+(g)$ is nondegenerate but clearly, $x_1^5 \in \mathcal{E}(g)$ and $x_1^5 \notin C(\overline{Jg})$.

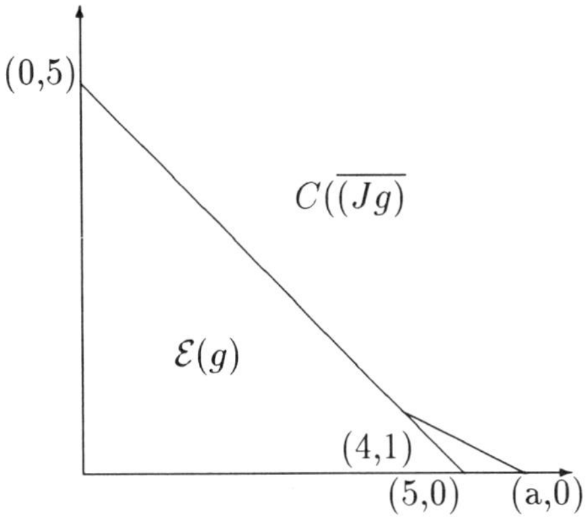

Notice that in this example, the vertices $(0,5)$ and $(4,1)$ define the polygon of the isolated singularity $f(x_1, x_2) = x_1^4 x_2 + x_2^5$, and the term x_1^a can be thought as a higher order term. We show in Proposition 5.4 that this is the general pattern for an equisingular deformation below the Newton boundary $\Gamma_+(g)$, when g is the germ of a plane curve.

We denote by $in(g)$ the part of the Taylor expansion of g whose terms have filtration equal to the filtration of g.

Proposition 5.4 *Let* g *be commode and nondegenerate and* $\theta(x_1, x_2) = x_1^c$, *with* $x_1^c \notin \Gamma_+(g)$. *Then* $G(x_1, x_2) = g(x_1, x_2) + t\theta(x_1, x_2)$ *is equisingular if and only if* $in(g) = x_1^a + f(x_1, x_2)$, *where*

$$f(x_1, x_2) = x_2^b + x_1 x_2^2 \lambda(x_1, x_2) + m x_1^d x_2, \ m \neq 0$$

defines an isolated singularity and $c \geq d + 1$.

Proof: Since g is commode, there are positive integers a and b, such that

$$in(g)(x_1, x_2) = x_1^a + x_2^b + x_1 x_2^2 \lambda(x_1, x_2) + m x_1^d x_2$$

Now, the equisingularity of the family of plane curves G is equivalent to the constancy of the Milnor number, i.e., $\mu(g_t) = \mu(g)$. Now, from Kouchnirenko's formula [K],

$$\mu(g) = \mu(in(g)) = 2S - (a+b) + 1,$$

where S is the area below $\Gamma_+(g)$.

Hence, $\mu(g_t) = \mu(g)$ if and only if $2(S - S_t) = a - c$. Since G cannot have mixed terms below $\Gamma_+(g)$, we must have $m \neq 0$.

□

References

[A] ALTMANN, K.,*Equisingular deformations below the Newton boundary*, Compositio Mathematica **80** 257–263 (1991).

[DG] DAMON, J.N., and GAFFNEY, T.,*Topological triviality of deformations of functions and Newton filtrations*, Inv. Math. **v.72** 335–358, (1983).

[G] GAFFNEY, T.,*Integral closure of modules and Whitney-equisingularity*, Inv. Math.**v.107** 301–322, (1992)

[K] KOUCHNIRENKO, A.,*Polyèdres de Newton et nombres de Milnor*, Inv. Math.**v.32** 1–31 (1976).

[S] SAIA, M. J.,*Integral Closure of Ideals and the Newton Filtration*. To appear in Journal of Algebraic Geometry.

[T1] TEISSIER, B.,*Multiplicities polaires, sections planes, et conditions de Whitney*, La Rabida Springer Lecture Notes #966 314–491 (1983).

[T2] ———,*Introduction to equisingularity problems*, Proc. A.M.S. Symp. in Pure Math. #29 Arcata (1974).

[T3] ———,*Variétés Polaires I. Invariants polaires des singularité d'hypersurfaces*, Inventiones Mathematicae. **40**, 267–292 (1977).

[Y] YOSHINAGA, E.,*Topological Principal part of analytic Functions*, Trans. of the A.M.S. Vol. 314, #2, (1989).

J. J. Nuño Ballesteros[1] & O. Saeki[2]

Singular surfaces in 3-manifolds, the tangent developable of a space curve and the dual of an immersed surface in 3-space

Abstract

In the first part of the paper, we prove that for a generic closed space curve, the number of pyramids (triple points of the tangent developable) is congruent modulo 2 to the sum of the indices of the torsion zero points. This index is defined as the number of trisecant lines of the curve passing through the torsion zero point. The result is deduced from the study of the singularities of the tangent developable surface of the curve. In the second part, we use analogous techniques to give a new proof of a theorem by Ozawa. This theorem says that if the Euler characteristic of a generic immersed surface in $\mathbf{R}^3$ is even, then the number of tritangent planes is congruent modulo 2 to the sum of the indices of the godron points (cusps of the Gauss map of the surface). In this case, the result is deduced from the study of the singularities of the dual surface.

1 Introduction

In this paper we study the number of triple points of two types of singular surfaces in 3-manifolds: the tangent developable of a space curve and the dual surface of an immersed surface in $\mathbf{R}^3$.

In the first case, the tangent developable of a space curve $\alpha : S^1 \to \mathbf{R}^3$ is the surface $\chi(\alpha)$ in $\mathbf{R}^3$ defined by the tangent lines of α. The local form of this surface was first studied by Cleave [**C**], and recently the first author [**N**] proved that it is a topologically stable surface when the curve is generic. He obtained as a consequence that if the curve has no torsion zero points (topological cross caps of $\chi(\alpha)$), then the number of pyramids of the curve (triple points of $\chi(\alpha)$) is even.

Here, we give one step more in this direction. The above mentioned result can be easily generalized by applying a theorem by Szücs [**Sz**], which gives the following congruence:

$$T(\alpha) \equiv \sum_{i=1}^{k} n(x_i, \alpha) \bmod 2.$$

[1]Partially supported by DGICYT Grant PB93–0707.

[2]Partially supported by CNPq, Brazil.

The number $T(\alpha)$ is the number of pyramids of α, $x_1, \cdots, x_k$ are the torsion zero points of α and $n(x_i, \alpha)$ is the index of each torsion zero point conveniently defined (see also [**NS**]). The problem is that this index, $n(x_i, \alpha)$, does not give a priori any information on the geometry of the curve. We will show, by using a new proof of the Szücs theorem given by the authors in [**NS**], that the index $n(x_i, \alpha)$ can be interpreted geometrically as the number of trisecant lines of α passing through x_i.

Note that this result can be considered dual to the congruence obtained by Banchoff, Gaffney and McCrory in [$\mathbf{BGM_2}$], where they show that for a generic space curve $\alpha : S^1 \to \mathbf{R}^3$ we have

$$\tilde{T}(\alpha) \equiv \sum_{i=1}^{k} \tilde{n}(x_i, \alpha) \bmod 2,$$

where now $\tilde{T}(\alpha)$ is the number of tritangent planes of α and the index $\tilde{n}(x_i, \alpha)$ is one half of the number of points in the intersection off the point x_i of the curve with the osculating plane at the torsion zero point x_i. Although their paper [$\mathbf{BGM_2}$] is previous to the Szücs one, they implicitly use the Szücs result for the dual surface of the curve. Remember that the dual surface of a regular space curve is the surface in $(\mathbf{R}P^3)^*$ defined by the tangent planes to α. When the curve is generic, the dual surface is again a topologically stable surface, the triple points corresponding to the tritangent planes to α and the cross caps to the torsion zero points (see [$\mathbf{BGM_2}$] for details and [**Sch**] for the duality between the dual surface and the tangent developable of a generic space curve).

In the second part of the paper, we show that the same techniques can be applied to the dual surface of an immersed surface $\varphi : M^2 \to \mathbf{R}^3$. Here the dual surface is defined as the map $\varphi^* : M \to (\mathbf{R}P^3)^*$ given by $\varphi^*(x) =$ the tangent plane to $\varphi(M)$ at $\varphi(x)$, for each point $x \in M$. Following Bruce [**Br**] we know that the dual surface is again a topologically stable map in the generic case. Then we apply the Szücs result and get a new proof of a theorem by Ozawa [**O**], which asserts that if the Euler characteristic of M is even,

$$T(\varphi) \equiv \sum_{i=1}^{k} \mathrm{Ind}(x_i) \bmod 2,$$

where $T(\varphi)$ is the number of tritangent planes of φ (or triple points of the dual surface), $x_1, \cdots, x_k$ are the godron points of φ (or topological cross caps of φ^*), and $\mathrm{Ind}(x_i)$ is the index of each godron point defined in the following way: remember that a godron point $p \in M$ is a point such that the tangent plane at $\varphi(p)$ has a contact of cusp type with the surface $\varphi(M)$, that is, the corresponding height function germ h is equivalent to the function germ $\pm x_1^2 + x_2^4$ at p; then $\mathrm{Ind}(p)$ is equal to the Euler characteristic of the set $M(p) = \{x \in M : h(x) > 0\}$.

2 The number of pyramids of a generic space curve

Suppose that $\alpha : S^1 \to \mathbf{R}^3$ is a smooth space curve satisfying the general position conditions (GP) given in [**N**]. We also assume the conditions (1) and (2) in [**N**, Lemma 7]:

(1) If the tangent lines at two distinct points of the curve are parallel, the osculating planes at these points are not parallel.

(2) There are no three distinct points such that their tangent lines are parallel.

Furthermore, we assume the additional conditions as follows:

(i) If $\tau(s) = 0$, then there are no quadrisecants of α passing through $\alpha(s)$, where τ is the torsion of α.

(ii) If there is a trisecant to α at $\alpha(s_0), \alpha(s_1), \alpha(s_2)$ with $\tau(s_0) = 0$, then the vectors $\alpha(s_1) - \alpha(s_0), \alpha'(s_1)$ and $\alpha'(s_2)$ are linearly independent.

(iii) The tritangent planes of α do not osculate at the tangency points.

(iv) The number of osculating planes of α containing a trisecant is finite.

(v) The subset $\mathcal{T}$ of $S^1 \times S^1 \times S^1 - \Delta$ defined by

$$\mathcal{T} = \{(s_1, s_2, s_3) : (\alpha(s_2) - \alpha(s_1)) \times (\alpha(s_3) - \alpha(s_1)) = 0\},$$

is a closed 1-dimensional submanifold of $S^1 \times S^1 \times S^1 - \Delta$, where

$$\Delta = \{(s_1, s_2, s_3) : s_i = s_j \text{ for some } i \neq j\}.$$

Lemma 2.1 *The set of smooth curves α satisfying the above general position conditions is residual in the space $C^\infty(S^1, \mathbf{R}^3)$, with the Whitney C^∞-topology.*

Proof. We consider the multijet space ${}_4J^3(S^1, \mathbf{R}^3)$ and take the coordinates

$$(s_1, s_2, s_3, s_4, r_1^0, r_2^0, r_3^0, r_4^0, \ldots, r_1^3, r_2^3, r_3^3, r_4^3),$$

where $s_i \in S^1$ and $r_j^i \in \mathbf{R}^3$. Then we define the subset $W_1 \subset {}_4J^3(S^1, \mathbf{R}^3)$ by the equations:

$$(r_2^0 - r_1^0) \times (r_3^0 - r_1^0) = (r_2^0 - r_1^0) \times (r_4^0 - r_1^0) = 0, \quad \det(r_1^1, r_1^2, r_1^3) = 0.$$

Clearly, W_1 is an algebraic subset of ${}_4J^3(S^1, \mathbf{R}^3)$ of codimension 5. By the multijet version of the Thom transversality theorem, the set of curves $\alpha \in C^\infty(S^1, \mathbf{R}^3)$ such that ${}_4j^3\alpha : (S^1)^{(4)} \to {}_4J^3(S^1, \mathbf{R}^3)$ is transversal to W_1 is residual in $C^\infty(S^1, \mathbf{R}^3)$. But since $(S^1)^{(4)}$ has dimension 4, it is obvious that the transversality condition is equivalent to condition (i).

Analogously, we see that the rest of conditions (ii),...,(v) give residual subsets in $C^\infty(S^1, \mathbf{R}^3)$. □

Lemma 2.2 *Let $p = \alpha(s_0)$ be a torsion zero point. If α satisfies the above general position conditions, then the number of trisecants of α passing through p is finite.*

The proof of Lemma 2.2 will be given later (see the paragraph just after the proof of Proposition 2.8).

Definition 2.3 For a torsion zero point $p = \alpha(s_0)$ of α, we define the *index* $n(p, \alpha)$ ($\in \mathbf{Z}$) to be the number of trisecants of α passing through p.

Recall that if α satisfies the condition (GP) as in [**N**] (condition 3), then the number of torsion zero points is finite.

The main purpose of this section is to prove the following.

Theorem 2.4 *Let $\alpha : S^1 \to \mathbf{R}^3$ be a smooth space curve satisfying the general position conditions as stated above. Then we have*

$$T(\alpha) \equiv \sum_{i=1}^{k} n(x_i, \alpha) \ mod\ 2,$$

where $T(\alpha)$ is the number of pyramids of α and $x_1, \cdots, x_k$ are the torsion zero points of α.

Lemma 2.5 *Let $f : (s_0 - \epsilon, s_0 + \epsilon) \to \mathbf{R}$ be a C^∞ function. Then there exists another C^∞ function $f_* : (s_0 - \epsilon, s_0 + \epsilon) \to \mathbf{R}$ such that if $s \neq s_0$, then*

$$f_*(s) = \frac{f(s) - f(s_0)}{s - s_0},$$

$$\frac{df_*}{ds}(s) = \frac{f'(s)(s - s_0) - (f(s) - f(s_0))}{(s - s_0)^2},$$

and if $s = s_0$,

$$f_*(s_0) = f'(s_0), \ \frac{df_*}{ds}(s_0) = \frac{1}{2} f''(s_0).$$

Proof. Just apply a classical argument of analysis. We see easily that

$$\begin{aligned} f(s) - f(s_0) &= \int_0^1 \frac{d}{dt}\{f(s_0 + t(s - s_0))\}dt \\ &= \int_0^1 f'(s_0 + t(s - s_0))(s - s_0)dt \\ &= (s - s_0)f_*(s), \end{aligned}$$

where $f_*(s) = \int_0^1 f'(s_0 + t(s - s_0))dt$. The rest is easy to check. □

Now let $\alpha : S^1 \to \mathbf{R}^3$ be a C^∞ curve which is regular and simple. Given a point $p \in \mathbf{R}^3$ we define $t_p : S^1 \to \mathbf{R}P^2$ by $t_p(s) = [\alpha(s) - p]$ if $\alpha(s) \neq p$, and $t_p(s) = [\alpha'(s)]$ if $\alpha(s) = p$. Note that t_p is a smooth map.

Lemma 2.6 *Suppose that $p = \alpha(s_0)$. Then t_p is an immersion at s_0 if and only if $\kappa(s_0) \neq 0$, where κ is the curvature of α.*

Proof. Suppose for instance that $\alpha_3'(s_0) \neq 0$, where $(\alpha_1(t), \alpha_2(t), \alpha_3(t)) = \alpha(t)$. We take coordinates in $\mathbf{R}P^2$ such that the homogeneous coordinate $[X, Y, Z]$ corresponds to $(X/Z, Y/Z)$. Then t_p gives the map $\tilde{t}_p$ in a neighbourhood of s_0 given by

$$\tilde{t}_p(s) = \left(\frac{\alpha_1(s) - \alpha_1(s_0)}{\alpha_3(s) - \alpha_3(s_0)}, \frac{\alpha_2(s) - \alpha_2(s_0)}{\alpha_3(s) - \alpha_3(s_0)}\right),$$

when $s \neq s_0$ and

$$\tilde{t}_p(s) = \left(\frac{\alpha_1'(s_0)}{\alpha_3'(s_0)}, \frac{\alpha_2'(s_0)}{\alpha_3'(s_0)}\right),$$

when $s = s_0$. By Lemma 2.5, this map is

$$\tilde{t}_p(s) = \left(\frac{\alpha_{1*}(s)}{\alpha_{3*}(s)}, \frac{\alpha_{2*}(s)}{\alpha_{3*}(s)}\right),$$

for s in a neighbourhood of s_0. Then $\tilde{t}_p$ is differentiable at s_0 and

$$\tilde{t}_p'(s_0) = \frac{1}{2\alpha_3'(s_0)^2}(\alpha_1''(s_0)\alpha_3'(s_0) - \alpha_1'(s_0)\alpha_3''(s_0), \alpha_2''(s_0)\alpha_3'(s_0) - \alpha_2'(s_0)\alpha_3''(s_0)).$$

Therefore $\tilde{t}_p'(s_0) = 0$ if and only if $\alpha'(s_0) \times \alpha''(s_0) = 0$; i.e., if and only if $\kappa(s_0) = 0$. □

Lemma 2.7 *Suppose that $p = \alpha(s_0)$. Then t_p is an immersion at $s \neq s_0$ if and only if $\alpha'(s), \alpha(s) - \alpha(s_0)$ are not collinear.*

Proof. Taking coordinates as in the proof of Lemma 2.6, we have

$$\tilde{t}_p(s) = \left(\frac{\alpha_1(s) - \alpha_1(s_0)}{\alpha_3(s) - \alpha_3(s_0)}, \frac{\alpha_2(s) - \alpha_2(s_0)}{\alpha_3(s) - \alpha_3(s_0)}\right).$$

Hence we have that $\tilde{t}_p'(s) = 0$ if and only if $\alpha'(s) \times (\alpha(s) - \alpha(s_0)) = 0$. □

Proposition 2.8 *If α is a smooth curve satisfying the general position conditions and $\tau(s_0) = 0$, then t_p is an immersion with normal crossings for $p = \alpha(s_0)$.*

Proof. Since $\kappa(s_0) > 0$ and there are no cross tangents passing through $\alpha(s_0)$ (see the condition 7 of [**N**]), t_p is an immersion.

On the other hand, a self-intersection of t_p happens when $t_p(s_1) = t_p(s_2)$ for $s_1 \neq s_2$. Again the fact that there are no cross tangents passing through $\alpha(s_0)$ implies that $s_1, s_2 \neq s_0$; therefore $[\alpha(s_1) - \alpha(s_0)] = [\alpha(s_2) - \alpha(s_0)]$, i.e., there is a trisecant to α passing through $\alpha(s_0), \alpha(s_1), \alpha(s_2)$. By our condition (i) t_p has no triple points. We prove that condition (ii) implies the normal crossing condition at a double point, that is, that $t'_p(s_1)$ and $t'_p(s_2)$ are not collinear.

Since $\alpha(s_1)-\alpha(s_0), \alpha(s_2)-\alpha(s_0)$ are collinear, we can choose a coordinate which is not zero for both vectors, for instance, $\alpha_3(s_i)-\alpha_3(s_0)\neq 0$, $i=1,2$. Then taking coordinates as in the proof of Lemma 2.6, t_p gives the map

$$\tilde{t}_p(s)=\left(\frac{\alpha_1(s)-\alpha_1(s_0)}{\alpha_3(s)-\alpha_3(s_0)}, \frac{\alpha_2(s)-\alpha_2(s_0)}{\alpha_3(s)-\alpha_3(s_0)}\right),$$

for $s\neq s_0$ in a neighbourhood of s_1, s_2 and hence

$$\tilde{t}'_p(s_i)=\frac{1}{(\alpha_3(s_i)-\alpha_3(s_0))^2}(\alpha'_1(s_i)(\alpha_3(s_i)-\alpha_3(s_0))-(\alpha_1(s_i)-\alpha_1(s_0))(\alpha'_3(s_i)),$$

$$\alpha'_2(s_i)(\alpha_3(s_i)-\alpha_3(s_0))-(\alpha_2(s_i)-\alpha_2(s_0))(\alpha'_3(s_i))).$$

Now, if we set

$$\alpha(s_1)-\alpha(s_0)=(a_{11},a_{12},a_{13}),$$
$$\alpha(s_2)-\alpha(s_0)=\lambda(\alpha(s_1)-\alpha(s_0))=\lambda(a_{11},a_{12},a_{13}),$$
$$\alpha'(s_1)=(a_{21},a_{22},a_{23}),$$
$$\alpha'(s_2)=(a_{31},a_{32},a_{33}),$$

then an easy (but tedious) computation gives

$$\det(\tilde{t}'_p(s_1),\tilde{t}'_p(s_2))=\frac{1}{a_{13}^3\lambda}\det(a_{ij})\neq 0.$$

This completes the proof. □

Note that Lemma 2.2 is a direct consequence of Proposition 2.8, since $n(\alpha(s_0),\alpha)$ is equal to the number of double points of the immersion with normal crossings $t_p: S^1\to \mathbf{R}P^2$.

Let $\alpha: S^1\to\mathbf{R}^3$ be a smooth space curve which satisfies our general position conditions as above. We define $\chi_\alpha: S^1\times\mathbf{R}\to\mathbf{R}^3$ by $\chi_\alpha(s,t)=\alpha(s)+t\alpha'(s)$. We denote by $\chi(\alpha)$ the tangent developable of α; i.e., $\chi(\alpha)=\chi_\alpha(S^1\times\mathbf{R})$. Furthermore, we define the smooth map $\Phi: S^1\times S^1\times\mathbf{R}\to\mathbf{R}^3$ by

$$\Phi(s_1,s_2,t)=\frac{\alpha(s_1)+\alpha(s_2)}{2}+t\int_0^1\alpha'(s_1+\tau(s_2-s_1))d\tau.$$

Note that, if $s_1\neq s_2$, we have

$$\Phi(s_1,s_2,t)=\frac{\alpha(s_1)+\alpha(s_2)}{2}+t\frac{\alpha(s_2)-\alpha(s_1)}{s_2-s_1}$$

and that, if $s_1 = s_2$, we have

$$\Phi(s_1, s_1, t) = \alpha(s_1) + t\alpha'(s_1).$$

Thus, for a point $q \in \mathbf{R}^3 - \chi(\alpha)$, t_q has a double point at s_1 and s_2 if and only if $\Phi(s_1, s_2, t) = q$ for some t.

Set $\mathcal{B}(\subset S^1 \times S^1)$ to be the bitangency set of α defined in [**NR**]. We know that $\mathcal{B}$ is a closed 1-dimensional submanifold of $S^1 \times S^1$. Define Ω to be $\Phi(\mathcal{B} \times \mathbf{R})$. Furthermore we define $T(\subset \mathbf{R}^3)$ to be the union of the trisecants of α. We see that T is of dimension 2 by our assumption (v). Note that Ω is also 2-dimensional and that the complements $\mathbf{R}^3 - \Omega$ and $\mathbf{R}^3 - T$ are open and dense in $\mathbf{R}^3$.

The following lemma can be proved by the same argument as in the proof of Proposition 2.8.

Lemma 2.9 *For* $p \in \mathbf{R}^3 - (\chi(\alpha) \cup \Omega \cup T)$ *the map* $t_p : S^1 \to \mathbf{R}P^2$ *is an immersion with normal crossings.*

Lemma 2.10 *Suppose that* $p, p' \in \mathbf{R}^3 - (\chi(\alpha) \cup \Omega \cup T)$ *and that they are in the same connected component of* $\mathbf{R}^3 - \chi(\alpha)$. *Then the number of double points of* t_p *has the same parity as that of* $t_{p'}$.

The above lemma is obvious, since t_p and $t_{p'} : S^1 \to \mathbf{R}P^2$ are regularly homotopic and the parity of the number of double points is an invariant of the regular homotopy class.

Remark. There are exactly four regular homotopy classes of immersions $f : S^1 \to \mathbf{R}P^2$, which are characterized by:
1) the $\mathbf{Z}_2$ class of f in $\pi_1(\mathbf{R}P^2)$;
2) the parity of the number of double points of $\tilde{f}$, where $\tilde{f}$ is an immersion with normal crossings that approximates f. Note that, for $p \in \alpha(S^1)$, t_p represents the nontrivial class of $\pi_1(\mathbf{R}P^2)$, while for $p \notin \alpha(S^1)$, the trivial class of $\pi_1(\mathbf{R}P^2)$.

Definition 2.11 Let C be a connected component of $\mathbf{R}^3 - \chi(\alpha)$. Take a point $p \in C - (\Omega \cup T)$, which is non-vacuous. We say that C is a *blue region* (resp. *red region*) if the number of double points of t_p is odd (resp. even). Note that this does not depend on the choice of the point p by Lemma 2.10. Furthermore, define B (resp. R) to be the union of all blue (resp. red) regions of $\mathbf{R}^3 - \chi(\alpha)$.

Lemma 2.12 $\bar{B} \cap \bar{R} = \partial B = \partial R = \chi(\alpha)$.

For the proof of Lemma 2.12, we need the following lemmas.

Lemma 2.13 $\chi(\alpha) - \Omega$ *is dense in* $\chi(\alpha)$.

Proof. Take a point $p \in \chi(\alpha) \cap \Omega$. We have only to show that there exists a point $q \in \chi(\alpha) - \Omega$ arbitrarily close to p. We may assume that p is a simple regular point of $\chi(\alpha)$. Suppose $p \in \chi(\alpha) \cap \Omega$. Note that $p = \Phi(s, s; t)$ for some $s \in S^1$ and $t \in \mathbf{R}^3$; in other words, $p = \chi_\alpha(s, t)$. First suppose that $(s, s) \in \mathcal{B}$. Then s is a torsion-zero point of α. Since α has only finitely many torsion-zero points, such a point p should lie in a 1-dimensional subspace of $\chi(\alpha)$. Thus we may assume that $p = \Phi(s_1, s_2; t')$ for some $(s_1, s_2) \in \mathcal{B}$ $(s_1 \neq s_2)$ and some $t' \in \mathbf{R}$. It is easily checked that the tangent space of Ω at p is spanned by $\{\alpha(s_2) - \alpha(s_1), \alpha'(s_i)\}$, where $i = 1$ or 2. On the other hand, the tangent space of $\chi(\alpha)$ at p is spanned by $\{\alpha'(s), \alpha''(s)\}$. Thus, if $\Phi|(\mathcal{B} \times \mathbf{R})$ and χ_α are not transverse at p, then there exists a plane P tangent to α at s_1, s_2 and s which osculates at $\alpha(s)$. Suppose that $s = s_i$. Then P is a bitangent osculating plane of α. Since there are only finitely many pairs $(t_1, t_2) \in \mathcal{B}$ $(t_1 \neq t_2)$ corresponding to bitangent osculating planes ([**NR**]), the point $p = \Phi(s_1, s_2; t')$ is in a 1-dimensional set. Hence, we may assume that s, s_1 and s_2 are all distinct. This contradicts to the condition (iii). Thus $\Phi|(\mathcal{B} \times \mathbf{R})$ and χ_α are transverse at p. Hence the intersection of Ω and $\chi(\alpha)$ at p is of 1-dimension. This completes the proof. □

Lemma 2.14 *$\chi(\alpha) - T$ is dense in $\chi(\alpha)$.*

Proof. Take a point p in $\chi(\alpha) \cap T$. We will find a point $q \in \chi(\alpha) - T$ which is arbitrarily close to p. We may assume that p is a simple regular point of $\chi(\alpha)$. Define

$$\Theta : \mathcal{T} \times \mathbf{R} \to \mathbf{R}^3$$

by

$$\Theta(s_1, s_2, s_3; t) = t\frac{\alpha(s_2) - \alpha(s_1)}{s_2 - s_1} \left(= t \int_0^1 \alpha'(s_1 + \sigma(s_2 - s_1)) d\sigma \right).$$

Note that Θ is a smooth map of a 2-dimensional manifold (see our condition (v)) and that $\Theta(\mathcal{T}) = T$.

Let V be the union of the tangent lines of α at the points where the osculating plane contains a trisecant. By our condition (iv), it is a finite union of lines. Thus we may assume that $p \in \chi(\alpha) - V$.

Now suppose that, for a point $(s_1, s_2, s_3; t) \in \mathcal{T} \times \mathbf{R}$, $p = \Theta(s_1, s_2, s_3; t)$ and that Θ is *not* transverse to $\chi(\alpha)$ at $(s_1, s_2, s_3; t)$. We assume that $p = \chi_\alpha(s', t')$ $(s' \in S^1, t' \neq 0)$. Then we see that the osculating plane of α at $\alpha(s')$ contains a trisecant. This contradicts to the fact that $p \notin V$. Hence, Θ is transverse to χ_α at p. Thus the intersection of $\Theta(\mathcal{T})$ and $\chi(\alpha)$ is of 1-dimension at p. This completes the proof. □

Proof of Lemma 2.12 Take a simple regular point $p \in \chi(\alpha)$ of the tangent developable. We have only to show that $p \in \bar{B} \cap \bar{R}$. By Lemmas 2.13 and 2.14, we may assume that $p \in \chi(\alpha) - (\Omega \cup T)$. Suppose $p = \alpha(s_0) + r_0\alpha'(s_0)$ $(s_0 \in S^1, r_0 \in \mathbf{R} - \{0\})$. Note that s_0 is not a torsion-zero point, since $p \notin \Omega$. By the proof of Proposition 2.8, the map $t_p : S^1 \to \mathbf{R}P^2$ is an immersion with normal crossings off s_0. If t_p has a double point at s_0, there exists a cross tangent passing through $\alpha(s_0)$. Since there are only finitely many cross tangents (see [**NR**]), we may assume that $t_p(s_0)$ is not a double point of t_p, changing p if necessary.

We may assume that $\alpha(s_0) = (0,0,0)$, $\alpha'(s_0) = (1,0,0)$, that the (x_1, x_2)-plane osculates at $\alpha(s_0)$, that $(\alpha'(s_0) \times \alpha''(s_0)) \cdot (0,0,1) > 0$ and that the torsion of α at s_0 is positive. Then we have $\alpha_2''(s_0) > 0$ and $\alpha_3'''(s_0) > 0$. Hence, for some small positive number θ, $\alpha(s) \in \{x_1 < 0, x_3 < 0\}$ for $s_0 - \theta < s < s_0$ and $\alpha(s) \in \{x_1 > 0, x_3 > 0\}$ for $s_0 < s < s_0 + \theta$, where we identify a neighborhood of s_0 in S^1 with an interval in $\mathbf{R}$.

For simplicity, we assume $r_0 > 0$. Take a point $q = p + (0,0,q_3) = (r_0, 0, q_3)$ close to p with $q_3 \neq 0$. Note that t_q is an immersion with normal crossings, since $q \notin \chi(\alpha) \cup \Omega \cup T$. Furthermore, the number of double points of t_q is equal to that of t_p off a small neighborhood of s_0. Recall that t_p has no double points in a neighborhood of s_0. Now we consider the number of double points of t_q in the neighborhood of s_0. We will prove that if $q_3 < 0$, then t_q does not have any double points in the neighborhood of s_0, while if $q_3 > 0$, t_q has exactly one double point in the neighborhood of s_0.

For a small open disk neighborhood D of p with $D \cap (\Omega \cup T) = \emptyset$, $D - \chi(\alpha)$ has exactly two connected components D_+ and D_-, where D_+ is the region which contains a point q with $(q-p) \cdot \alpha'''(s_0) > 0$ (i.e., $q_3 > 0$).

Lemma 2.15 *If D is sufficiently small, t_q has exactly one double point in a neighborhood of s_0 if $q \in D_+$, and it has no double point in the neighborhood if $q \in D_-$.*

Proof. We identify a neighborhood of $s_0 \in S^1$ with $(-\varepsilon, \varepsilon)$, s_0 being identified with 0, and we set $W = \{(s_1, s_2, t) \in (-\varepsilon, \varepsilon) \times (-\varepsilon, \varepsilon) \times (r_0 - \delta, r_0 + \delta) : s_1 > s_2\}$ $(\delta > 0)$. Recall that the image of $W' = \{(s_1, s_2, t) : s_1 = s_2\}$ by Φ is precisely the tangent developable of α. We have only to show that Φ maps W injectively onto D_+ for some ε and δ. For this, we calculate the differential of Φ at $u = (0,0,r_0)$. As a basis of $T_u(S^1 \times S^1 \times \mathbf{R})$, we take $\{(\partial/\partial s_1) + (\partial/\partial s_2), (\partial/\partial s_1) - (\partial/\partial s_2), (\partial/\partial t)\}$. Note that $\{(\partial/\partial s_1) + (\partial/\partial s_2), (\partial/\partial t)\}$ constitutes a basis of $T_u W'$ and that $(\partial/\partial s_1) - (\partial/\partial s_2)$ is the direction normal to W' in $S^1 \times S^1 \times \mathbf{R}$ toward W. Then we have

$$\begin{aligned} d\Phi_u((\partial/\partial s_1) + (\partial/\partial s_2)) &= \left.\frac{\partial}{\partial s}\Phi(s,s,t)\right|_{(s,s,t)=u} \\ &= \left.\frac{\partial}{\partial s}(\alpha(s) + t\alpha'(s))\right|_{(s,s,t)=u} \\ &= \alpha'(s_0) + r_0\alpha''(s_0), \end{aligned}$$

$$\begin{aligned} d\Phi_u(\partial/\partial t) &= \left.\frac{\partial}{\partial t}\Phi(s_1,s_2,t)\right|_{(s_1,s_2,t)=u} \\ &= \left.\left(\int_0^1 \alpha'(s_1 + \tau(s_2 - s_1))d\tau\right)\right|_{(s_1,s_2,t)=u} = \alpha'(s_0), \end{aligned}$$

and

$$
\begin{aligned}
d\Phi_u&((\partial/\partial s_1) - (\partial/\partial s_2)) \\
&= \frac{\partial}{\partial s}\Phi(s, -s, t)\Big|_{(s,s,t)=u} \\
&= \frac{\partial}{\partial s}\left(\frac{\alpha(s) + \alpha(-s)}{2} + t\int_0^1 \alpha'(s - 2s\tau)d\tau\right)\Big|_{(s,-s,t)=u} \\
&= \left(\frac{\alpha'(s) - \alpha'(-s)}{2} + t\int_0^1 (1 - 2\tau)\alpha''(s - 2s\tau)d\tau\right)\Big|_{(s,-s,t)=u} = 0.
\end{aligned}
$$

Thus the rank of $d\Phi_u$ is equal to 2. Note also that $d\Phi_u(T_u(S^1 \times S^1 \times \mathbf{R})) = d(\Phi|W')_u(T_uW')$ and that $\Phi|W'$ is a local diffeomorphism around u onto an open neighborhood of p in $\chi(\alpha)$. Furthermore, we have

$$
\begin{aligned}
\frac{\partial^2}{\partial s^2}&\Phi(s, -s, t)\Big|_{(s,s,t)=u} \\
&= \frac{\partial}{\partial s}\left(\frac{\alpha'(s) - \alpha'(-s)}{2} + t\int_0^1 (1 - 2\tau)\alpha''(s - 2s\tau)d\tau\right)\Big|_{(s,-s,t)=u} \\
&= \left(\frac{\alpha''(s) + \alpha''(-s)}{2} + t\int_0^1 (1 - 2\tau)^2\alpha'''(s - 2s\tau)d\tau\right)\Big|_{(s,-s,t)=u} \\
&= \alpha''(s_0) + \frac{r_0}{3}\alpha'''(s_0).
\end{aligned}
$$

Thus, if we take the derivative of second order along $(\partial/\partial s_1) - (\partial/\partial s_2)$, it does not lie on the image of $d\Phi_u$. Hence we can change the system of coordinates of $\mathbf{R}^3$ around p in a homeomorphic way so that $d(\Phi|W \cup W')_u$ has rank 3 with respect to this new C^0 coordinates. Hence, Φ maps W injectively onto D_+ for some small W. (In other words, Φ has a fold singularity (or $\Sigma_{1,0}$-singularity) along W' in a neighborhood of u, the discriminant set being the tangent developable. See [**Mo**].) This completes the proof of Lemma 2.15. □

Thus we see that p is in the closure of both B and R. This completes the proof of Lemma 2.12. □

Remark. As a digression, it would be interesting to study the behavior of the map Φ. For a generic curve, is Φ stable? What are the singularities and the critical values? Is it related to the theory of self-translation surfaces [**MNR**]?

By the above lemma, the decomposition $\mathbf{R}^3 - \chi(\alpha) = B \cup R$ coincides with the decomposition guaranteed by the 2-color theorem ([**NS**, Lemma 2.1]) applied to the topologically stable map $\chi_\alpha : S^1 \times \mathbf{R} \to \mathbf{R}^3$. (More precisely, we have to compactify the map as is done in [**N**] in order to apply the result of [**NS**] and then we restrict to

$S^1 \times \mathbf{R}$ and $\mathbf{R}^3$.) Recall that, in [**NS**], we have made the convention that the index $n(p, \chi(\alpha))$ of a cross cap point p of $\chi(\alpha)$ with respect to the 2-color theorem is defined to be 1 if the outside region of $\mathbf{R}^3 - \chi(\alpha)$ in a neighborhood of p is red and 0 if it is blue. Note also that the torsion-zero points of α coincides exactly to the cross cap points of $\chi(\alpha)$ ([**N**]).

Proposition 2.16 *The two definitions of the index of a torsion-zero point p — the index with respect to the number of trisecants passing through p, $n(p, \alpha)$, and that with respect to the 2-color theorem, $n(p, \chi(\alpha))$ — coincide with each other modulo 2.*

Proof. Let $p = \alpha(s_0)$ ($s_0 \in S^1$) be a torsion-zero point. Take a point $q \in \mathbf{R}^3 - \chi(\alpha)$ close to p. We may assume that $q \notin \chi(\alpha) \cup \Omega \cup T$. Then the map t_q and t_p are immersions with normal crossings. Although t_q and t_p are not even homotopic, yet they have the same number of double points off a neighborhood J of s_0. This is because the maps $t_p|(S^1 - J)$ and $t_q|(S^1 - J)$ are sufficiently "close" to each other as immersions into $\mathbf{R}P^2$ (note that t_p has no double points in J). By Lemma 2.10, the parity of the number of double points of t_q in the neighborhood of s_0 depends only on the region to which q belongs; more precisely it is odd when q is in the "outside region" of the cross-cap point p of $\chi(\alpha)$, and it is even when q is in the "inside region". This is because we can take an appropriate point q in the "outside region" such that t_q has exactly one double point in the neighborhood J of s_0. In fact, given a pair of bitangent points $\alpha(s_1), \alpha(s_2)$ in J, the segment joining $\alpha(s_1), \alpha(s_2)$ is included in the local convex hull of the curve, which lies in the "outside region" of the cross-cap point. Thus, for any point q in this segment, t_q has exactly one double point in J (for example, see Figure 4 of [**C**]). Now consider the case where the index $n(p, \alpha)$ with respect to trisecants is odd. Then the outside region is the red region and the inner region is the blue region. Hence the index $n(p, \chi(\alpha))$ of p with respect to the 2-color theorem is equal to 1 and coincides with the index with respect to trisecants. The other case is similar. Hence the index with respect to the number of trisecants coincides with the index with respect to the 2-color theorem modulo 2. This completes the proof. □

Now our Theorem 2.4 follows from [**NS**] (or [**Sz**]) and Proposition 2.16, by using the compactification $\tilde{\chi}_\alpha : S^1 \times \mathbf{R}^* \to \mathbf{R}P^3$ of χ_α as in [**N**]. Note that $(\tilde{\chi}_\alpha)_*[S^1 \times \mathbf{R}^*] \in H_2(\mathbf{R}P^3; \mathbf{Z}_2)$ vanishes, since $S^1 \times \mathbf{R}^* \cong S^1 \times S^1$ is orientable, where $[S^1 \times \mathbf{R}^*] \in H_2(S^1 \times \mathbf{R}^*; \mathbf{Z}_2)$ is the fundamental class.

Remark. Theorem 2.4 implies that the number of pyramids of a generic space curve is congruent modulo 2 to the number of trisecants passing through the torsion zero points, where such trisecants are counted with multiplicities which are defined to be the number of torsion zero points they pass through. We note that we could add the condition that a trisecant passes through at most one torsion zero point in the general position conditions from the beginning. This condition is generic (i.e., even after adding this condition, we have Lemma 2.1), and then we have that the number of pyramids is congruent modulo 2 to the number (counted without multiplicities) of trisecants passing through the torsion zero points.

Corollary 2.17 *Let $\alpha : S^1 \to \mathbf{R}^3$ be a smooth curve satisfying the general position conditions, such that the number of pyramids $T(\alpha)$ is odd. Then α has at least two torsion zero points.*

A special case of Theorem 2.4 is when the curve α is convex. By using a result by Sedykh [**Se**], we can look at the structure of the convex envelope of a generic curve and deduce that a convex generic curve has no trisecants. Then we get the following immediate consequence.

Corollary 2.18 *Let $\alpha : S^1 \to \mathbf{R}^3$ be a convex curve satisfying our general position conditions and the general position conditions given in* [**Se**]. *Then the number of pyramids of α is even.*

3 The number of tritangent planes of an immersed surface in $\mathbf{R}^3$

Let M be a smooth closed surface and $\varphi : M \to \mathbf{R}^3$ a smooth immersion. We denote by $\varphi^* : M \to (\mathbf{R}P^3)^*$ the dual surface of M; i.e., $(\mathbf{R}P^3)^*$ is the dual projective space, which is the natural compactification of the set of all planes in $\mathbf{R}^3$, and φ^* is the map given by $\varphi^*(x) =$ the tangent plane of $\varphi(M)$ at $\varphi(x)$, for each $x \in M$.

The following lemma is due to Bruce [**Br**].

Lemma 3.1 *For a residual subset in the space of smooth immersions of M into $\mathbf{R}^3$, the dual surface is topologically stable.*

In the following, we assume that $\varphi : M \to \mathbf{R}^3$ is such a generic smooth immersion.

Lemma 3.2 *$(\varphi^*)_*[M] = 0$ in $H_2((\mathbf{R}P^3)^*; \mathbf{Z}_2)$ if and only if the Euler characteristic $Eul(M)$ of M is even, where $[M] \in H_2(M; \mathbf{Z}_2)$ is the fundamental class of M.*

Proof. Since M is compact, we may assume that $\varphi(M)$ is contained in the space $\{(x, y, z) \in \mathbf{R}^3 : x < 0, |z| < a\}$ for some $a > 0$ and that $h = \pi \circ \varphi : M \to \mathbf{R}$ is a Morse function, where $\pi : \mathbf{R}^3 \to \mathbf{R}$ is the height function defined by $\pi(x, y, z) = z$. Suppose that M has an even Euler characteristic. Then the number of critical points of h is even. This shows that the number of planes which are tangent to $\varphi(M)$ and which are perpendicular to the z-axis is even. Let $\alpha : S^1 \to (\mathbf{R}P^3)^*$ be the piecewise smooth curve which corresponds to the planes $z = t$ ($|t| \leq a$), $-x \sin\theta + (z + a)\cos\theta = 0$ ($0 \leq \theta \leq \pi/2$) and $x \cos\theta + (z - a)\sin\theta = 0$ ($0 \leq \theta \leq \pi/2$). Note that $\alpha(S^1)$ represents the generator of $H_1((\mathbf{R}P^3)^*; \mathbf{Z}_2) \cong \mathbf{Z}_2$, since the orientation of the plane is reversed when we traverse α once around. Furthermore, it is not difficult to show that α

intersects $\varphi^*(M)$ transversely. Then the above argument shows that the intersection number modulo 2 of $\alpha(S^1)$ and $\varphi^*(M)$ in $(\mathbf{R}P^3)^*$ is zero. By the ring structure of $H^*((\mathbf{R}P^3)^*; \mathbf{Z}_2)$, we see that $[\varphi^*(M)]$ is zero in $H_2((\mathbf{R}P^3)^*; \mathbf{Z}_2)$. Note that this argument shows that if the Euler characteristic of M is odd, then $[\varphi^*(M)]$ is not zero in $H_2((\mathbf{R}P^3)^*; \mathbf{Z}_2)$. This completes the proof. □

Suppose that $\mathrm{Eul}(M)$ is even. By the above lemma, we can apply [**NS**, Lemma 2.1], which shows that $(\mathbf{R}P^3)^* - \varphi^*(M) = R \cup B$, where R and B are disjoint nonempty open sets with common boundary $\varphi^*(M)$. In fact, the proof of the above lemma suggests the following observation. Let T be a plane in $\mathbf{R}^3$ which is not tangent to $\varphi(M)$. For simplicity we assume that $T = \pi^{-1}(c)$ for some $c \in \mathbf{R}$. If $T \cap \varphi(M) \neq \emptyset$, then T divides M (not $\varphi(M)$) into two pieces M_+ and M_-, both of which are compact surfaces with boundary, where $M_+ = \{x \in M : \pi \circ \varphi(x) \geq c\}$ and $M_- = \{\pi \circ \varphi(x) \leq c\}$. Define R (or B) to be the subset of $(\mathbf{R}P^3)^* - \varphi^*(M)$ which consists of the planes for which the Euler characteristic of one of the two pieces $M_\pm$ above is odd (resp. even). Note that since the Euler characteristic of M is even, the parity of the Euler characteristic of the piece does not depend on the choice of the piece and that these two sets R and B serve exactly as the open sets mentioned above.

Then, for each cross cap $p \in M$ of φ^*, we define the index $n(p, \varphi^*) \in \{0, 1\}$ as in [**NS**, Definition 2.2]. More precisely, $n(p, \varphi^*) = 1$ if the inside components of $\mathbf{R}P^3 - \varphi^*(M)$ in a neighborhood of $\varphi^*(p)$ are blue and $n(p, \varphi^*) = 0$ if they are red.

Recall that the cross cap points of φ^* correspond exactly to the godron points (i.e., the cusp points of the Gauss mapping) of $\varphi(M)$ (see, for example, [**BGM$_1$**]). Ozawa [**O**] defined the index $\mathrm{Ind}(p) \in \mathbf{Z}$ of a godron point $p \in M$ as the Euler characteristic $\mathrm{Eul}(M(p))$ of $M(p)$, where $M(p) = \{x \in M : h(x) > 0\}$, $h(x) = (\varphi(x) - \varphi(p)) \cdot \nu$, and ν is a unit vector at $\varphi(p)$ normal to $\varphi(M)$ such that h is equivalent to the function germ $\pm x_1^2 + x_2^4$ at p. Recall also that a godron point p is called *elliptic* if h is equivalent to $x_1^2 + x_2^4$ at p and *hyperbolic* if $-x_1^2 + x_2^4$ (for details, see [**O**]).

Proposition 3.3 *For each godron point $p \in M$ of φ, we have*

$$n(p, \varphi^*) \equiv \mathit{Ind}(p) \quad \mathit{mod}\ 2.$$

Proof. First suppose that $p \in M$ is an elliptic godron point. Take a unit vector ν at $\varphi(p)$ normal to $\varphi(M)$ as mentioned above. Let $T = h^{-1}(0)$ be the plane tangent to $\varphi(M)$ at $\varphi(p)$. Recall that T corresponds to a cross cap point of the dual surface φ^* of φ and that it is not a double point of φ^*, since φ^* is a topologically stable map. Hence T is transverse to $\varphi(M - \{p\})$. Take a bitangent plane $T' \in (\mathbf{R}P^3)^*$ of $\varphi(M)$ which is close to T; in other words, T is the cross cap of φ^* corresponding to the godron point p of φ and T' is a double point of φ^* close to the cross cap. Note that there are 4 regions of $(\mathbf{R}P^3)^* - \varphi^*(M)$ in a neighborhood of T'. The four planes T_i $(i = 1, 2, 3, 4)$ corresponding to the four regions cut $\varphi(M)$ in a neighborhood of $\varphi(p)$ as in Figure 1.

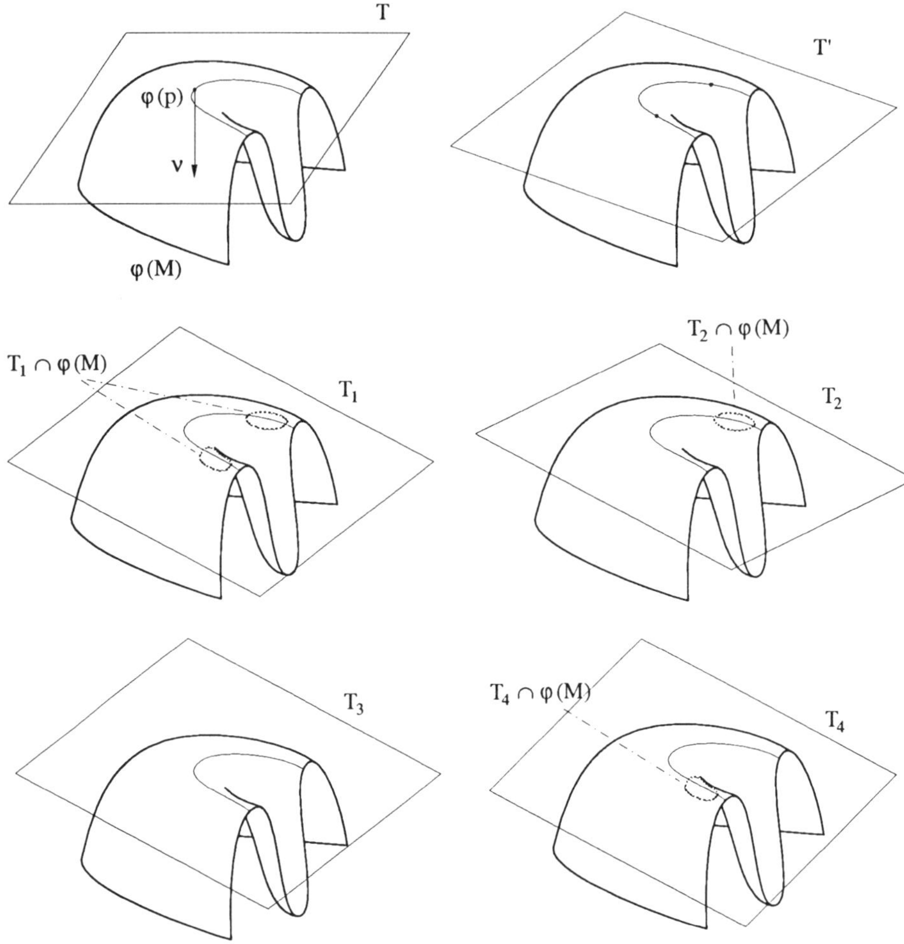

Figure 1

Note that, away from the neighborhood of $\varphi(p)$, T_i cuts $\varphi(M)$ in a way similar to T, since T is transverse to $\varphi(M-\{p\})$ and T_i is close to T. Then we see that T_2 and T_4 can be connected by a path in $(\mathbf{R}P^3)^* - \varphi^*(M)$ in a neighborhood of $T = \varphi^*(p)$. This shows that T_1 and T_3 are in the inside components as mentioned in [**NS**, Figure 2]. Hence the index $n(p, \varphi^*)$ is equal to 1 if and only if $T_1, T_3 \in B$. Recall that $T_1 \in B$ if and only if the Euler characteristic of one of the cut pieces of M by T_1 is even. Since the Euler characteristic of the cut piece by T_1 is equal to the Euler characteristic of $M(p) = \{x \in M : h(x) > 0\}$ plus one modulo 2, we see that $n(p, \varphi^*) = 1$ if and only if the Euler characteristic $\mathrm{Eul}(M(p)) = \mathrm{Ind}(p)$ of $M(p)$ is odd. Hence we have $n(p, \varphi^*) \equiv \mathrm{Ind}(p) \bmod 2$.

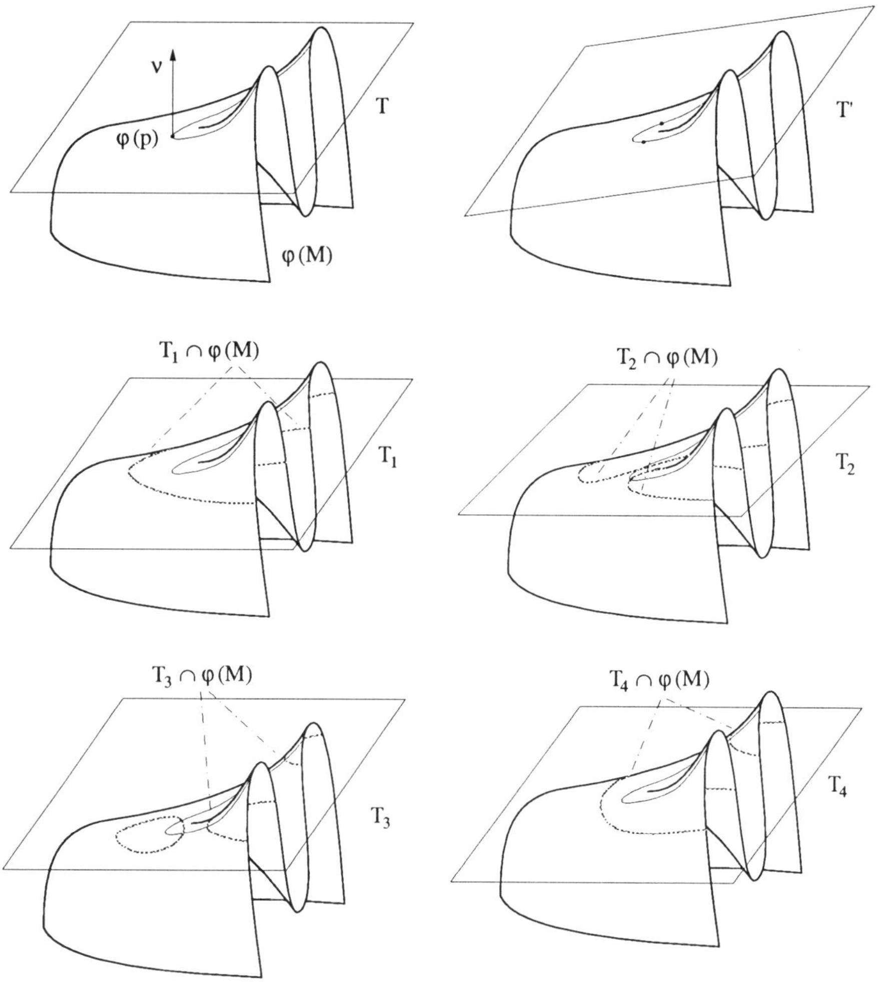

Figure 2

The case where p is a hyperbolic godron point is treated similarly (see Figure 2). This completes the proof. □

Now the following theorem is an easy consequence of [**NS**, Theorem 1.1], Lemma 3.2 and Proposition 3.3 above.

Theorem 3.4 (Ozawa [O]) *Let $\varphi : M \to \mathbf{R}^3$ be a generic smooth immersion of a closed surface M into $\mathbf{R}^3$. Suppose that the Euler characteristic $Eul(M)$ of M is even. Then the number of tritangent planes of $\varphi(M)$ is congruent modulo 2 to the sum $\sum_x Ind(x)$ of the indices of all godron points $x \in M$ of φ.*

References

[$\mathbf{BGM_1}$] T. Banchoff, T. Gaffney and C. McCrory, "Cusps of Gauss mappings," Research Notes in Math. no.55, Pitman Advanced Publ. Program, Boston, London, Melbourne, 1982.

[$\mathbf{BGM_2}$] T. Banchoff, T. Gaffney and C. McCrory, *Counting tritangent planes of space curves,* Topology **24** (1985), 15-23.

[**Br**] J. W. Bruce, *The duals of generic hypersurfaces,* Math. Scand. **49** (1981), 36–60.

[**C**] J. Cleave, *The form of the tangent developable at points of zero torsion on space curves,* Math. Proc. Camb. Phil. Soc. **88** (1980), 403–407.

[**MNR**] C. McCrory, J. J. Nuño Ballesteros and M. C. Romero Fuster, *Self-translation surfaces,* Matemática Contemporânea **5** (1993), 61–76.

[**Mo**] B. Morin, *Formes canoniques des singularités d'une application différentiable,* C. R. Acad. Sci. Paris **260** (1965), 5662–5665, 6503–6506.

[**N**] J. J. Nuño Ballesteros, *On the number of triple points of the tangent developable,* Geom. Dedicata **47** (1993), 241–254.

[**NR**] J. J. Nuño Ballesteros and M. C. Romero Fuster, *Global bitangency properties of generic space curves,* Math. Proc. Camb. Phil. Soc. **112** (1992), 519–526.

[**NS**] J. J. Nuño Ballesteros and O. Saeki, *On the number of singularities of a stable surface with boundary in a 3-manifold,* preprint (1994).

[**O**] T. Ozawa, *On the number of tritangencies of a surface in* $\mathbf{R}^3$, in "Global Differential Geometry and Global Analysis 1984," Edited by D. Ferus, R. B. Gardner, S. Helgason and U. Simon, Lecture Notes in Math. no.1156, Springer-Verlag, Berlin, Heidelberg, New York, Tokyo, 1985 pp. 240–253.

[**Sch**] O. P. Scherback, *Projectively dual space curves and Legendre singularities,* Sel. Math. Sov. **5** (no. 4) (1986), 391–421.

[**Se**] V. D. Sedykh, *The structure of the convex envelope of a space curve,* Trudy Sem. Petrovsk. **6** (1981), 239–256.

[**Sz**] A. Szücs, *Surfaces in* $\mathbf{R}^3$, Bull. London Math. Soc. **18** (1986), 60–66.

D. Mond & D. Ratcliffe

An infinite number of counter-examples to a conjecture on stems

1 Introduction

The notion of a *stem* in singularity theory was introduced in an attempt to characterise those branches in the classification tree from which there is only one subsequent ramification. That is, the notion of stem is intended as an analogue of the notion of *genus* in biology.

Definition 1 *Let $\mathcal{G}$ be any one of the groups $\mathcal{R}, \mathcal{L}, \mathcal{C}, \mathcal{K}$ or $\mathcal{A}$ (or any other) and let $f : \mathbb{C}^n, S \to \mathbb{C}^p, 0$ be an analytic map-germ. Then f is a $\mathcal{G}$-stem if*

1. *it is not finitely determined, and*
2. *there exists an integer k such that if $g : \mathbb{C}^n, 0 \to \mathbb{C}^p, 0$ is analytic and $j^k g = j^k f$ then either g is finitely $\mathcal{G}$-determined or g is $\mathcal{G}$-equivalent to f. The lowest k for which this holds is the* order *of the stem.*

Thus, a stem is a non-finitely determined germ which is finitely determined within the class of non-finitely determined germs.

For example, the function $f : \mathbb{C}^2, 0 \to \mathbb{C}, 0$ defined by $f(x, y) = x^2 y$ is an $\mathcal{R}$-stem. One can almost prove this by inspection of the classification tree: for each integer $k > 4$, if g is a function with $j^{k-1} g = j^{k-1} f$, then either $j^k g$ is equivalent to the (sufficient) k-jet $x^2 y + y^k$, and g is of type D_{k+1}, or $j^k g$ is equivalent to $j^k f$. In fact what this observation shows is that if $j^4 g = j^4 f$ and g is not finitely determined, then for every $k \geq 4$, $j^k g$ is equivalent to $j^k f$. Germs f with this property are known as *weak stems*; in [7], Theorem 2.1, the second author used Artin approximation to show that in all reasonable circumstances a weak stem is indeed a stem.

The example we have given suggests a relation between stems and series, and indeed all single-index series are associated to a stem, which can be thought of as the branch on which all the members of the series are twigs, or as the limit of the series. In [6], the first paper published on this topic, Ruud Pellikaan showed that a germ $f : \mathbb{C}^n, 0 \to \mathbb{C}, 0$ is an $\mathcal{R}$-stem if and only if its singular locus Σ is an irreducible curve, and at any point of $\Sigma - 0$, the restriction of f to a smooth transverse slice to Σ is $\mathcal{R}$-equivalent to $x_1^2 + \cdots + x_{n-1}^2$. That is, along its singular locus f is a trivial deformation of an $\mathcal{R}_e$-codimension 1 singularity.

Motivated by Pellikaan's result, C.T.C. Wall remarked to the first author that it was natural to conjecture that a map-germ $f : \mathbb{C}^2, 0 \to \mathbb{C}^3, 0$ should be an $\mathcal{A}$-stem if and only if its unstable locus (in $\mathbb{C}^3, 0$) is an irreducible curve along which f is a trivial unfolding of one of the $\mathcal{A}_e$-codimension 1 singularities of maps $\mathbb{C} \to \mathbb{C}^2$: the natural parametrisations of the first-order cusp A_2, the tacnode A_3 and the plane curve germ consisting of three concurrent lines meeting two-by-two transversely, D_4. This conjecture was borne out rather nicely by the examples known at that time: the first three series of singularities to appear in the classification ([4]) of singularities of maps from surfaces to 3-space were

$$\begin{array}{ll} S_k & (x,y) \mapsto (x, y^2, y^3 + x^{k+1}y) \\ B_k & (x,y) \mapsto (x, y^2, x^2y + y^{2k+1}) \\ H_k & (x,y) \mapsto (x, y^3, xy + y^{3k-1}) \end{array}$$

and it is easy to check that the non-finitely determined germs

$$\begin{array}{ll} S_\infty & (x,y) \mapsto (x, y^2, y^3) \\ B_\infty & (x,y) \mapsto (x, y^2, x^2y) \\ H_\infty & (x,y) \mapsto (x, y^3, xy) \end{array}$$

which give rise to these series (or can be viewed as their limits) do indeed have irreducible (even smooth) unstable loci, along which the transverse section of the image is A_2, A_3 and D_4 respectively.

In [7], the second author proved one half of this conjecture, for weighted homogenenous germs:

Theorem 1 *Suppose that $f : \mathbb{C}^2, 0 \to \mathbb{C}^3, 0$ is weighted homogeneous, and that the unstable locus Σ_f is an irreducible curve, a long which the image of f has transversal type A_2, A_3 or D_4. Then f is an $\mathcal{A}$- stem.* □

This was a corollary of a much more general theorem, which is the natural generalisation (for weighted homogenenous germs) of Wall's original conjecture:

Theorem 2 *Suppose that $f : \mathbb{C}^n, 0 \to \mathbb{C}^p, 0$ is weighted homogenenous, that $n > 1, p > 2$, and that the unstable locus Σ of f is an irreducible curve. If $\ell(\theta(f)/T\mathcal{A}_e f)_\Sigma) = 1$ then f is an $\mathcal{A}$-stem.* □

Here $(\theta(f)/T\mathcal{A}_e f)_\Sigma$ is the localisation of the $\mathcal{O}_{\mathbb{C}^p,0}$-module $\theta(f)/T\mathcal{A}_e f$ at the prime ideal defining the curve Σ, and ℓ is its length over $(\mathcal{O}_{\mathbb{C}^3,0})_\Sigma$. However, she also showed with an example ([7, 3.8]) that the "only if" part of the conjecture is false:

Example 1 *Let $f(x,y) = (x, xy + xy^3, y^4)$. Then*

1. *f is an $\mathcal{A}$- stem of order 5.*
2. *The unstable locus of f is the third coordinate axis in the target, along which the transverse section to the image consists of four concurrent lines meeting two by two transversely.*

The nature of the example was mysterious; the proof that it was a stem was purely classification-theoretic, and involved listing the equivalence classes of map-germs having the same 5-jet as f. In this paper we give a geometrical explanation, and in the process we find an infinite series of counterexamples to the original conjecture (Theorem 3 and Corollary 4 below). Clearly, the structure of classification for $\mathcal{A}$-equivalence is considerably more complicated than it is for $\mathcal{R}$-equivalence.

2 Examples and calculations

From now on in this paper we will be concerned exclusively with $\mathcal{A}$ -stems and we shall drop reference to the group.

Define $f_n : (\mathbb{C}^2, 0) \to (\mathbb{C}^3, 0)$ by $f_n(x, y) = (x, xy, y^n)$. In this section we will look at the geometry of map-germs whose k-jet is equal to $j^k f_n$ for some k, n.

Recall that a map-germ $(\mathbb{C}^2, 0) \to (\mathbb{C}^3, 0)$ is $\mathcal{A}$-finite if and only if its image is non-singular off the origin except for curves of transverse self-intersection [5].

The image of f_n is non-singular off the third coordinate axis. However each point $(0, 0, Z)$ on this axis has n preimages, the points $(0, y)$ with $y^n = Z$; f_n maps the y-axis n to 1 onto the Z-axis. A section of the image transverse to the Z-axis consists of n concurrent lines meeting two by two transversely. Thus, f_n is not $\mathcal{A}$-finite for $n \geq 3$. If $j^k f = j^k f_n$ then f may or may not have an n to 1 curve. For example, if $n = pq$ and $f(x, y) = (x, xy + y^{rn+p}, y^n)$ for some $r \in \mathbb{N}$, then f is p to 1 on the image of the y-axis (and will have other curves of self-intersection). If $p \geq 3$ then f is not $\mathcal{A}$-finite. But there is no such choice of p if $n = 4$ or n is prime. In fact we shall prove:

Theorem 3 *If n is prime or $n = 4$, and $k \geq 2n - 3$ then every map-germ f satisfying $j^k f = j^k f_n$ is either $\mathcal{A}$-finite or has an n-to-one curve.*

As a consequence, we have

Corollary 4 *For each integer $s > 0$, the map-germ f_s defined by*

$$f_s(x, y) = (x, xy + xy^{4s+3}, y^4)$$

is a stem, and therefore a counterexample to the conjecture, since it has a curve of 4-to-one points.

Our method of proof will be to consider the triple point scheme of f as defined in [3]: suppose that $g : (\mathbb{C}^2, 0) \to (\mathbb{C}^3, 0)$ has corank 1, so that in appropriate coordinates it has the form $g(x, y) = (x, g_2(x, y), g_3(x, y))$. Then since points with the same image must have the same x-coordinate, for each k its (non-reduced) k-tuple point scheme D^k can be embedded in $\mathbb{C} \times \mathbb{C}^k$. Taking coordinates x, y, y_1, y_2 on $\mathbb{C} \times \mathbb{C}^3$, $D^3(g)$ is then defined by the ideal generated by

$$\left|\begin{array}{lll} 1 & y & g_2(x,y) \\ 1 & y_1 & g_2(x,y_1) \\ 1 & y_2 & g_2(x,y_2) \end{array}\right| \Big/ A, \qquad \left|\begin{array}{lll} 1 & y & g_3(x,y) \\ 1 & y_1 & g_3(x,y_1) \\ 1 & y_2 & g_3(x,y_2) \end{array}\right| \Big/ A,$$

$$\left|\begin{array}{lll} 1 & g_2(x,y) & y^2 \\ 1 & g_2(x,y_1) & y_1^2 \\ 1 & g_2(x,y_2) & y_2^2 \end{array}\right| \Big/ A, \qquad \left|\begin{array}{lll} 1 & g_3(x,y) & y^2 \\ 1 & g_3(x,y_1) & y_1^2 \\ 1 & g_3(x,y_2) & y_2^2 \end{array}\right| \Big/ A,$$

where A is the Vandermode determinant

$$A(x,y,y_1,y_2) = \left|\begin{array}{lll} 1 & y & y^2 \\ 1 & y_1 & y_1^2 \\ 1 & y_2 & y_2^2 \end{array}\right| = (y-y_1)(y_2-y)(y_1-y_2).$$

Denote these generators by $v_1(g_i)(x,y,y_1,y_2), v_2(g_i)(x,y,y_1,y_2)$ respectively, for $i = 1,2$. We need to simplify the f of Theorem 3 before calculating the generators for $\mathcal{I}^3 f$.

Lemma 5 *For any $n \geq 3$, if $k \geq 2n-3$ and f is such that $j^k f = (x, xy, y^n)$ then f is $\mathcal{A}$-equivalent to a map-germ of the form*

$$(x,y) \mapsto (x, xy + P(x,y), y^n),$$

where $P \in y(m_{\mathbb{C}^2,0})^2$.

Proof. As f has corank 1 it is certainly $\mathcal{A}$-equivalent to a map-germ f' of the form $f'(x,y) = (x, xy+f_2, y^n+f_3)$, where $f_2(x,0) = f_3(x,0) = 0$. We can rewrite f' as

$$f'(x,y) = (x, (1+\alpha(x))xy + y^2 f_2', xy\beta(x) + y^2 f_3' + y^n).$$

Then the diffeomorphism $\Phi_1 : (\mathbb{C}^3,0) \to (\mathbb{C}^3,0)$ defined by

$$\Phi_1(X,Y,Z) = (X, Y, Z - \beta(X)Y/(1+\alpha(X)))$$

transforms f' to

$$f''(x,y) = (x, (1+\alpha(x))xy + y^2 f_2, y^2 f_3'' + y^n).$$

If $k \geq 3$ then we can write f'' as

$$f''(x,y) = (x, (1+\alpha(x))xy + y^2 f_2', x^2y^2\beta'(x) + y^3 f_3^{(3)} + y^n).$$

Then the diffeomorphism $\Phi_2 : (\mathbb{C}^3,0) \to (\mathbb{C}^3,0)$ defined by

$$\Phi_2(X,Y,Z) = (X, Y, Z - \beta'(X)Y^2/(1+\alpha(X))^2$$

transforms f'' to

$$f'''(x,y) = (x, (1+\alpha(x))xy + y^2 f_2, y^3 f_3^{(4)} + y^n).$$

If $k \geq 5$ then we can write f''' as

$$f'''(x,y) = (x, (1+\alpha(x))xy + y^2 f_2', x^3 y^3 \beta''(x) + y^4 f_3^{(5)} + y^n).$$

Continuing in this way, provided that $k \geq 2n-3$, we see that f is $\mathcal{A}$-equivalent to a map-germ of the form

$$(x,y) \mapsto (x, xy + f_2, y^n(1+\gamma(x,y)))$$

for some function-germ $\gamma \in (m_{\mathbb{C}^2,0})_2$. The coordinate change $x' = x$, $y' = y(1+\gamma)^{1/n}$ now brings this to the form

$$(x', y') \mapsto (x', x'y' + P(x', y'), y'^n)$$

as required. □

NB The final map-germ might no longer have f_n as its k-jet, but this is not relevant to what follows.

Lemma 6 *If $f(x,y) = (x, xy + P(x,y), y^n)$ for some $P \in y(m_{\mathbb{C}^2,0})^2$ then f is $\mathcal{A}$-finite if and only if $D^3 f$ has dimension 0.*

Proof. The map-germ f is $\mathcal{A}$-finite if and only if $D^2 f$ is an ICIS of dimension 1 and $D^3 f$ is of dimension 0 [3, 2.14].

Now $D^2 f \subseteq \mathbb{C}^3$ is defined by means of the ideal $\mathcal{I}^2 f$ generated by

$$\frac{xy + P(x,y) - xy_1 - P(x,y_1)}{y - y_1} \text{ and } \frac{y^n - y_1^n}{y - y_1},$$

where $\mathbb{C}^3$ has coordinates (x, y, y_1). The first generator is

$$x + \frac{P(x,y) - P(x,y_1)}{y - y_1}.$$

Let $x' = x + \frac{P(x,y)-P(x,y_1)}{y-y_1}$, $y' = y$ and $y_1' = y_1$. Then $\mathcal{I}^2 f$ becomes $(x', \prod_{i=1}^{n-1}(y' - \zeta^i y_1'))$, where ζ is a primitive nth root of unity. Thus $D^2 f$ is always an ICIS (and $\mu(D^2 f) = (n-2)^2$). This proves the lemma. □

Now that we have a simpler expression for f, we shall calculate $\mathcal{I}^3 f$. We find that

$$v_1(y^n)(x, y, y_1, y_2) = \frac{y^n(y_2 - y_1) + y_1^n(y - y_2) + y_2^n(y_1 - y)}{A},$$

$$v_2(y^n)(x,y,y_1,y_2) = \frac{y^n(y_2^2-y_1^2)+y_1^n(y^2-y_2^2)+y_2^n(y_1^2-y^2)}{A}.$$

Subtracting $(y_2+y_1)v_1(y^n)(x,y,y_1,y_2)$ from $v_2(y^n)(x,y,y_1,y_2)$ and rearranging, we obtain

$$(y_1^n - y_2^n)/(y_2-y_1),$$

and similarly

$$(y^n - y_1^n)/(y-y_1) \in \mathcal{I}^3 f.$$

Define $P_{r,s} \subseteq \mathbb{C}^4$ by

$$P_{r,s} = \{(x,y,y_1,y_2) : y_1 = e^{2\pi i r/n}y, y_2 = e^{2\pi i s/n}y\}.$$

Then $D^3 f \subseteq \bigcup_{0<r<s<n} P_{r,s}$, and any irreducible component of $D^3 f$ lies in one of the $P_{r,s}$. Thus we have proved

Lemma 7 *The map-germ $f(x,y) = (x, xy+P(x,y), y^n)$ is $\mathcal{A}$-finite if and only if for every pair r and s satisfying $0<r<s<n$, $D^3 f \cap P_{r,s}$ has dimension* 0. □

Write $g(x,y) = xy + P(x,y)$. We have shown that it is enough to look at the ideals

$$(y_1 - \alpha_r y, y_2 - \alpha_s y, v_1(g), v_2(g)),$$

where $\alpha_r = e^{2\pi i r/n}$. Equivalently, we calculate the ideals

$$\mathcal{I}_{r,s} = (Q_1^{r,s}, Q_2^{r,s})$$

where

$$Q_1^{r,s}(x,y) = v_1(g)(x,y,\alpha_r y, \alpha_s y)$$

and

$$Q_2^{r,s}(x,y) = v_2(g)(x,y,\alpha_r y, \alpha_s y).$$

To make these calculations easier, we shall write $g(x,y) = xy + P(x,y)$ as

$$(x+P_1)y + P_2 y^2 + \cdots + P_n y^n,$$

where each function P_i is of the form

$$P_i(x,y) = \sum_{l,m} a_{l,m} x^l y^{nm}$$

for some $a_{l,m} \in \mathbb{C}$. Note that for each r and i, $P_i(x,y) = P_i(x,\alpha_r y)$. Note also that without loss of generality, we may assume that $P_n = 0$ (by composing with the diffeomorphism $\Phi(X,Y,Z) = (X, Y - P_n(X,Z), Z)$). So

$$
\begin{aligned}
Q_1^{r,s} &= \begin{vmatrix} 1 & y & xy & +P(x,y) \\ 1 & y_1 & xy_1 & +P(x,y_1) \\ 1 & y_2 & xy_2 & +P(x,y_2) \end{vmatrix} \Big/ A_{r,s} \\
&= \begin{vmatrix} 1 & y & (x+P_1(x,y))y & +P_2(x,y)y^2 & +\cdots+P_{n-1}(x,y)y^{n-1} \\ 1 & \alpha_r y & (x+P_1(x,y))\alpha_r y & +P_2(x,y)\alpha_r^2 y^2 & +\cdots+P_{n-1}(x,y)\alpha_r^{n-1}y^{n-1} \\ 1 & \alpha_s y & (x+P_1(x,y))\alpha_s y & +P_2(x,y)\alpha_s^2 y^2 & +\cdots+P_{n-1}(x,y)\alpha_s^{n-1}y^{n-1} \end{vmatrix} \Big/ A_{r,s} \\
&= (x+P_1)y^2 \begin{vmatrix} 1 & 1 & 1 \\ 1 & \alpha_r & \alpha_r \\ 1 & \alpha_s & \alpha_s \end{vmatrix} \Big/ A_{r,s} + P_2 y^3 \begin{vmatrix} 1 & 1 & 1 \\ 1 & \alpha_r & \alpha_r^2 \\ 1 & \alpha_s & \alpha_s^2 \end{vmatrix} \Big/ A_{r,s} \\
&\quad + \cdots + P_{n-1}y^n \begin{vmatrix} 1 & 1 & 1 \\ 1 & \alpha_r & \alpha_r^{n-1} \\ 1 & \alpha_s & \alpha_s^{n-1} \end{vmatrix} \Big/ A_{r,s} \\
&= P_2 + P_3 y^4 \begin{vmatrix} 1 & 1 & 1 \\ 1 & \alpha_r & \alpha_r^3 \\ 1 & \alpha_s & \alpha_s^3 \end{vmatrix} \Big/ A_{r,s} + \cdots + P_{n-1}y^n \begin{vmatrix} 1 & 1 & 1 \\ 1 & \alpha_r & \alpha_r^{n-1} \\ 1 & \alpha_s & \alpha_s^{n-1} \end{vmatrix} \Big/ A_{r,s} \qquad (1)
\end{aligned}
$$

and similarly

$$
Q_2^{r,s} = (x+P_1) + P_3 y^5 \begin{vmatrix} 1 & 1 & 1 \\ 1 & \alpha_r^3 & \alpha_r^2 \\ 1 & \alpha_s^3 & \alpha_s^2 \end{vmatrix} \Big/ A_{r,s} + \cdots + P_{n-1}y^{n+1} \begin{vmatrix} 1 & 1 & 1 \\ 1 & \alpha_r^{n-1} & \alpha_r^2 \\ 1 & \alpha_s^{n-1} & \alpha_s^2 \end{vmatrix} \Big/ A_{r,s} \qquad (2)
$$

where

$$
A_{r,s}(y) = y^3 \begin{vmatrix} 1 & 1 & 1 \\ 1 & \alpha_r & \alpha_r^2 \\ 1 & \alpha_s & \alpha_s^2 \end{vmatrix} = y^3(1-\alpha_r)(\alpha_s-1)(\alpha_r-\alpha_s).
$$

Define $R_{r,s}$ by

$$
R_{r,s}(y) = Q_1^{r,s}(0,y). \qquad (3)
$$

Lemma 8 *If $R_{r,s} \neq 0$ then $D^3 f \cap P_{r,s}$ is of dimension 0.*

Proof If $R_{r,s} \neq 0$ then it has a first term ay^m, say. We claim that in this case,

$$
(Q_1^{r,s}, Q_2^{r,s}) = (x, y^m).
$$

By Nakayama's lemma, this holds if and only if

$$
(Q_1^{r,s}, Q_2^{r,s}) + (x, y^m)m_{\mathbb{C}^2,0} = (x, y^m).
$$

But it is clear that

$$
Q_1^{r,s} = ay^m \bmod (x, y^m)m_{\mathbb{C}^2,0}
$$

and that

$$Q_2^{r,s} = x \bmod 2(x, y^m)m_{\mathbb{C}^2,0},$$

which proves the lemma. □

Note that $R_{r,s} = 0$ if and only if $P_2(0,y) = 0$ and for all i, $(2 < i < n)$,

$$P_i(0,y) \begin{vmatrix} 1 & 1 & 1 \\ 1 & \alpha_r & \alpha_r^i \\ 1 & \alpha_s & \alpha_s^i \end{vmatrix} \Big/ A_{r,s}(1) = 0.$$

Lemma 9 *If $n = 4$ or n is prime then for all i with $2 < i < n$ and all r and s $(0 < r < s < n)$,*

$$\begin{vmatrix} 1 & 1 & 1 \\ 1 & \alpha_r & \alpha_r^i \\ 1 & \alpha_s & \alpha_s^i \end{vmatrix} \Big/ A_{r,s}(1) \neq 0.$$

Proof. If $n = 4$ and $i = 3$ then α_r and α_s must be chosen from i, -1 and $-i$, and it is easy to check that the determinants do not vanish for any of the choices. Otherwise

$$\begin{aligned} \begin{vmatrix} 1 & 1 & 1 \\ 1 & \alpha_r & \alpha_r^i \\ 1 & \alpha_s & \alpha_s^i \end{vmatrix} \Big/ A_{r,s}(1) &= \frac{(\alpha_s^i - 1)(\alpha_r - 1) - (\alpha_r^i - 1)(\alpha_s - 1)}{(1 - \alpha_r)(\alpha_s - 1)(\alpha_r - \alpha_s)} \\ &= -\frac{(\alpha_s^{i-1} + \cdots + 1) - (\alpha_r^{i-1} + \cdots + 1)}{\alpha_r - \alpha_s} \\ &= (\alpha_s^{i-2} + \alpha_s^{i-3}\alpha_r + \cdots + \alpha_r^{i-2} + \cdots + \alpha_s + \alpha_r + 1). \end{aligned}$$

Suppose now that $n > 2$ and $n \neq 4$ (but might not be prime). Then $\alpha_r = \xi^\rho$ and $\alpha_s = \xi^\sigma$ for an nth root of unity ξ, which is not necessarily primitive, but permits 1, α_r and α_s to be all different (so is not a root of $t^2 - 1 = 0$). Then ξ is a root of both

$$t^n - 1 = 0$$

and

$$S_{i,\rho,\sigma}(t) = t^{\sigma(i-2)} + t^{\sigma(i-3)+\rho} + \cdots + t^{\rho(i-2)} + \cdots + t^\sigma + t^\rho + 1 = 0. \tag{4}$$

The *dth Cyclotomic Polynomial* Φ_d is defined by

$$\Phi_d(t) = \prod_{1 \leq l \leq m} (t - \zeta_l)$$

where $\zeta_1, \zeta_2, \ldots, \zeta_m$ are all the primitive dth roots of unity [2, p. 94]. Each cyclotomic polynomial is irreducible over $\mathbb{Q}$ [2, 7.4]. Then

$$t^n - 1 = \prod_{d|n} \Phi_d(t),$$

so ξ is a root of the equation $\Phi_d(t) = 0$ for some d dividing n. But Φ_d is irreducible, and ξ is a common root of Φ_d and $S_{i,\rho,\sigma}$, so Φ_d must be a factor of $S_{i,\rho,\sigma}$ [1, p.167, Ex.(4)]. If a polynomial p is irreducible over $\mathbb{Z}$ then it is irreducible over $\mathbb{Q}$ [1, p. 166]. So there exists some polynomial g over $\mathbb{Z}$ such that

$$S_{i,\rho,\sigma}(t) = g(t)\Phi_d(t).$$

Hence

$$S_{i,\rho,\sigma}(1) = \sum_{a=1}^{i-1} a = i(i-1)/2 = g(1)\Phi_d(1).$$

Now suppose that d is prime. Then

$$\Phi_d(t) = t^{d-1} + t^{d-2} + \cdots + t + 1,$$

and $\Phi_d(1) = d$, implying that d divides $i(i-1)/2$, and hence d divides i or $i-1$. If n is prime then we must have $d = n$ and this cannot occur, since $i < n$. This concludes the proof of the lemma. □

Proof of Theorem 3. Suppose that $j^k f = (x, xy, y^n)$, that f is not $\mathcal{A}$-finite and that $k \geq 2n-3$. Then by lemma 5, after a change of coordinates, we may assume that

$$f(x,y) = (x, xy + P(x,y), y^n).$$

By lemmas 7 and 8, if $R_{r,s} \neq 0$ for all r and s $(0 < r < s < n)$, then f is $\mathcal{A}$-finite, where $R_{r,s}(y) = Q_1^{r,s}(0,y)$ (eqn 3) and $Q_1^{r,s}$ is given by eqn 1. Thus f is $\mathcal{A}$-finite unless there is some choice of r and s such that $R_{r,s} = 0$.

By lemma 9, if n is prime or $n = 4$, $R_{r,s} = 0$ if and only if $P_i(0,y) = 0$ for all i $(2 \leq i < n)$. But then $P(x,0) = 0$ and f is clearly n-to-1 on the y-axis. This concludes the proof of Theorem 3. □

Corollary 10 *Suppose that f is of the form $f(x,y) = (x, xy + P(x,y), y^n)$. If $n = 4$ or n is prime, then either $P(0,y) = 0$ and f has an n-to-1 curve or $P(0,y) \neq 0$ and f is $\mathcal{A}$-finite.*

Proof of Corollary 4 It is straightforward to check, using Nakayama's lemma, that

$$tf((m_{\mathbb{C}^2,0})^2\theta_{\mathbb{C}^2,0}) + \omega f((m_{\mathbb{C}^3,0})^2\theta_{\mathbb{C}^3,0}) \supset (x^2, xy^{4s+4})\theta(f) +$$

$$+(xy^4, y^{4s+7})\mathcal{O}_2\partial/\partial Z + \mathbb{C}\{(y^{4j},0,0),(0,y^{4j},0),(y^j, y^{j+1} + y^{4s+3+j}, 0) : j \geq 2\}.$$

It follows from [7, 3.1] that if $f \in \mathcal{A}^k f_s$ for some $k \geq 4s+7$, then $f \in \mathcal{A}^{k+1} f_s$ or

$$f \in \mathcal{A}^{k+1}(f_s + (0, ay^{k+1}, 0)),$$

for some $a \in \mathbb{C}$ (and $k+1 = 2,3 \bmod 4$). It follows that either $f \in \bigcap_{k\in\mathbb{N}} \mathcal{A}^k f_s$, or that $P(0,y) \neq 0$ and f is $\mathcal{A}$-finite. But, by [7, 2.1], $\bigcap_{k\in\mathbb{N}} \mathcal{A}^k f_s = \mathcal{A} f_s$. □

In view of Theorem 3, one might think that from the f_n with $n \geq 5$ and prime one could construct further examples of stems whose transverse section is not A_2, A_3, or D_4. However, this is not the case, since functional moduli appear in the classification of germs with curves of quintuple points or worse. The moduli, of course, come from the cross-ratios of the tangents to the distinct branches of a transverse section to the image at a point of the quintuple curve. Suppose that each point of the curve Σ in the image of f has k preimages. Then (once an ordering of the branches has been chosen), each point of Σ determines $k-3$ independent cros-ratios. Thus, to f is associated (ambiguously) a map $c_k : \Sigma \to \mathbb{C}^{k-3}$. The germ of the image of this map is an invariant of the $\mathcal{A}$-equivalence class of f. If $k > 4$ then it is possible to modify arbitrarily high-order terms in the power series expansion of f, in such a way that f continues to have a k-tuple curve Σ', but the germs of $c_k(\Sigma')$ and $c_k(\Sigma)$ are distinct.

Consider, for example, the family of map-germs g_q defined by $g_q(x,y) = (x, xy + xy^2 + xy^4 q(y), y^5)$, where $q(y) = \sum_i a_i y^{5i}$. It appears that if $f \in \mathcal{A}^k g_q$, for some q, k large, then f is either $\mathcal{A}$-finite or $\mathcal{A}$-equivalent to $g_{q'}$ for some q', but that none of the g_q's are $\mathcal{A}$-stems, since adding a term λy^{5r} to q ($r \in \mathbb{N}$) will always change the $\mathcal{A}$-class of g_q.

On the other hand, it seems very likely that germs such a g_q constructed from the f_n for $n > 4$ may be stems for *topological* $\mathcal{A}$-equivalence.

References

[1] P. M. Cohn, *Algebra, Volume 1, (Second Ed.)*, J. Wiley & Sons (1982)

[2] P. M. Cohn, *Algebra, Volume 2, (Second Ed.)*, J. Wiley & Sons (1989)

[3] W. L. Marar and D. Mond, *Multiple point schemes for corank 1 maps*, J. London Math. Soc.(2) 39 (1989), pp. 553–567.

[4] D.Mond, *On the classification of germs of maps from* $\mathbb{R}^2$ *to* $\mathbb{R}^3$, Proc.London Math. Soc. 50 (1985), 333-369

[5] D. Mond, *Some remarks on the geometry an classification of germs of maps from surfaces to 3-space*, Topology 26 (3) (1987), pp. 361–383.

[6] R. Pellikaan, *On hypersurface singularities which are stems*, Compositio Math. 71 (1989), 229-240

[7] D. Ratcliffe *Stems and Series in $\mathcal{A}$-classification*, to appear in Proc. London Math. Soc. 69 (3) 1994.

T. Kawakami

Nash structures of equivariant smooth manifolds and equivariant smooth vector bundles

1. Introduction.

H. Seifert [16] showed that any closed C^∞ stably parallelizable manifold is C^∞ diffeomorphic to a union of connected components of a nonsingular real algebraic set. J. Nash [13] conjectured the following in 1952.

Nash conjecture. *Every closed C^∞ manifold is C^∞ diffeomorphic to a nonsingular algebraic set.*

A. Tognoli [22] solved affirmatively this conjecture in 1973 and the proof consists of two parts. One is a cobordism process and the other is an approximation process. Using only the approximation one, one can prove any closed C^∞ manifold is C^∞ diffeomorphic to a union of nonsingular connected components of some algebraic set (cf. [21]), however, one cannot get rid of extra connected components [1]. R. Benedetti and A. Tognoli [3] proved a stronger result which asserts that each closed C^∞ manifold admits a structure Y of nonsingular algebraic set such that any topological vector bundle over Y admits one and only one structure of strongly algebraic vector bundle. J. Bochnak and W. Kucharz [5] proved that any closed C^∞ manifold admits a continuous family of structures of nonsingular algebraic sets.

Remark that this conjecture is not always true for an open manifold. We are concerned with the Nash conjecture for an affine Nash manifold and a compactifiable C^∞ manifold (See Theorem 3.1).

In 1990 K.H. Doverman, M. Masuda and T. Petrie [6] posed the following conjecture.

Equivariant Nash conjecture. *Let G be a compact Lie group. Any closed $C^\infty G$ manifold is $C^\infty G$ diffeomorphic to a nonsingular algebraic G set. Here a nonsingular algebraic G set means a G invariant nonsingular algebraic one in some representation of G.*

They gave some partial solutions, in particular, for any closed $C^\infty G$ manifold X, the disjoint union $X \coprod X$ is $C^\infty G$ diffeomorphic to some nonsingular algebraic G set. In the equivariant cases, the approximation process works, however, we do not know whether the cobordism one works or not. Moreover it is known that if G is a finite group and that a closed $C^\infty G$ manifold admits a structure of nonsingular algebraic G set, then it admits a continuous family of such structures [7]. But the equivariant Nash conjecture remains open.

In general we have the following inclusions as sets:

$$\{ \text{ nonsingular algebraic } G \text{ sets } \} \subset \{ \text{ affine Nash } G \text{ manifolds } \}$$
$$\subset \{ \text{ Nash } G \text{ manifolds } \} \subset \{ C^\infty G \text{ manifolds } \} .$$

Therefore it is reasonable to consider the following problem.

Problem. Let G be a compact affine Nash group.
(1) For a closed $C^\infty G$ manifold X, does there exist an affine Nash G manifold Y such that X is $C^\infty G$ diffeomorphic to Y ?
(2) How many Nash structures does X admit ?
(3) For an open $C^\infty G$ manifold X, is (1) true ?
(4) Does a $C^\infty G$ vector bundle admit a Nash structure ?
(5) How many Nash structures do they admit ?

Our main purpose of this paper is an exposition of the author's results concerning this problem.

This paper is organized as follows. We recall some definitions and useful facts in section 2. We state our results in section 3 and sketch the proof of them in section 4. In the present paper all Nash manifolds and Nash groups are of class C^ω.

2. Nash G manifolds and Nash G vector bundles.

Recall the definition of Nash manifolds. A *semialgebraic subset of* $\mathbb{R}^n$ is defined to be a finite union of sets of the following form:

$$\{x \in \mathbb{R}^n | f_1(x) = \cdots = f_k(x) = 0, g_1(x) > 0, \ldots, g_l(x) > 0\},$$

where $f_1, \ldots, f_k, g_1, \ldots, g_l$ are polynomial functions on $\mathbb{R}^n$. Let U and V be semialgebraic sets. We call a map $f : U \longrightarrow V$ *semialgebraic* if the graph of f is semialgebraic. The following theorem is well known (cf. 2.2.7 [4]).

Theorem 2.1. *Let X and Y be semialgebraic sets and let $f : X \longrightarrow Y$ be a semialgebraic map. Then $f(A)$ is semialgebraic if $A \subset X$ is semialgebraic.* □

Let U and V be open semialgebraic sets. A map $f : U \longrightarrow V$ is called *a Nash map* if f is semialgebraic and of class C^∞. Notice that a Nash map is a C^ω one. A Nash map $f : U \longrightarrow V$ is called *a Nash diffeomorphism* if there exists a Nash map $h : V \longrightarrow U$ such that $f \circ h = id, h \circ f = id$.

Definition 2.2. (1) A semialgebraic subset X in $\mathbb{R}^n$ is called *a Nash submanifold of dimension* d if for any $x \in X$ there exists a Nash diffeomorphism ϕ from some semialgebraic open neighborhood U of the origin in $\mathbb{R}^n$ onto some semialgebraic open neighborhood V of x in $\mathbb{R}^n$ such that $\phi(0) = x, \phi(\mathbb{R}^d \cap U) = X \cap V$. Here $\mathbb{R}^d$ denotes the subset of $\mathbb{R}^n$ those which the last $(n - d)$ components are zero.
(2) *A Nash manifold of dimension* d is a C^∞ manifold with a finite system of charts $\{\phi_i : U_i \longrightarrow \mathbb{R}^d\}$ such that for each i and j $\phi_i(U_i \cap U_j)$ is an open semialgebraic subset of $\mathbb{R}^d$ and the map $\phi_j \circ \phi_i^{-1} | \phi_i(U_i \cap U_j) : \phi_i(U_i \cap U_j) \longrightarrow \phi_j(U_i \cap U_j)$ is a

Nash diffeomorphism. We call these charts *Nash*. A Nash submanifold is of course a Nash manifold (cf. 1.3.9 [20]).
(3) Let X (resp. Y) be a Nash manifold with Nash charts $\{\phi_i : U_i \longrightarrow \mathbb{R}^n\}_i$ (resp. $\{\psi_j : V_j \longrightarrow \mathbb{R}^m\}_j$). A map $f : X \longrightarrow Y$ is said to be *a Nash map* if for any i and j $\phi_i(f^{-1}(V_j) \cap U_i)$ is open and semialgebraic in $\mathbb{R}^n$ and the map $\psi_j \circ f \circ \phi_i^{-1} : \phi_i(f^{-1}(V_j) \cap U_i) \longrightarrow \mathbb{R}^m$ is a Nash map.
(4) Let X and Y be Nash manifolds. We say that X is *Nash diffeomorphic to* Y if one can find Nash maps $f : X \longrightarrow Y, h : Y \longrightarrow X$ so that $f \circ h = id, h \circ f = id$.
(5) A Nash manifold is said to be *affine* if it is Nash diffeomorphic to a Nash submanifold in some Euclidean space $\mathbb{R}^l$.
(6) A topological vector bundle (E, p, X) with rank k is called *a Nash vector bundle* if the following three conditions are satisfied:

(a) The total space E and the base space X are Nash manifolds.
(b) The projection p is a Nash map.
(c) There exists a family of finitely many local trivializations $\{U_i, \phi_i : U_i \times \mathbb{R}^k \longrightarrow p^{-1}(U_i)\}_i$ such that $\{U_i\}_i$ is a semialgebraic open covering of X and that for any i and j the map $\phi_i^{-1} \circ \phi_j|(U_i \cap U_j) \times \mathbb{R}^k : (U_i \cap U_j) \times \mathbb{R}^k \longrightarrow (U_i \cap U_j) \times \mathbb{R}^k$ is a Nash map.

We call these local trivializations *Nash*.
(7) Let $\eta = (E, p, X)$ (resp. $\zeta = (F, q, X)$) be a Nash vector bundle with rank n (resp. m). Let $\{U_i, \phi_i : U_i \times \mathbb{R}^n \longrightarrow p^{-1}(U_i)\}_i$ (resp. $\{V_j, \psi_j : V_j \times \mathbb{R}^m \longrightarrow q^{-1}(V_j)\}_j$) be Nash local trivializations of η (resp. ζ). A vector bundle map $f : \eta \longrightarrow \zeta$ is said to be *a Nash vector bundle map* if for any i and j the map $(\psi_j)^{-1} \circ f \circ \phi_i|(U_i \cap V_j) \times \mathbb{R}^n : (U_i \cap V_j) \times \mathbb{R}^n \longrightarrow (U_i \cap V_j) \times \mathbb{R}^m$ is a Nash map.

It is known that a number of nonaffine Nash manifolds exist [17]. Remark that the finiteness condition in Definition 2.2 (2) is important. This induces satisfactory results [17], [18], [19], [20]. We say that a C^∞ manifold which admits a system of countably infinitely many Nash charts is *a locally Nash manifold*.

Definition 2.3. A group G is called *a Nash group* (resp. *an affine Nash group*, *a locally Nash group*) if G is a Nash manifold (resp. an affine Nash manifold, a locally Nash manifold) and that the multiplication $G \times G \longrightarrow G$ and the inversion $G \longrightarrow G$ are Nash maps.

Remark that one-dimensional connected (resp. locally) Nash groups are classified by J.J Madden and C.H Stanton [12] (resp. M. Shoita [19]).

Definition 2.4. Let G be a Nash group.
(1) A Nash submanifold in a representation of G is called *a Nash G submanifold* if it is G invariant.
(2) A Nash manifold X is said to be *a Nash G manifold* if X admits a G action whose action map $G \times X \longrightarrow X$ is a Nash map.

We can define Nash G maps, Nash G diffeomorphisms, and affine Nash G manifolds in the similar way. Recall the definition of Nash G vector bundles.

Definition 2.5. Let G be a Nash group.
(1) A Nash vector bundle (E, p, X) is said to be a *Nash G vector bundle* if the following three conditions are satisfied:

(a) The total space E and the base space X are Nash G manifolds.
(b) The projection p is a Nash G map.
(c) For any $x \in X$ and $g \in G$, the map $p^{-1}(x) \longrightarrow p^{-1}(gx)$ is linear.

(2) Let η and ζ be two Nash G vector bundles. A Nash vector bundle map $f : \eta \longrightarrow \zeta$ is called a *Nash G vector bundle map* if f is a G map.
(3) Two Nash G vector bundles η and ζ are said to be *Nash G vector bundle isomorphic* if there exist Nash G vector bundle maps $f : \eta \longrightarrow \zeta$ and $h : \zeta \longrightarrow \eta$ such that $f \circ h = id$ and $h \circ f = id$.

Definition 2.6. Let Ω be an n-dimensional representation of G and B the representation map $G \longrightarrow GL_n(\mathbb{R})$ of Ω. Suppose that $M(\Omega)$ denotes the vector space of $n \times n$-matrices with the action $(g, A) \in G \times M(\Omega) \longrightarrow B(g)^{-1}AB(g) \in M(\Omega)$. For any positive integer k, we define the vector bundle $\gamma(\Omega, k) = (E(\Omega, k), u, G(\Omega, k))$ as follows:

$$G(\Omega, k) = \{A \in M(\Omega) | A^2 = A, A = A', TrA = k\},$$

$$E(\Omega, k) = \{(A, v) \in G(\Omega, k) \times \Omega | Av = v\},$$

$$u : E(\Omega, k) \longrightarrow G(\Omega, k) : u((A, v)) = A,$$

where A' denotes the transposed matrix of A. Then $\gamma(\Omega, k)$ is an algebraic one. Since the action on $\gamma(\Omega, k)$ is algebraic, it is an algebraic G vector bundle. We call it *the universal G vector bundle associated with Ω and k*. Since $G(\Omega, k)$ and $E(\Omega, k)$ are nonsingular [14], $\gamma(\Omega, k)$ is a Nash G vector bundle.

Definition 2.7. A Nash G vector bundle $\eta = (E, p, X)$ with rank k is said to be *strongly Nash* if the base space X is affine and that there exist some representation Ω of G and a Nash G map $f : X \longrightarrow G(\Omega, k)$ such that η is Nash G vector bundle isomorphic to $f^*(\gamma(\Omega, k))$.

3. Results. We have the next result concerning the Nash conjecture for an affine Nash manifold and a compactifiable C^∞ manifold.

Theorem 3.1 [11]. *Every affine Nash manifold (resp. compactifiable C^∞ manifold) X with dim $X \geq 1$ has infinitely many nonsingular algebraic subsets $\{Y_n\}_{n \in \mathbf{N}}$ of some Euclidean space such that each Y_n is Nash diffeomorphic (resp. C^∞ diffeomorphic) to X and that Y_n is not birationally equivalent to Y_m for $n \neq m$.* □

Let G be a Nash group. We say that $C^\infty G$ manifold X *admits a Nash G manifold structure* (resp. *an affine Nash G manifold structure*) if there exists a Nash G manifold (resp. an affine Nash G manifold) Y such that X is $C^\infty G$ diffeomorphic to Y.

We proved in [10] the following three results.

Theorem 3.2 [10]. *Let G be a compact affine Nash group and let X be a compact $C^\infty G$ manifold with dim $X \geq 1$.*
(1) X admits exactly one affine Nash G manifold structure Y up to Nash G diffeomorphism.
(2) If G acts on X transitively, then for any other Nash G manifold structure Z (not necessarily nonaffine), Z is Nash G diffeomorphic to Y.
(3) If X is connected and the action is not transitive, then X admits a continuous family of nonaffine Nash G manifold structures. □

A $C^\infty G$ manifold X is called *compactifiable as a $C^\infty G$ manifold* if there exists a compact $C^\infty G$ manifold Y with boundary such that X is $C^\infty G$ diffeomorphic to the interior of Y.

Theorem 3.3 [10]. *Let G be a compact affine Nash group and let X be a noncompact compactifiable $C^\infty G$ manifold with dim $X \geq 1$.*
(1) X admits an affine Nash G manifold structure.
(2) X admits uncountably many nonaffine Nash G manifold structures. □

Theorem 3.4 [10]. *Let G be a compact affine Nash group and X a $C^\infty G$ manifold. Then X is compactifiable as a $C^\infty G$ manifold if and only if X admits an affine Nash G manifold structure.* □

We obtain the following concerning Nash structures of $C^\infty G$ vector bundles.

Theorem 3.5 [9]. *Let G be a compact affine Nash group and X a compact affine Nash G manifold. For any $C^\infty G$ vector bundle η over X, there exists a strongly Nash G vector bundle ζ over X such that η and ζ are $C^\infty G$ vector bundle isomorphic, and that ζ is unique up to Nash G vector bundle isomorphism.* □

We get the following as a weak equivariant version of the Artin-Mazur theorem, and prove it in section 5.

Proposition 3.6. *Suppose that G is a compact affine Nash group and that X is an affine Nash G manifold with free action. Let $f : X \longrightarrow \mathbb{R}^k$ be a G invariant Nash map. Then there exist an affine Nash G manifold Z and a Nash G diffeomorphism $\phi : Z \longrightarrow X$ such that $f \circ \phi$ is a G invariant polynomial map.*

4. Sketches of proofs.

In this section we give sketches of proofs of our results.

The simplest case of Theorem 3.1. Let X be the unit circle $\{(x,y) \in \mathbb{R}^2 | x^2 + y^2 = 1\} \subset \mathbb{R}^2 \subset \mathbf{P}^2(\mathbb{R})$. For $n \geq 2$, let $Y_n = \{(x,y) \in \mathbb{R}^2 | x^{2n} + y^{2n} = 1\} \subset \mathbb{R}^2 \subset \mathbf{P}^2(\mathbb{R})$. Then each Y_n is C^∞ diffeomorphic to X. Thus they are Nash diffeomorphic. If Y_n is birationally equivalent to Y_m then the complexifications of them are birationally equivalent. Birational invariance of arithmetic genus shows Y_n is not birationally equivalent to Y_m for $n \neq m$.

In the general case, we use birational invariance of Hodge number to distinguish Y_n. □

Sketch of proof of Theorem 3.2. Let G be a compact affine Nash group and let X be a compact $C^\infty G$ manifold. By [6] the disjoint union $X \coprod X$ is $C^\infty G$ diffeomorphic to some nonsingular algebraic G set. Thus X is $C^\infty G$ diffeomorphic to some affine Nash G manifold. Uniqueness follows from the Weierstrass approximation theorem and Proposition 4.1 below.

One can find a Nash G tubular neighborhood of an affine Nash G manifold [10].

Proposition 4.1 [10]. *Let G be a compact affine Nash group and let X be an affine Nash G submanifold in a representation Ω of G. Then there exists a Nash G tubular neighborhood (U, p) of X in Ω, namely, U is an affine Nash G submanifold in Ω and the orthogonal projection $p : U \longrightarrow X$ is a Nash G map.* □

If X is connected, dim $X \geq 1$, and that the action is not transitive, then we can find a Nash G tubular neighborhood U of some orbit such that $U \neq X$. We construct a G invariant open covering $\{U_i\}_{i=1}^3$ of X using U. We paste them with a family of Nash G diffeomorphisms $f_\lambda^i, \lambda \in \Lambda$, where Λ is an uncountable set. Let Y_λ be the resulting Nash G manifold. Showing that for any $\lambda \in \Lambda$ the set of μ such that Y_μ is Nash G diffeomorphic to Y_λ is at most countable, we get (3). □

Sketch of proof of Theorem 3.3. Since X is non-compact and compactifiable as a $C^\infty G$ manifold, there exists a compact $C^\infty G$ manifold X' with boundary so that X is $C^\infty G$ diffeomorphic to the interior of X'. Let Y be the double of X'. We use the following to get (1).

Theorem 4.2 [10]. *Let G be a compact affine Nash group, X a compact $C^\infty G$ manifold, and X' a compact $C^\infty G$ submanifold of X. Then there exist an affine Nash G manifold Y, an affine Nash G submanifold Y' of Y, and a $C^\infty G$ diffeomorphism $\phi : X \longrightarrow Y$ such that $\phi(X') = Y'$.* □

By Theorem 4.2 there exist an affine Nash G manifold Z and a $C^\infty G$ diffeomorphism $f : Y \longrightarrow Z$ so that $f(\partial X')$ is an affine Nash G manifold. Therefore X is $C^\infty G$ diffeomorphic to an affine Nash G manifold.

The similar argument in the proof of Theorem 3.2 (3) proves (2). □

Sketch of proof of Theorem 3.4. Theorem 3.2 and Theorem 3.3 show one direction. The other follows from an improvement of [15]. □

Sketch of proof of Theorem 3.5. Let Ω and Ξ be representations of G, and f a map $\Omega \longrightarrow \Xi$. Denote the Haar measure of G by dg, and let x be a point in Ω. Then

$$A(f)(x) = \int_G g^{-1} f(gx) dg.$$

Lemma 4.3 [6]. *Let G be a compact Lie group.*
(1) $A(f)$ is equivariant, and $A(f) = f$ if f is equivariant.
(2) If f is a polynomial map then so is $A(f)$.

(3) Let $C^r(\Omega, \Xi)$ be the set of all r times continuously differentiable maps from Ω to Ξ. The self map of $C^r(\Omega, \Xi)$ induced by A is continuous with respect to the C^k topology for any $k \le r$ and $r < \infty$. □

Recall the following theorem due to [23].

Theorem 4.4 [23]. *Let G be a compact Lie group and X, Y $C^\infty G$ manifolds. Suppose that η is a $C^\infty G$ vector bundle over Y and that f and h are $C^\infty G$ maps from X to Y. If f is $C^\infty G$ homotopic to h then $f^*(\eta)$ are $C^\infty G$ vector bundle isomorphic to $h^*(\eta)$.* □

By the Weierstrass approximation theorem, Proposition 4.1, Lemma 4.3, and Theorem 4.4, η admits a strongly Nash G vector bundle structure. Uniqueness follows from the following two propositions.

Proposition 4.5 [9]. *Let G be a Nash group. If η and ζ are strongly Nash G vector bundles then $Hom(\eta, \zeta)$ is strongly Nash.* □

Proposition 4.6 [9]. *Let G be a compact affine Nash group and let η be a strongly Nash G vector bundle over a compact affine Nash G manifold. Then every C^0G section of η can be approximated by a Nash one.* □ □

5. Artin-Mazur Theorem.

We recall the Artin-Mazur theorem.

Theorem 5.1 (Artin-Mazur Theorem) [2]. *Let X be an affine Nash manifold and let $f : X \longrightarrow \mathbb{R}^n$ be a Nash map. Then there exist a union Y of connected components of some nonsingular algebraic set and a Nash diffeomorphism $\phi : Y \longrightarrow X$ such that $f \circ \phi$ is a polynomial map.* □

We do not know whether we can replace a union Y of connected components of some nonsingular algebraic set by a nonsingular algebraic set.

Proposition 5.2. *Let G be a compact affine Nash group, X an affine Nash G manifold with free action in an n-dimensional representation Ω of G. Then the quotient space X/G admits a unique affine Nash manifold structure such that :*
(1) The projection $p : X \longrightarrow X/G$ is a Nash map.
(2) For a map f from X/G to some affine Nash manifold Y in $\mathbb{R}^m$, f is a Nash map if and only if $p \circ f$ is a Nash one.

Moreover $p : X \longrightarrow X/G$ is a Nash fiber bundle.

Proof. The algebra of G invariant polynomial functions on Ω is finitely generated [24]. Let l_i $(1 \le i \le k)$ be a system of its generators. Set

$$p : \Omega \longrightarrow \mathbb{R}^k, p(x) = (l_1(x), \dots, l_k(x)).$$

By Theorem 2.1 $p(X)$ is a semialgebraic set, and its structure is unique up to semialgebraic isomorphism [8]. It is well known that X/G admits a unique C^∞ manifold structure with the corresponding properties. Since a semialgebraic set with C^∞ manifold structure is an affine Nash manifold (cf. 1.3.9 [20]), X/G is an affine Nash manifold. Clearly the projection p is a Nash map.

We now prove (2). Assume that $p \circ f$ is a Nash map. For a map k, let $\Gamma(k)$ denote the graph of k. Let F be the projection $(\mathbb{R}^n \times \mathbb{R}^k) \times (\mathbb{R}^n \times \mathbb{R}^m) \longrightarrow \mathbb{R}^k \times \mathbb{R}^m$. Then $\Gamma(f) = F(\Gamma(p) \times \Gamma(p \circ f))$. Hence f is semialgebraic because of Theorem 2.1. Therefore f is a Nash map. The remaining assertion is obvious. □

Proof of Proposition 3.6. Let p be the orbit map $X \longrightarrow X/G$ constructed in Proposition 5.2. Then X/G is an affine Nash manifold and $\eta = (X, p, X/G)$ is a Nash fiber bundle. Since f is G invariant, we can find a Nash map $f/G : X/G \longrightarrow \mathbb{R}^k$ such that $f/G \circ p = f$. Applying $f/G : X/G \longrightarrow \mathbb{R}^k$ to the Artin-Mazur theorem, we have a union X' of connected components of a nonsingular algebraic set and a Nash diffeomorphism $f' : X' \longrightarrow X/G$ such that $f/G \circ f'$ is a polynomial map. Then $Z := (f')^*(\eta)$ is an affine Nash G manifold and f' induces the Nash G diffeomorphism ϕ from Z to X. Therefore Z and $\phi : Z \longrightarrow X$ fulfill the requirements. □

REFERENCES

1. S. Akbulut and H. King, *Submanifolds and homology of nonsingular real algebraic varieties*, Amer. J. Math **107** (1985), 45-83.
2. M. Artin and B. Mazur, *On periodic points*, Ann. of Math **81** (1965), 82-99.
3. R. Benedetti and A. Tognoli, *On real algebraic vector bundles*, Bull. Sci. Math. (2) (1980), 89-112.
4. J. Bochnak, M. Coste and M.F. Roy, *Géométie algébrique réelle*, Ergebnisse der Mathematik, und ihrer Grenzebiete, Springer verlag, 1987.
5. J. Bochnak and W. Kucharz, *Nonisomorphic algebraic models of a smooth manifold*, Math. Ann. **290** (1991), 1-2.
6. K.H. Dovermann, M. Masuda, and T. Petrie, *Fixed point free algebraic actions on varieties diffeomorphic to* $\mathbb{R}^n$, Progress in Math. 80, Birkhäuser (1990), 49-80.
7. K.H.Doverman, M. Masuda, and D.Y. Suh, *Nonisomorphic models of a smooth manifold with group action*, to appear.
8. S. Illman, *The isomorphism classes of a representation of a compact Lie group is determined by the equivariant simple homotopy type of the representation*, Transformation Groups, Lecture Note in Mathematics **1375** (1989), 98-110.
9. T. Kawakami, *Algebraic G vector bundles and Nash G vector bundles*, Chinese J. Math **22** (1994), 275-289.
10. T. Kawakami, *Nash G manifold structures of compact or compactifiable* $C^\infty G$ *manifolds*, to appear in J. Math. Soc. Japan.
11. T. Kawakami, *Nonisomorphic algebraic models of Nash manifolds and compactifiable* C^∞ *manifolds*, Osaka J. Math **31**, 831-835.
12. J.J. Madden and C.H. Stanton, *One-dimensional Nash groups*, Pacific J. Math. **154** (1992), 331-344.
13. J. Nash, *Real algebraic manifolds*, Ann. of Math. **56** (1952), 405-421.
14. R. Palais, *Equivariant real algebraic differential topology, Part 1, Smoothness categories and Nash manifolds, Notes Brandeis Univ.* (1972).
15. T. Petrie, *Smooth conjugacy of algebraic actions on affine varieties*, Topology **27** (1988), 473-477.
16. H. Seifert, *Algebarische approximation von mannigfalitigkeiten*, Math. Z. **41** (1936), 1-17.

17. M. Shiota, *Abstract Nash manifolds*, Proc. Amer. Math. Soc. **96** (1986), 155-162.
18. M. Shiota, *Approximation theorems for Nash mappings and Nash manifolds*, Trans. Amer. Math .Soc **293** (1986), 319-337.
19. M. Shiota, *Nash functions and manifolds (Chapter 2. of a preparing book)*, preprint.
20. M. Shiota, *Nash manifolds*, Lecture Note in Mathematics vol 1269, Springer-Verlag, 1987.
21. A. Tognoli, *Algebraic geometry and Nash functions*, Academic Press, 1978.
22. A. Tognoli, *Su una congettura di Nash*, Annali Sc. Norm. Sup. Pisa **27** (1973), 167-185.
23. A.G. Wasserman, *Equivariant differential topology*, Topology **8** (1969), 127-150.
24. H. Weyl, *The classical groups, 2nd ed*, Princeton Univ. Press, Princeton, N.J., 1946.

S. Izumiya

Local classifications of multi-valued solutions of quasilinear first order partial differential equations

1 Introduction

We consider the following partial differential equation which is called *a quasilinear first order partial differential equation* (briefly, a quasilinear equation):

$$\sum_{i=1}^{n} a_i(x,y)\frac{\partial y}{\partial x_i} - b(x,y) = 0$$

where $a_i(x,y)$ and $b(x,y)$ are C^∞-functions. This equation is well studied in several articles ([3],[5],[6],[7],[9],[11], etc.).

One of the most interesting qusilinear equations are scalar conservation laws. The notion of entropy solutions has provided the right weak setting for the study of the Cauchy problem for scalar conservation laws. Existence and uniqueness of the entropy solution have been established in [7].

On the other hand, if we solve the eqation by the classical method of characteristics, in generally, we have multi-valued solutions. The multi-valued solution is classical solution near the Cauchy data and it coinicdes with the entropy solution, however, around *the singular point* of the multi-valued solution, the entropy solution has *shock waves* (i.e., surfaces across which the entropy solution is discontinuous). In order to understand the process how shock waves appear, interact and bifurcate, we need to classify sinuglar points of multi-valued solutions. After that we can solve Riemann problem for each normal form of the classification. In [5] we have studied a classification of multi-valued solutions of time-dependent quasilinear equations in order to study shock waves of the Cauchy problem for scalar conservation laws. Most of the techniques we use here have been developed in [5], however, the situation here is slightly different. For time-dependent equations, we distinguish the time parameter and the space parameters, so the bifurcations of singularities of multi-valued solutions along the time parameter is most interesting. Here we consider stationary quasilinear equations, so that there are no reason to distinguish a special parameter from the space parameters. There is another motivation to consider the stationary quasilinear equations. Recently, Tsuji[10] studied the singularities of solutions for hyperbolic Monge-Ampère equations, in which the notion of intermediate integral is quite important. One of the examples of such first order partial differential equations are quasilinear equations. We can show that each quasilinear equation can be realized as an intermediate integral of a Monge-Ampère equation (cf., [10]). So our result gives a classification of multi-valued solutions of hyperbolic Monge-Ampère equations whose intermediate integral are quasilinear equations.

All maps are differentiable of class C^∞ unless stated otherwise.

2 Geometry of quasilinear equations

We construct the geometric framework of equations in the projective cotangent bundle $\pi : PT^*(\mathbf{R}^n \times I\!R) \to I\!R^n \times I\!R$. Consider the tangent bundle $\tau : TPT^*(I\!R^n \times I\!R) \to PT^*(I\!R^n \times I\!R)$ and the differential map $d\pi : TPT^*(I\!R^n \times I\!R) \to T(I\!R^n \times I\!R)$ of π.

For any $X \in TPT^*(I\!R^n \times I\!R)$, there exists an element $\alpha \in T^*_{(x,y)}(I\!R^n \times I\!R)$ such that $\tau(X) = [\alpha]$. For an element $V \in T_{(x,y)}(I\!R^n \times I\!R)$, the property $\alpha(V) = 0$ does not depend on the choice of representative of the class $[\alpha]$. Thus we can define the canonical contact structure on $PT^*(I\!R^n \times I\!R)$ by

$$K = \{X \in TPT^*(I\!R^n \times I\!R) | \tau(X)(d\pi(X)) = 0\}.$$

Because of the trivialization $PT^*(I\!R^n \times I\!R) \cong (I\!R^n \times I\!R) \times P(I\!R^n \times I\!R)^*$, we call

$$((x_1, \ldots, x_n, y), [\xi_1; \ldots; \xi_n; \eta])$$

a homogeneous coordinate, where $[\xi_1; \ldots; \xi_n; \eta]$ is the homogeneous coordinate of the dual projective space $P(I\!R^n \times I\!R)^*$.

It is easy to show that $X \in K_{((x,y),[\xi;\eta])}$ if and only if $\sum_{i=1}^n \mu_i \xi_i + \lambda\eta = 0$, where $d\tilde{\pi}(X) = \sum_{i=1}^n \mu_i \frac{\partial}{\partial x_i} + \lambda \frac{\partial}{\partial y}$.

We remark that $PT^*(I\!R^n \times I\!R)$ is a fibrewise compactification of the 1-jet space $J^1(I\!R^n, I\!R)$ as follows : We consider an open subset $U_\eta = \{((x,y),[\xi;\eta]) | \eta \neq 0\}$ of $PT^*(I\!R^n \times I\!R)$. For any $((x,y),[\xi;\eta]) \in U_\eta$, we have

$$((x_1, \ldots, x_n, y), [\xi_1; \ldots; \xi_n; \eta]) = ((x_1, \ldots, x_n, y), [-\frac{\xi_1}{\eta}; \ldots; -\frac{\xi_n}{\eta}; -1])$$

so that we may adopt the corresponding *affine coordinates*

$$((x_1, \ldots, x_n, y), (p_1, \ldots, p_n)),$$

where $p_i = -\frac{\xi_i}{\eta}$. On U_η we can easily show that $\theta^{-1}(0) = K|U_\eta$, where $\theta = dy - \sum_{i=1}^n p_i dx_i$. This means that U_η may be identified with the 1-jet space $J^1(I\!R^n, I\!R)$. We call the above coordinate *a system of canonical coordinates.* Throughout the remainder of this paper, we use this identification so that we have

$$J^1(I\!R^n, I\!R) \subset PT^*(I\!R^n \times I\!R).$$

A quasilinear equation is defined to be a hypersurface

$$E(a_1, \ldots, a_n, b) = \{((x,y),[\xi;\eta]) \in PT^*(\mathbb{R}^n \times \mathbb{R}) | \sum_{i=1}^{n} a_i(x,y)\xi_i + b(x,y)\eta = 0\}$$

where $a_i(x,y), b(x,y)$ are C^∞ -function.

We usually assume that $(a_1(x,y), \ldots, a_n(x,y)) \neq (0, \ldots, 0)$.

We remark that $\sum_{i=1}^{n} a_i(x,y)\xi + b(x,y)\eta = 0$ if and only if $\sum_{i=1}^{n} a_i(x,y)p_i - b(x,y) = 0$ under the canonical coordinate of $U_\eta = J^1(\mathbb{R}^n, \mathbb{R})$. So the above definition is consistent with the classical definition of quasilinear equations.

In accordance with the philosophy of Lie, we may define the notion of geometric solutions as follows: An immersion $i : L \to PT^*(\mathbb{R}^n \times \mathbb{R})$ is *a Legendrian immersion* if $\dim \mathrm{L} = \mathrm{n}$ and $di_q(T_qL) \subset K_{i(q)}$ for any $q \in L$. *A geometric solution* of $E(a_1, \ldots, a_n, b)$ is defined to be a Legendrian immersion $i : L \to PT^*(\mathbb{R}^n \times \mathbb{R})$ such that Image$(L) \subset E(a_1, \ldots, a_n, b)$. We also say that the geometric solution is *smooth* if $\pi|L : L \to \mathbb{R}^n \times \mathbb{R}$ is an immersion.

We consider the meaning of the notion of geometric solutions. Let S be a smooth hypersurface in $\mathbb{R}^n \times \mathbb{R}$, then we have a unique Legendrian submanifold $\hat{S}$ in $PT^*(\mathbb{R}^n \times \mathbb{R})$ such that $\pi(\hat{S}) = S$ which is given as follows :

$$\hat{S} = \{((x,y),[\xi;\eta]) | \sum_{i=1}^{n} \xi_i \frac{\partial}{\partial x_i} + \eta \frac{\partial}{\partial y} \in T_{(x,y)}S^\perp\}.$$

It follows that if L is a smooth geometric solution of $E(a_1, \ldots, a_n, b)$, then we have $L = \widehat{\pi(L)}$.

We consider the condition on a smooth hypersurface S that $\hat{S} \subset E(a_1, \ldots, a_n, b)$. For any $(x_0, y_0) \in S$, there exists a smooth submersion germ $f : (\mathbb{R}^n \times \mathbb{R}, (x_0, y_0)) \to (\mathbb{R}, 0)$ such that $(f^{-1}(0), (x_0, y_0)) = (S, (x_0, y_0))$. A vector $\sum_{i=1}^{n} \mu_i \frac{\partial}{\partial x_i} + \lambda \frac{\partial}{\partial y}$ is tangent to S at $(x,y) \in (S, (x_0, y_0))$ if and only if it satisfies $\sum_{i=1}^{n} \mu_i \frac{\partial f}{\partial x_i} + \lambda \frac{\partial f}{\partial y} = 0$ at (x,y). Then we have the following representation of $\hat{S}$:

$$(\hat{S}, ((x_0, y_0); [\xi_0; \eta_0])) = \{((x,y), [\frac{\partial f}{\partial x}; \frac{\partial f}{\partial y}]) | (x,y) \in (S, (x_0, y_0))\}.$$

We now consider the following vector field on $\mathbb{R}^n \times \mathbb{R}$ associated with the quasilinear equation $E(a_1, \ldots, a_n, b)$:

$$X(a_1, \ldots, a_n, b) = \sum_{i=1}^{n} a_i(x,y) \frac{\partial}{\partial x_i} + b(x,y) \frac{\partial}{\partial y}.$$

Then we have the following proposition.

Proposition 2.1. *Let S be a smooth hypersurface in $\mathbb{R}^n \times \mathbb{R}$. Then $\hat{S}$ is a smooth geometric solution of $E(a_1, \ldots, a_n, b)$ if and only if the vector field $X(a_1, \ldots, a_n, b)$ is tangent to S.*

Proof. We have a local representation

$$(\hat{S}, ((x_0, y_0); [\xi_0; \eta_0])) = \{((x, y), [\frac{\partial f}{\partial x}; \frac{\partial f}{\partial y}])|(x, y) \in (S, (x_0, y_0))\},$$

where $f : (\mathbb{R}^n \times \mathbb{R}, (x_0, y_0)) \to (\mathbb{R}, 0)$ is a submersion germ with $f^{-1}(0) = S$. It follows that $\hat{S} \subset E(a_1, \ldots, a_n, b)$ if and only if

$$\sum_{i=1}^{n} a_i(x, y)\frac{\partial f}{\partial x_i} + b(x, y)\frac{\partial f}{\partial y} = 0$$

for $(x, y) \in (S, (x_0, y_0))$. The last condition is equivalent to the condition that $X(a_1, \ldots, a_n, b)$ is tangent to S at (x, y). □

The Cauchy problem can be solved by the method of characteristics. We say that *a geometric Cauchy problem (GCP)* for an equation $E(a_1, \ldots, a_n, b)$ is a given $(n-1)$-dimensional submanifold $i' : S' \subset \mathbb{R}^n \times \mathbb{R}$ such that the characteristic vector field $X(a_1, \ldots, a_n, b)$ is not tangent to S'. The following result is well-known (cf. [1]), however it clarifies the geometric situation in this paper.

Theorem 2.2 (Local existence theorem). *A GCP $i' : S' \subset \mathbb{R}^n \times \mathbb{R}$ has a unique smooth geometric solution, that is, there exists a smooth hypersurface $S \subset \mathbb{R}^n \times \mathbb{R}$, $S' \subset S$ such that $\hat{S} \subset E(a_1, \ldots, a_n, b)$, and any two such smooth hypersurfaces coincide in a neighborhood of S'.*

Proof. Consider the embedding $i : S \subset \mathbb{R}^n \times \mathbb{R}$, where $S \subset (-\varepsilon, \varepsilon) \times S'$ is a neighborhood of $0 \times S'$, $\varepsilon > 0$; here $i(t, q) = T_t(q)$, $q \in S'$, $(t, q) \in S$, and T_t is an one-parameter group of translations along $X(a_1, \ldots, a_n, b)$. By the construction, we have $X(a_1, \ldots, a_n, b) \in TS$,so that $\hat{S} \subset E(a_1, \ldots, a_n, b)$ by Proposition 2.1.

On the other hand, for any geometric solution L of $E(a_1, \ldots, a_n, b)$, $\pi(L)$ must be invariant under $X(a_1, \ldots, a_n, b)$, so that it coincides with S in some neighborhood. □

We also consider the notion of classical solutions of $E(a_1, \ldots, a_n, b)$. We say that a C^∞-function $y = g(x_1, \ldots, x_n)$ is *a classical solution* of $E(a_1, \ldots, a_n, b)$ if it satisfies

$$\sum_{i=1}^{n} a_i(x, g(x))\frac{\partial g}{\partial x_i} - b(x, g(x)) = 0.$$

Let $g : \mathbb{R}^n \to \mathbb{R}$ be a smooth function, then we have a smooth hypersurface $S_g = \{(x, g(x))|x \in \mathbb{R}^n\}$ in $\mathbb{R}^n \times \mathbb{R}$. Since $S_g = f^{-1}(0)$, where $f(x, y) = g(x) - y$, it follows from the previous arguments that $\widehat{S_g}$ is a smooth geometric solution of $E(a_1, \ldots, a_n, b)$ if and only if $y = g(x_1, \ldots, x_n)$ is a classical solution.

On the other hand, we consider the canonical projection $\pi_1 : \mathbb{R}^n \times \mathbb{R} \to \mathbb{R}^n$ given by $\pi_1(x, y) = x$. Let L be a smooth geometric solution of $E(a_1, \ldots, a_n, b)$. We say that

$q \in L$ is *a singular point* if rank $d(\pi_1 \circ \pi|L) < n$. We remark that $q \in L$ is not a singular point if and only if there exists a smooth function germ $g : (I\!R^n, x_0) \to I\!R$ such that $(S, (x_0, y_0)) = (S_g, (x_0, y_0))$. Our problem in this paper is summarized as follows:

Problem. Classify the generic singularities of smooth geometric solutions of GCP for a quasilinear equation which is solved by the characteristic method.

3 Projections of smooth hypersurfaces

The smooth geometric solution of GCP for a quasilinear equation is a smooth hypersurface and the multi-valuedness of geometric solutions is caused by sinuglarities of the projection on to $I\!R^n$, so that we consider the classification of singularities of the projection of hypersurfaces. Since any smooth hypersurface is locally given as a zero point set of a smooth submersion, we consider the following. Let $F : (I\!R^n \times I\!R, 0) \to (I\!R, 0)$ be a submersion germ, then $F^{-1}(0)$ is a smooth hypersurface in $(I\!R^n \times I\!R, 0)$. For the study of the singularities of the projection $\pi_F : (F^{-1}(0), 0) \to I\!R^n$, we define the following equivalence relation. Let

$$F_i : (I\!R^n \times I\!R, 0) \to (I\!R, 0), \ (i = 0, 1)$$

be submersion germs. We say that F_0 and F_1 are *P-$\mathcal{K}$-equivalent* if there exists a diffeomorphism germ

$$\Phi : (I\!R^n \times I\!R, 0) \to (I\!R^n \times I\!R, 0)$$

of the form

$$\Phi(x, y) = (\phi_1(x), \phi_2(x, y))$$

such that

$$\langle F_1 \circ \Phi \rangle_{\mathcal{E}_{(x,y)}} = \langle F_0 \rangle_{\mathcal{E}_{(x,y)}},$$

where $\langle F_0 \rangle_{\mathcal{E}_{(x,y)}}$ denotes the ideal generated by F_0 in the ring $\mathcal{E}_{(x,y)}$ of function germs of (x, y)-variables at the origin.

Our purpose in this section is to classify submersions F under the P-$\mathcal{K}$-equivalence. For this purpose we need some notions from singularity theory (cf., [4], [8]). For a submersion germ $F : (I\!R^n \times I\!R, 0) \to (I\!R, 0)$, we define

$$T_e(\mathcal{K})(f) = \langle \frac{\partial f}{\partial y}, f \rangle_{\mathcal{E}_y},$$

where $f = F|0 \times I\!R$. We say that F is *a $\mathcal{K}$-versal deformation of f* if it satisfies

$$\mathcal{E}_y = \langle \frac{\partial F}{\partial x_1}|0 \times I\!R, \dots, \frac{\partial F}{\partial x_n}|0 \times I\!R \rangle_{I\!R} + T_e(\mathcal{K})(f).$$

The above definiton is an infinitesimal version of the definition of the $\mathcal{K}$-versality, however it is known that these are equivalent (cf., [8]), so that we adopt the above definiton.

It is well known that the following classification theorem holds (cf.,[2],[8]).

Theorem 3.1. *Let* $F : (\mathbb{R}^n \times \mathbb{R}, 0) \to (\mathbb{R}, 0)$ *be a* $\mathcal{K}$*-versal deformation of* f*, then* F *is P-*$\mathcal{K}$*-equivalent to one of the following* ${}^0A_k$*-type germs* $(0 \leq k \leq n)$*:*

$$y^{k+1} + \sum_{i=1}^{k} x_i y^{i-1}.$$

Remarks. (1) The above theorem gives a "generic" classification of subersion germs under the P-$\mathcal{K}$-equivalence.
(2) If $F : (\mathbb{R}^n \times \mathbb{R}, 0) \to (\mathbb{R}, 0)$ is a $\mathcal{K}$-versal deformation of f, then π_F is an $\mathcal{A}$-stable map germ in the sense of Mather (cf., [9]).

The following theorem is the classification theorem of Mather (cf., [8]) and it is quite useful in the latter sections.

Theorem 3.2 (J. Mather). *Let* $F_i : (\mathbb{R}^n \times \mathbb{R}, 0) \to (\mathbb{R}, 0)$ *be* $\mathcal{K}$*-versal deformations of* $f_i = F_i|0 \times \mathbb{R}$, $i = 0, 1$*. Then* f_0 *and* f_1 *are* $\mathcal{K}$*-equivalent if and only if* F_0 *and* F_1 *are* P-$\mathcal{K}$*-equivalent.*

The definiton of the $\mathcal{K}$*-equivalence can be found in* ([8]).

4 Realization theorems

In this section we identify geometric solutions of a (GCP) introduced in Section 2 with smooth hypersurfaces.

We have already observed that the local geometric solution of the geometric Cauchy problem for a quasilinear equation $E(a_1, \ldots, a_n, b)$ is a smooth surface germ in Theorem 2.2. Here, we prove that the converse is also true.

Theorem 4.1. *Let* $(S, (x_0, y_0))$ *be a smooth hypersurface germ. Then there exist function germs* $a_i(x, y)$ $(i = 1, \ldots, n)$, $b(x, y)$ *and codimension one submanifold germ* $(S', (x_0, y_0)) \subset (S, (x_0, y_0))$ *such that* $(\widehat{S}, (x_0, y_0))$ *is a local geometric solution of the geometric Cauchy problem for the quasilinear equation* $E(a_1, \ldots, a_n, b)$ *with the Cauchy data* $(S', (x_0, y_0))$.

Proof. Let $\phi : (\mathbb{R}^n, 0) \to (S, (x_0, y_0))$ be a local parametrization of the smooth hyper surface $(S, (x_0, y_0))$ (i.e., $\phi : (\mathbb{R}^n, 0) \to (\mathbb{R}^n \times \mathbb{R}, (x_0, y_0))$ is an immersion germ with $(\mathrm{Image}(\phi), (x_0, y_0)) = (S, (x_0, y_0)))$. Then $d\phi(\frac{\partial}{\partial u_1})$ is a non-singular vector field on $(S, (x_0, y_0))$, where $(u_1, \ldots, u_n)$ are the canonical coordinates of $(\mathbb{R}^n, 0)$. Since ϕ is an immersion germ, there exists a vector field germ X on $(\mathbb{R}^n \times \mathbb{R}, (x_0, y_0))$ such that

$X \circ \phi = d\phi(\frac{\partial}{\partial u_1})$. So there exist function germs $a_i(x, y)$ ($i = 1, \ldots, n$ and $b(x, y)$ such that

$$X = \sum_{i=1}^{n} a_i(x, y)\frac{\partial}{\partial x_i} + b(x, y)\frac{\partial}{\partial y}.$$

By Proposition 2.1, $(\widehat{S}, (x_0, y_0))$ is a local geometric solution of the quasilinear equation $E(a_1, \ldots, a_n, b)$. We consider the smooth submanifold $(S', (x_0, y_0))$ of codimension one in $(S, (x_0, y_0))$ which is transverse to the vector field X in $(S, (x_0, y_0))$, $(\widehat{S}, (x_0, y_0))$ is the geometric solution of the geometric Cauchy problem where the Cauchy data is $(S', (x_0, y_0))$ □

The above theorem guarantees that the class of smooth hypersurfaces supplies the correct class to describe the geometric solutions of GCP for quasilinear equations. We must also concern ourselves with which types of singularities the geometric solution to a given quasilinear equation may exhibit. Representation Theorems 4.2 and 4.3 address this question.

Theorem 4.2. *Let $E(a_1, \ldots, a_n, b)$ be a quasilinear equation. If ${}^0A_k$ $k < n$, then there exists a submersion germ $G : (I\!R^n \times I\!R, (x_0, y_0) \to (I\!R, 0)$ such that $\widehat{G^{-1}(0)}$ is a geometric solution of $E(a_1, \ldots, a_n, b)$ and G is*P-$\mathcal{K}$*-equivalent to the germ ${}^0A_k$.*

Proof. Without loss of generality, we may assume that $(x_0, y_0) = (0, 0)$ and $a_n(0, 0) \neq 0$.

We now consider the germ

$$H(x_1, \ldots, x_n, y) = y^{k+1} + \sum_{i=1}^{k} x_i y^{i-1} \quad (0 \leq k < n),$$

then $S' = (H|_{\{x_n=0\}})^{-1}(0))$ is a smooth hypersurface in $I\!R^{n-1} \times 0 \times I\!R = \{x_n = 0\}$. Since $a_n(0, 0) \neq 0$, $X_{(a_1, \ldots, a_n, b)}$ is transverse to S', so that we have a hypersurface S such that $\hat{S}$ is a geometric solution of $E(a_1, \ldots, a_n, b)$ by Theorem 2.2.

Let $G : (I\!R^n \times I\!R, (x_0, y_0)) \to (I\!R, 0)$ a submersion with $G^{-1}(0) = S$. By the uniqueness of the geometric solution, we have $(G|I\!R^{n-1} \times 0 \times I\!R)^{-1}(0) = S'$, so that $G|I\!R^{n-1} \times 0 \times I\!R$ is P-$\mathcal{K}$-equivalent to $H|I\!R^{n-1} \times 0 \times I\!R$. It follows that $g = G|0 \times 0 \times I\!R)$ is $\mathcal{K}$-equivalent to y^{k+1}. Since $k < n$, $H|I\!R^{n-1} \times 0 \times I\!R$ is a $\mathcal{K}$-versal deformation of y^{k+1}. Therefore $G|I\!R^{n-1} \times 0 \times I\!R$ is a $\mathcal{K}$-versal deformation of g. It follows that G is also a $\mathcal{K}$-versal deformation of g. By Theorem 3.2, G is P-$\mathcal{K}$-equivalent to H. This completes the proof. □

We remark that ${}^0A_k$-type germs in Theorem 3.1 can always be realized as geometric solutions for any quasilinear equation if $k < n$. However, if $k = n$, the situation is different. We fix $i \in \{1, \ldots, n\}$, let $E(a_1, \ldots, a_n, b)$ be a quasilinear equation with

$a_i(x_0, y_0) \neq 0$. We say that it is *k-non-degenerated at* (x_0, y_0) $(1 \leq k \leq n)$ if

$$\frac{\partial^k \tilde{a}}{\partial y^k}(x_0, y_0) \neq 0,$$

where $\frac{\partial^k \tilde{a}}{\partial y^k}(x_0, y_0) = (\frac{\partial^k a_1}{\partial y^k}(x_0, y_0), \ldots, \frac{\partial^k a_{i-1}}{\partial y^k}(x_0, y_0), \frac{\partial^k a_{i+1}}{\partial y^k}(x_0, y_0), \frac{\partial^k a_n}{\partial y^k}(x_0, y_0))$. we simply say that $E(a_1, \ldots, a_n, b)$ is *non-degenerated* if it is 1-non-degenerated. Then we have the following realization theorem.

Theorem 4.3. *Let $E(a_1, \ldots, a_n, b)$ be a quasilinear equation and $F : (\mathbb{R}^n \times \mathbb{R}, 0) \to (\mathbb{R}, 0)$ be a submersion germ such that F is a $\mathcal{K}$-versal deformation of f_0. If the equation is non-degenerated at (x_0, y_0), then there exists a submersion germ $G : (\mathbb{R}^n \times \mathbb{R}, (x_0, y_0)) \to (\mathbb{R}, 0)$ such that $\widehat{G^{-1}(0)}$ is a local geometric solution of $E(a_1, \ldots, a_n.b)$ and F, G are P-$\mathcal{K}$-equivalent.*

Proof. Without the loss of generality, we may assume that $(x_0, y_0) = (0, 0)$, $a_n(0,0) \neq 0$ and $F(x, y) = y^{k+1} + \sum_{i=1}^{k} x_i y^{i-1}$. It follows that $S' = f^{-1}(0)$ is a smooth hypersurface in $\mathbb{R}^{n-1} \times 0 \times \mathbb{R} = \{x_n = 0\}$, where $f = F|\mathbb{R}^{n-1} \times 0 \times \mathbb{R}$. Since $a_n(0,0) \neq 0$, $X(a_1, \ldots, a_n, b)$ is transverse to S', so that we have a hypersurface S such that $\hat{S}$ is a geometric solution of $E(a_1, \ldots, a_n, b)$ by Theorem 1.2. Let $G : (\mathbb{R}^n \times \mathbb{R}, 0) \to (\mathbb{R}, 0)$ be a submersion with $G^{-1}(0) = S$, then $g = G|\mathbb{R}^{n-1} \times 0 \times \mathbb{R}$ and f are $\mathcal{C}$-equivalent (i.e., there exists a function germ $\lambda(x_1, \ldots, x_{n-1}, y)$ with $\lambda(0) \neq 0$ such that $g = \lambda \cdot f$). If $k \leq n-1$, then f is already a $\mathcal{K}$-versal deformation of $f_0 = f|0 \times \mathbb{R}$. Hence, by the same reason as in the proof of Theorem 3.2, F is P-$\mathcal{K}$-equivalent to G. We now assume that $k = n$.

We now consider a vector space

$$T(g_0) = \langle \frac{\partial g}{\partial x_1}|0 \times \mathbb{R}, \ldots, \frac{\partial g}{\partial x_{n-1}}|0 \times \mathbb{R} \rangle_{\mathbb{R}} + T_e(\mathcal{K})(g_0).$$

If $T(g_0) = \mathcal{E}_y$, then g is already a $\mathcal{K}$-versal deformation of g_0, so that we can get the required assertion by Theorem 2.2 as in the previous case.

Suppose that $T(g_0) \neq \mathcal{E}_y$ and $\frac{\partial G}{\partial x_n}|0 \times \mathbb{R} \in T(g_0)$ for any submersion germ G with $G^{-1}(0) = S$. Since $G^{-1}(0)$ is a geometric solution of $E(a_1, \ldots, a_n, b)$, we have a relation

$$\sum_{i=1}^{n} a_i(x, y) \frac{\partial G}{\partial x_i} + b(x, y) \frac{\partial G}{\partial y} = 0 \quad \text{on } \mathrm{G}^{-1}(0),$$

so that

$$-a_n(x, y) \frac{\partial G}{\partial x_n} \equiv \sum_{i=1}^{n-1} a_i(x, y) \frac{\partial G}{\partial x_i} + b(t, x, y) \frac{\partial G}{\partial y} \text{ mod } \langle \mathrm{G} \rangle_{\mathcal{E}_{(x,y)}}.$$

It follows that that

$$-a_n(0, y) \frac{\partial G}{\partial x_n}|0 \times \mathbb{R} \equiv \sum_{i=1}^{n-1} a_i(0, y) \frac{\partial g}{\partial x_i}(0, y) + b(0, 0, y) \frac{\partial g_0}{\partial y} \text{ mod } \langle \mathrm{g}_0 \rangle_{\mathcal{E}_y}.$$

We may assume that $g_0 \in \mathcal{M}_y^k$ for $k \geq 3$, where $\mathcal{M}_y$ is the unique maximal ideal of $\mathcal{E}_y$.

We now consider the Taylor polynomial of $a_i(x,y)$ for sufficiently higher order at $(x,0)$ with respect to y-variables as follows :

$$a_i(x,y) = a_i(x,0) + \frac{\partial a_i}{\partial y}(x,0)y + \frac{1}{2}\frac{\partial^2 a_i}{\partial y^2}(x,0)y^2 + \text{higher terms.}$$

On the other hand, we have

$$\frac{\partial G}{\partial x_n}|0 \times I\!R \in T(g_0),$$

so that

$$\begin{aligned}\sum_{i=1}^{n-1} a_i(0,y)\frac{\partial g}{\partial x_i}(0,y) \quad &+ \quad b(0,y)\frac{\partial g_0}{\partial y}(y) \\ &\in \quad \langle g_0(y), \frac{\partial g_0}{\partial y}(y)\rangle_{\mathcal{E}_y} + \langle \frac{\partial g}{\partial x_1}(0,y), \ldots, \frac{\partial g}{\partial x_{n-1}}(0,y)\rangle_{I\!R},\end{aligned}$$

and

$$\begin{aligned}\sum_{i=1}^{n-1}(a_i(0,0) + \frac{\partial a_i}{\partial y}(0,0)y \quad &+ \quad \frac{1}{2}\frac{\partial^2 a_i}{\partial y^2}(0,0)y^2 + \text{higher terms})\frac{\partial g}{\partial x_i}(0,y) \\ &\in \quad \langle g_0(y), \frac{\partial g_0}{\partial y}(y)\rangle_{\mathcal{E}_y} + \langle \frac{\partial g}{\partial x_1}(0,y), \ldots, \frac{\partial g}{\partial x_{n-1}}(0,y)\rangle_{I\!R}.\end{aligned}$$

For any linear isomorphism $A : I\!R^{n-1} \to I\!R^{n-1}$, we have a relation

$$\frac{\partial f(Ax',0,y)}{\partial x_i} = \sum_{j=1}^{n-1} A_{ij}\frac{\partial f}{\partial x_j}(Ax',0,y),$$

where $x' = (x_1, \ldots, x_{n-1})$. We remark that the vector space

$$\langle g_0, \frac{\partial g_0}{\partial y}\rangle_{\mathcal{E}_y} + \langle \frac{\partial g}{\partial x_1}(0,y), \ldots, \frac{\partial g}{\partial x_{n-1}}(0,y)\rangle_{I\!R}$$

is an invariant under the action of the linear isomorphism A. Since

$$\sum_{i=1}^{n-1} a_i(0,0)\frac{\partial g}{\partial x_i}(0,y) \in \langle \frac{\partial g}{\partial x_1}(0,y), \ldots, \frac{\partial g}{\partial x_{n-1}}(0,y)\rangle_{I\!R},$$

we have

$$\begin{aligned}(\sum_{i=1}^{n-1} \frac{\partial a_i}{\partial y}(0,0)\frac{\partial g}{\partial x_i}(0,y))y \quad &+ \quad \text{higher terms} \\ &\in \quad \langle g_0, \frac{\partial g_0}{\partial y}\rangle_{\mathcal{E}_y} + \langle \frac{\partial g}{\partial x_1}(0,y), \ldots, \frac{\partial g}{\partial x_{n-1}}(0,y)\rangle_{I\!R}.\end{aligned}$$

By assumptions, $\frac{\partial a_i}{\partial y}(0,0) \neq 0$ and $\frac{\partial g}{\partial x_j}(0,0) \neq 0$ for some $i, j = 1, \ldots, n-1$. If necessary, by applying some linear isomorphism A, we have

$$\sum_{i=1}^{n-1} \frac{\partial a_i}{\partial y}(0,0)\frac{\partial g}{\partial x_i}(0,0) \neq 0.$$

It follows that $\mathcal{M}_y \subset \langle g_0, \frac{\partial g_0}{\partial y}\rangle_{\mathcal{E}_y} + \langle \frac{\partial g}{\partial x_1}(0,y), \ldots, \frac{\partial g}{\partial x_{n-1}}(0,y)\rangle_{\mathbb{R}} \bmod \mathcal{M}_y^2$.

Since $\mathcal{K}-\mathrm{cod}\,(g_0)$ is finite (for the definition of $\mathcal{K}-$finiteness, see [18,22]), then there exists $r \in \mathbb{N}$ such that $\mathcal{M}_y^r \subset \langle g_0, \frac{\partial g_0}{\partial y}\rangle_{\mathcal{E}_y} \subset \mathcal{M}_y^{r-1}$. If $y \in \langle g_0, \frac{\partial g_0}{\partial y}\rangle_{\mathcal{E}_y}$, we have $\mathcal{M}_y = \langle g_0, \frac{\partial g_0}{\partial y}\rangle_{\mathcal{E}_y}$. Then this case is corresponding to the case $r = 1$.

We may assume that $y \in \langle \frac{\partial g}{\partial x_1}(0,y), \ldots, \frac{\partial g}{\partial x_{n-1}}(0,y)\rangle_{\mathbb{R}} \bmod \mathcal{M}_y^2$. It follows that there exist real numbers λ_i $(i = 1, \ldots, n-1)$ such that $y = \sum_{i=1}^{n-1} \lambda_i \frac{\partial g}{\partial x_i}(0,y)$. If necessary, applying a linear isomorphism A, we may assume that $y = \frac{\partial g}{\partial x_i}(0,y)$ for some $i = 1, \ldots, n-1$. By the same arguments as those of previous paragraphs, we can asserts that $y^2 \in \langle g_0, \frac{\partial g_0}{\partial y}\rangle_{\mathcal{E}_y} + \langle \frac{\partial g}{\partial x_1}(0,y), \ldots, \frac{\partial g}{\partial x_{n-1}}(0,y)\rangle_{\mathbb{R}} \bmod \mathcal{M}_y^3$. We can continue this procedure up to degree $r-1$. Eventually, every polynomial of degree $r-1$ is contained in the vector space, so that we ahve

$$\mathcal{M}_y \subset \langle g_0, \frac{\partial g_0}{\partial y}\rangle_{\mathcal{E}_y} + \langle \frac{\partial g}{\partial x_1}(0,y), \ldots, \frac{\partial g}{\partial x_{n-1}}(0,y)\rangle_{\mathbb{R}} \bmod \mathcal{M}_y^r.$$

Since $\mathcal{M}_y^r \subset \langle g_0, \frac{\partial g_0}{\partial y}\rangle_{\mathcal{E}_y}$ and $\frac{\partial g}{\partial x_j}(0,0) \neq 0$ for some $i = 1, \ldots n-1$, we have

$$\mathcal{E}_y = \langle g_0, \frac{\partial g_0}{\partial y}\rangle_{\mathcal{E}_y} + \langle \frac{\partial g}{\partial x_1}(0,y), \ldots, \frac{\partial g}{\partial x_{n-1}}(0,y)\rangle_{\mathbb{R}} = T(g_0).$$

This contradicts to the assumption that $T(g_0) \neq \mathcal{E}_y$, so that $\frac{\partial G}{\partial x_n}|0 \times \mathbb{R} \notin T(g_0)$ Since f and g are $\mathcal{C}$-equivalent, it is P-$\mathcal{K}$-equivalent, so that

$$\dim_{\mathbb{R}} \frac{\mathcal{E}_{(x',y)}}{\langle g, \frac{\partial g}{\partial y}\rangle_{\mathcal{E}_{(x',y)}} + \langle \frac{\partial g}{\partial x_1}, \cdots \frac{\partial g}{\partial x_{n-1}}\rangle_{\mathcal{E}_{x'}}} = \dim_{\mathbb{R}} \frac{\mathcal{E}_{(x',y)}}{\langle f, \frac{\partial f}{\partial y}\rangle_{\mathcal{E}_{(x',y)}} + \langle \frac{\partial f}{\partial x_1}, \cdots \frac{\partial f}{\partial x_{n-1}}\rangle_{\mathcal{E}_{x'}}}.$$

It is easy to calculate that the right hand side of the above equality is equal to 1.

We now consider the canonical inclusion $i : (\mathbb{R}, 0) \to (\mathbb{R}^{n-1} \times \mathbb{R}, 0)$ given by $i(y) = (0, y)$ and the induced epimorphism $i^* : \mathcal{E}_{(x',y)} \to \mathcal{E}_y$ Then we have $\mathrm{Ker}(i^*) = \mathcal{M}_{x'}\mathcal{E}_{(x',y)}$ and

$$i^*(\langle \frac{\partial g}{\partial x_1}, \cdots \frac{\partial g}{\partial x_{n-1}}\rangle_{\mathcal{E}_{x'}}) = T(g_0).$$

It follows that

$$\dim_{\mathbb{R}} \frac{\mathcal{E}_y}{T(g_0)} = \dim_{\mathbb{R}} \frac{\mathcal{E}_{(x',y)}}{\langle g, \frac{\partial g}{\partial y}\rangle_{\mathcal{E}_{(x',y)}} + \langle \frac{\partial g}{\partial x_1}, \cdots \frac{\partial g}{\partial x_{n-1}}\rangle_{\mathcal{E}_{x'}} + \mathcal{M}_{x'}\mathcal{E}_{(x',y)}} \leq 1.$$

Since $\frac{\partial G}{\partial x_n}|0 \times I\!R \notin T(g_0)$, we have

$$\mathcal{E}_y = \frac{\partial G}{\partial x_n}|0 \times I\!R + T(g_0).$$

This means that G is a $\mathcal{K}$-versal deformation of g_0. Since F is a $\mathcal{K}$-versal deformation of f_0 and the fact that f_0 and g_0 are $\mathcal{C}$-equivalent, F and G are P-$\mathcal{K}$-equivalent by Theorem 2.2. This completes the proof. □

References

[1] V. I. Arnol'd, *Geometric Methods in the Theory of Ordinary Differential Equations,* Springer-Verlag, 1983

[2] C. G. Gibson, *Singular points of Smooth Mappings,* Pitman, London 1979

[3] J. Guckenheimer, *Solving a single conservation law,* Lecture notes in Mathematics **468** Springer Verlag, New York, 1975, 108-134

[4] S. Izumiya, *Generic bifurcations of varieties,* manuscripta math. **46** (1984), 137–164.

[5] S. Izumiya and G. Kossioris, *Geometric singularities for solutions of single conservation laws,* preprint, (1993).

[6] G. Jennings, *Piecewise smooth solutions of a single conservation law exist,* Adv. in Math **33** (1979), 192-205.

[7] S. N. Kruzkov, *First order quasilinear equations in several independent variables,* Math. USSR Sb. **10** (1970), 217-243.

[8] J. Martinet, *Singularities of Smooth Functions and Mappings,* Cambridge Univ. Press, Cambridge, 1982.

[9] S. Nakane, *Formation of shocks for a single conservation law,* SIAM J. Math. Anal. **19** (1988), 1391–1408.

[10] M. Tsuji, *Formation of singularities for Monge-Ampère equations,* Bull. des sciences mathématiques, **119** (1995), to appear.

[11] D. Wagner, *The Riemann problem in two space dimensions for a single conservation law,* SIAM J. Math. Ann. **14** (1983), 534–559.

T. Gaffney

Equisingularity of plane sections, t_1 condition and the integral closure of modules

The theory of the integral closure of modules provides a powerful tool for studying stratification conditions which are connected with limits of linear spaces. It gives an expression which is both algebraic and geometric for these conditions. This connection allows one to control many geometric phenomena through associated numerical invariants. This paper will illustrate these points by examining the t_r conditions which were introduced by Thom and Trotman. We will show how to apply these conditions to the study of certain families of plane sections of an analytic space. This is a report on joint work by the author with David J. A. Trotman and Leslie C. Wilson of the Université de Provence and University of Hawaii. Full details will appear elsewhere.

To date the t_r conditions have been developed only for the case of real sets. Our formulation works for real and complex analytic sets. In the real case it agrees with the currently used definition. Our formulation gives a new simple proof of the change of the conditions under Grassman modification; this is used in conjunction with the principle of specialization of integral dependence to give numerical criteria for familes of plane sections of complete intersections to be Whitney equisingular. The following theorem is a sample of what can be proved:

Corollary 3.8. *Suppose X is the germ of a complex analytic hypersurface at a point z, with smooth singular locus Y. Suppose that for every plane P of dimension complementary to Y and transverse to Y at z, the Milnor number of $\{X \cap P\}$ is independent of P. Then the family $\{X \cap P\}$ is Whitney Equisingular.*

The author thanks Lê Dung Trang, and his collaborators David Trotman and Leslie Wilson for helpful comments, and Robert Gassler for reading over the manuscript carefully.

§1. Background on the theory of integral closure of modules.

In this section we review some basic facts about the integral closure of a module. We begin with a definition.

Definition 1.1. *Suppose X, x is a complex analytic germ, M a submodule of $\mathcal{O}^p_{X,x}$. Then $h \in \mathcal{O}^p_{X,x}$ is in the integral closure of M, denoted $\overline{M}$ iff for all $\phi : \mathbf{C}, 0 \to X, x$, $h \circ \phi \in (\phi^* M)\mathcal{O}_1$.*

An algebraic definition of the integral closure of a module has been given by David Rees in [R2]. In the case of ideals, this notion has been used by Zariski-Samuel, Hironaka, and Bernard Teissier in algebraic geometry and in particular in the theory of equisingularity of sets. If we are working in the real analytic case, we simply use real analytic curves instead of complex analytic curves. The set of germs satisfying the condition of Definition 1.1 in this case is called the real analytic closure of M and are denoted by $\overline{M}_{\mathbf{R}}$. To see that these notions are different, it is easy to check that the integral closure in $\mathcal{O}_{\mathbf{C}^2}$ of the ideal (x^2+y^2) is (x^2+y^2), while the integral closure in $\mathcal{O}_{\mathbf{R}^2}$ of the ideal (x^2+y^2) is (x^2, y^2, xy). In the first case, it is clear that h must vanish on the curve defined by (x^2+y^2), hence must be divisible by (x^2+y^2). In the second case, it is clear that $\phi^*((x^2+y^2))\mathcal{O}_1$ is just (t^{2k}) where k is the minimum of the orders of the first non-vanishing term in the components of ϕ. It is clear that the same equality holds for $\phi^*(x^2, y^2, xy)\mathcal{O}_1$. An algebraic definition of the real integral closure of an ideal has been given by Brumfiel ([B]).

The connection between the integral closure of ideals and modules is given by the next proposition (which in fact proves that our definition agrees with that of Rees). We denote the $(p-k)$-th fitting ideal of $\mathcal{O}^p_{X,x}/M$ by $J_k(M)$; if h is an element of $\mathcal{O}^p_{X,x}$, we denote by (h, M) the module generated by M and h.

Proposition 1.2. *Suppose M is a submodule of $\mathcal{O}^p_{X,x}$, $h \in \mathcal{O}^p_{X,x}$ and the rank of (h, M) is k on each component of X, x. Then $h \in \bar{M}$ iff $J_k(h, M) \subset \overline{J_k(M)}$.*

Proof. Cf [1.7] and [1.8] of [G1]. □

In the applications we make, we are interested in an equi-dimensional complex analytic set X defined by $F : \mathbf{C}^N, 0 \to \mathbf{C}^p, 0$. The modules we are interested in are the jacobian module, denoted $JM(F)$, which is the submodule of $\mathcal{O}^p_{X,x}$ generated by the partial derivatives of F, and its reductions. (A submodule N of M is a reduction of M if $\overline{N} = \overline{M}$.)

As a corollary of 1.2 we obtain:

Corollary 1.3. *Suppose $X^d, 0$ is an equidimensional complex analytic space defined by $F : \mathbf{C}^N, 0 \to \mathbf{C}^p, 0$. Then N is a reduction of $JM(F)$ iff for all $\phi : \mathbf{C}, 0 \to X, 0$ such that $\phi(\mathbf{C} - \{0\}) \subset X - S(X)$, we have $(\phi^* N)\mathcal{O}_1 = (\phi^* JM(F))\mathcal{O}_1$.*

Proof. Since $X^d, 0$ is equidimensional, the rank of $JM(F)$ is $N-d$ on each component of $X^d, 0$. The singular set of X, denoted $S(X)$, is exactly where the ideal $J_{N-d}(JM(F))$ vanishes. The implication $\Rightarrow$ is trivial, while the implication $\Leftarrow$ is now clear, since if $\phi(\mathbf{C}) \subset S(X)$ both $J_{N-d}(JM(F))$ and $J_{N-d}(N)$ vanish. □

Roughly speaking an element h is in the integral closure of an ideal I if the order of vanishing of h on the zero set of I is the same order as the vanishing of I. Sometimes we require the order of vanishing of h to be greater. The notion of strict dependence makes this precise. Suppose M is submodule of $\mathcal{O}^p_{X,x}$, $h \in \mathcal{O}^p_{X,x}$. Then h is *strictly dependent* on M if for all $\phi : \mathbf{C}, 0 \to X, x$ we have $h \circ \phi \in m_1 \phi^* M$, where m_1 is the maximal ideal in $\mathcal{O}_1$. We denote the set of elements strictly dependent on M by $\overline{M}^+$.

Note that M in general neither contains nor is contained in $\overline{M}^+$. (For example if $M = (x^3, y^3)$ in $\mathcal{O}_2$, then $\overline{M}^+ = m_2^4$.) If M is an $\mathcal{O}_n$ module then $\overline{M}^+$ clearly contains $m_n\overline{M}$ but this inclusion may be strict.

By reviewing the proofs of 1.2 and 1.3 we can check that in the setup of 1.3, and N is a reduction of $JM(F)$, then whether or not $h \in JM(F)$ is strictly dependent on N can be checked using curves ϕ such that ϕ maps $\mathbf{C} - \{0\}$ to the smooth part of X. The precise analogue of 1.2 is:

Corollary 1.4. *Suppose M is a submodule of $\mathcal{O}^p_{X,x}$, $h \in \mathcal{O}^p_{X,x}$ and the rank of (h, M) is k on each component of X, x. Then $h \in \overline{M}^+$ iff each minor in $J_k(h, M)$ which depends on h lies in $\overline{J_k(M)}^+$.*

To see the connection between integral closure and geometry, suppose that $F : \mathbf{C}^{n+1} \to \mathbf{C}$ defines the germ of a hypersurface X at the origin in $\mathbf{C}^{n+1}, 0$. Suppose X contains the t-axis, $\{0\} \times \mathbf{C}$. When does every sequence of limiting tangent planes to X at the origin contain the t-axis? In [G3], it is shown that for each limiting tangent plane H to a hypersurface X, you can find a curve $c(s)$, such that the image of c lies in the smooth set of X except for c(0), and the limit of the tangent planes to X at c(s) is H. This means that

$$\lim_{s \to 0}(1/s^k)(DF(c(s)) = (a_1, \ldots, a_{n+1})$$

where $a_1 z_1 + \cdots + a_n z_n + a_{n+1} t = 0$ defines H, and k is the minimum of the orders of the first non-vanishing term in $\frac{\partial F}{\partial t} \circ \phi, \frac{\partial F}{\partial z_1} \circ \phi, \ldots, \frac{\partial F}{\partial z_n} \circ \phi$. The condition that H contains the t-axis is just that $a_{n+1} = 0$. This is equivalent to asking that $\frac{\partial F}{\partial t} \circ c \in m_1 c^*(\frac{\partial F}{\partial z_1}, \ldots, \frac{\partial F}{\partial z_n})$. This means that the condition that every limit of tangent planes to X at the origin contain the t-axis is equivalent to $\frac{\partial F}{\partial t} \in \overline{(\frac{\partial F}{\partial z_1}, \ldots, \frac{\partial F}{\partial z_n})}^+$.

We can also describe the limiting tangent planes in terms of integral closure. If H is a limiting tangent plane, and $v = (v_1, \ldots, v_{n+1})$ is any vector in H, then $\sum a_i v_i = 0$, so the order in s of $DF(c(s))(v)$ must be greater than k. Let $JM(F)_H$ denote the submodule of $JM(F)$ generated by $\frac{\partial F}{\partial v}$ for all $v \in H$. Then we see that H is a limiting tangent plane iff $JM(F)_H$ is not a reduction of $JM(F)$. Both of these results hold more generally than for hypersurfaces, with an important change. In the hypersurface case, the Jacobian matrix of F had only one row, and this determined the tangent plane to X at a smooth point of X. In general, the rows of the the Jacobian matrix of F will each determine a tangent hyperplane to X at a smooth point, and the row space of the Jacobian matrix of F at a smooth point will determine all of the tangent hyperplanes. Thus our results in the general case will be stated in terms of tangent hyperplanes. Here are the corresponding results.

Proposition 1.5. *([G3] (Theorem 2.2)) Suppose X^d is an equidimensional complex analytic set, defined by a map germ F. A hyperplane H is a limiting tangent hyperplane to X^d at 0 iff $JM(F)_H$ is not a reduction of $JM(F)$.*

Proposition 1.6. *([G3] (Theorem 2.3)) Suppose F, X^d as above. Then every limiting tangent hyperplane contains v iff $DF(v)$ is in $\overline{JM(F)}^+$.*

Integral closure can also be used to describe the tangent cones of curves. If $c : \mathbf{C} \to \mathbf{C}^n$, then the tangent cone to the image at c at the origin will be a line, and the next proposition gives a test for a plane to contain this line.

Proposition 1.7. *Suppose c is as above, and P is a plane. Then the tangent line to c at the origin lies in P iff $c^*(I(P)) \subset m_1 c^* m_n$*

Proof. Let k be the minimum of the orders of the first non-vanishing term in the components of c. Then $c^* m_n = (s^k)$. We have that

$$\lim_{s \to 0} 1/s^k(c(s)) = (v_1, \ldots, v_n).$$

If $h = a_1 z_1 + \ldots + a_n z_n$ is a generator of I(P), then

$$\lim_{s \to 0} 1/s^k(h(c(s))) = a_1 v_1 + \ldots + a_n v_n = 0.$$

So the order of vanishing of every element of $c^*(I(P))$ is greater than k. □

Two important conditions in the study of stratifications are Whitney's conditions A and B. Whitney's condition A holds for a pair of strata (X,Y) at a point y if every limit of tangent hyperplanes to X at y contains the tangent space to Y at y. This condition seems to be necessary for any reasonable equisingularity condition. (For example, if $\{X_t\}$ is any topologically trivial family of complex analytic hypersurfaces with isolated singularities, and X is the total space of the family, with T the singular set, then (X-T,T) satisfies Whitney A.([L-S]))

Whitney's condition B holds at y if for any sequence of pairs of points (x_i, y_i) which converge to (y,y), such that the secant line joining the pair converges to a line l, and the tangent plane to X at x_i converges to a plane T, then T contains l. This condition implies local topological triviality but it also preserves some important local infinitesimal structure, the aureole ([L-T]). For a further discussion of these conditions see [Gi],[T]. We are also interested in a third regularity condition, Verdier's condition W. Roughly speaking, this condition says that as you approach y from X, the distance between the tangent space to X at x and the tangent space to Y at y goes to zero as fast as the distance between x and Y goes to zero. In the complex analytic case W is the same as the Whitney conditions ([T]), while in the real analytic case it implies the Whitney conditions, but there are examples due to Brodersen and Trotman which show that the Whitney conditions can hold even though W fails. (Cf. [Tr1] for a discussion of the implications between the various conditions.) We are interested in W because integral closure methods connect with W rather than the Whitney conditions in the real analytic case. The precise definitions are as follows:

Definition 1.8. *Suppose A, B are linear subspaces at the origin in* $\mathbf{C}^N$, *then*

$$\text{dist}\,(A,B) = \sup_{\substack{u \in B^{\perp} - \{0\} \\ v \in A - \{0\}}} \frac{|(u,v)|}{\|u\|\,\|v\|}$$

In the applications B is the "big" space and A the "small" space. Note that dist(A,B) is not in general the same as dist(B,A). This allows us to talk about the Whitney conditions holding with a certain exponent.

Definition 1.9. *Suppose* $Y \subset \bar{X}$, X, Y *strata in a stratification of an analytic space, and* $\text{dist}\,(TY_0, TX_x) \leq \text{dist}\,(x, Y)^e$. *Then* (X, Y) *satisfy Whitney A with exponent* e *at* $0 \in Y$. *Verdier's condition W is Whitney A with exponent 1.*

To date, the cases of most interest are those when $e = 1$ (Condition W), and when $e > 0$ (Whitney A), but indeterminate. If one is working with the Whitney A condition the value of e is usually not relevant.

In the statements that follow, let X_0 denote the smooth points of X. The theory of the integral closure of modules allows us to show:

Theorem 1.10. *Suppose* $X^d, 0 \subset \mathbf{C}^N$ *is an equidimensional complex analytic set,* $X = F^{-1}(0)$, S *a smooth subset of* X. *Then Whitney A holds for the pair* (X_0, S) *at the origin iff* $\frac{\partial F}{\partial s}$ *is strictly dependent on* $JM(F)$ *for all tangent vectors* $\frac{\partial}{\partial s}$ *to* S *at the origin.*

proof. Cf. [G3] Cor. 2.4. □

Theorem 1.11. *Let* X, Y *be as above,* $F : \mathbf{C}^N \to \mathbf{C}^p$ *coordinates chosen so that* $\mathbf{C}^t J \times J0 = Y$, $m_n = (z_1, \ldots, z_n)$ *denoting the ideal defining* Y, F *defines* X *with reduced structure. Then* $\frac{\partial F}{\partial s} \in \overline{m_n(\frac{\partial F}{\partial z_1}, \ldots \frac{\partial F}{\partial z_n})}$ *for all tangent vectors* $\frac{\partial}{\partial s}$ *to* Y *iff W holds for* (X_0, Y).

proof. Cf. [G1] Theorem 2.5. □

The analogous results hold in the real analytic case using the real integral closure instead.

§2. The t_1 condition and the Grassman modification.

Definition 2.1. *Suppose* X *and* Y *are disjoint smooth submanifolds of* $\mathbf{R}^n$ *in a neighborhood of the origin. Then* X *is* t_s *regular over* Y *at 0 if every* C^s *submanifold* Z *transverse to* Y *at 0 and of complementary dimension to* Y *is transverse to* X *near 0.*

This concept was first introduced by Thom; Thom allowed Z to be of dimension greater than the complementary dimension of Y. Transversals Z of dimension complementary to Y are called direct transversals. Trotman realized that the degree of smoothness of the transversal was important; t_r implies t_{r+1}, but the converse does not hold in general. Further Trotman showed that if you work with direct transversals, Whitney A implies t_1, but the converse doesn't hold except when the dimension of Y is 1 ([Tr2]).

A little thought reveals the connection between these ideas and the family of sections of X by planes which are direct transversals. If, as we move through the family of slices, one of the planes has higher order of contact with X than the others, than that slice should not be equisigular with the other slices. The failure of t_1 for a plane is a way of detecting this higher contact.

In this section we will give a formulation for the t_r conditions for the real and complex analytic cases, using integral closure theory. We will show that in the case of t_1, in the real analytic case it agrees with Trotman's definition. We will also make clear the connection between the t_1 condition and the family of sections of X by planes which are direct transversals. To do this, we will need to discuss the notion of the Grassman modification of X.

Definition 2.2. *Let $G_{n,d}$ denote the Grassmanian of d-planes in $\mathbf{k}^n$, $E_{n,d}$, the canonical bundle associated to $G_{n,d}$; the fiber of $E_{n,d}$ over a point P is just the points of $\mathbf{k}^n$ in P. Denote the projection of $E_{n,d}$ to $\mathbf{k}^n$ by $\beta_{n,d}$. If X is a subset of $\mathbf{k}^n$, we call $\tilde{X} = {\beta_{n,d}}^{-1}(X)$, the $G_{n,d}$ modification of X. When n and d are clear from the context, we will simply refer to the Grassman modification of X.*

This notion was introduced in connection with the t_r conditions by Kuo and Trotman ([K-T]). Note that $G_{n,d}$ is embedded in $E_{n,d}$ as the zero section of $E_{n,d}$. This means that we can think of $0 \times G_{n,d}$ as a stratum of $\tilde{X}$; note that the projection to $0 \times G_{n,d}$ makes $\tilde{X}$ a family of analytic sets with $0 \times G_{n,d}$ as the parameter space. The members of this family are just $\{P \cap X\}$ as P varies through the points of $G_{n,d}$. This means that if we want to study the equisingularity of the family $\{P \cap X\}$, then we should study the regularity conditions that the pair $(\tilde{X}_0, 0 \times G_{n,d})$ satisfy. In this paper, we will be interested only in the behavior of this pair at points of $G_{n,d}$ which correspond to direct transversals. *We denote by $\tilde{Y}$ the set of points of* $0 \times G_{n,d}$ *where the plane corresponding to a point of* $0 \times G_{n,d}$ *is a direct transversal.* Kuo and Trotman proved a remarkable result in [K-T] about the behavior of regularity conditions under Grassman modification. In the following result, $t_0 = W$.

Theorem 2.3. *(Kuo-Trotman) Suppose X and Y are disjoint smooth submanifolds of $\mathbf{R}^n$ in a neighborhood of the origin, Y is analytic, and X is subanalytic; $s \geq 0$. Then X is t_{s+1} regular over Y at 0 iff $\tilde{X}$ is t_s regular over $\tilde{Y}$ at every point of $\tilde{Y}$.*

proof. Cf. [K-T] Theorem 1. □

This result shows that if t_1 holds at the origin for the pair (X, Y), and $X \cup Y$ is a closed set, with Y the singular set of $X \cup Y$, then the family $\{P \cap X\}$ is Verdier regular, hence topologically trivial.

To arrive at an integral closure formulation of the t_1 condition, we must understand the connection between limits of hyperplanes and the t_1 condition, because, as we have seen, integral closure describes limits of hyperplanes. This connection can be seen as follows. We know from the definition that t_1 fails iff there exists a direct C^1 transversal Z to Y at the origin, which fails to be transverse to X in any neighborhood of the origin. Trotman in [Tr3] shows that if X is subanalytic, then Z can be taken to subanalytic as well. This means that you can find a curve ϕ on $Z \cap X$ such that a tangent hyperplane to X at $\phi(s)$ contains the tangent space to Z as well. The tangent plane to Z at the origin will be a direct transversal, which will be tangent to ϕ at the origin, and which will be contained in a limit of tangent hyperplanes to X along ϕ. This suggests that we look for an integral closure condition that will prevent this from happening. In the definition that follows, if P is a submanifold which is an analytic direct transversal, let $I(P)$ denote the ideal of functions vanishing on P. We let $JM(F)_P$ denote the submodule of $JM(F)$ obtained by applying vectorfields tangent to the fibers of some submersion defining P to the map F. (Note that if P is a plane this module will be larger than the module we denoted by $JM(F)_P$ previously.) If $(z_1, ..., z_n, t_1, ..., t_k)$ are coordinates on $\mathbf{k}^{n+k}$, let $JM_t(F)$ denote the submodule of $JM(F)$ generated by the partials of f with respect to t.

Definition 2.4. *Suppose $X \subset \mathbf{k}^{n+k}$ is the germ of an analytic space at the origin defined by F, Y a submanifold embedded as $\{0\} \times \mathbf{k}^k$, P an analytic direct transversal, then t_r holds for the transversal P iff*

$$m_n{}^r JM_t(F) \subset \overline{m_n JM(F)_P + I(P) JM_t(F)}.$$

Condition t_r holds for the pair (X_0, Y) at the origin iff t_r holds for all direct transversals P. We say $t_{r\_}$ holds for P iff

$$m_n{}^r JM_t(F) \subset \overline{m_n JM(F)_P + I(P) JM_t(F)}^{+}.$$

Condition $t_{r\_}$ holds for the pair (X_0, Y) at the origin iff $t_{r\_}$ holds for all direct transversals P. In all of the above cases we take the integral closure inside $\mathcal{O}_{X,0}{}^p$, where p is the number of components of F; in the real analytic situation we take real integral closure.

In the above definition, it turns out that to compute $m_n JM(F)_P + I(P) JM_t(F)$, it suffices to fix a submersion defining P. This is because the modules obtained by using two different submersions differ by elements of $I(P) JM_t(F)$. This means that if P is a plane we can use either of the two modules we have denoted by $JM(F)_P$ in checking the condition of 2.4.

It is important to note that this definition has two arguments—in applications we can vary both P and F. This is why the same condition can be used to study the stratification of a complex analytic space, (take F to be the defining equations for X,

P a "probing set") or the order of V-determinacy of a map-germ $f : \mathbf{C}^n \to \mathbf{C}^p$. To do this, work on the ambient space of the graph of f, take P to be the graph of f, and take $F = (t_1, ...t_p)$; X then becomes $\mathbf{C}^n$, while Y is still given by $(z_1 = ... = z_n = 0)$. If we restrict $I(P)$ to X we get $f^*(m_p)$, while $JM_t(F)$ becomes $\mathcal{O}_n{}^p$. The condition of Defintion 2.4 then becomes

$$m_n{}^r \mathcal{O}_n{}^p \subset \overline{m_n JM(f) + f^*(m_p)\mathcal{O}_n{}^p}.$$

In [G1], in Corollary 3.6, it is shown that if this condition holds for f, then any deformation of f which fixes the r jet of f gives a Verdier equisingular deformation of $f^{-1}(0)$, hence f is r V-determined.

We will show shortly that at least for t_1 this definition is equivalent to the one given by Trotman. The condition $t_{r\_}$ first appeared in the paper of Trotman and Wilson ([T-W]) in the real case. For the rest of the paper we will restrict our attention mainly to the case of t_0, t_1 and $t_{1\_}$ for simplicity. As we will see in the proof of 2.7, in the above definition, the $I(P)JM_t(F)$ term ensures that we only have to check the condition for curves which are tangent to P at the origin, while the $JM(F)_P$ term ensures that no limit of tangent hyperplanes to X along such a curve can contain P. Here are some basic results about these conditions.

Proposition 2.5. *Suppose X and Y are as in 2.4. Then t_1 holds for the pair (X_0, Y) at the origin if it holds for all linear direct transversals, and t_0 and $t_{1\_}$ hold for the pair if they hold for the plane defined by $\{t_1 = ... = t_k = 0\}$*

proof. Suppose t_1 holds for all linear direct transversals; let P be a direct transversal, and P_0 the linear direct transversal determined by the tangent plane to P at the origin. If ϕ is a curve on X, then the two modules $M(P) =$
$= \phi^*(m_n JM(F)_P + I(P)JM_t(F))$ and $M(P_0) = \phi^*(m_n JM(F)_{P_0} + I(P_0)JM_t(F))$ differ by $\phi^* m_n{}^2 JM_t(F)$. The inclusion

$$\phi^* m_n JM_t(F) \subset M(P_0)$$

implies that

$$M(P) + \phi^* m_n M(P_0) = M(P_0).$$

By Nakayama's lemma we know that $M(P) = M(P_0)$, so that the inclusion

$$\phi^* m_n JM_t(F) \subset M(P)$$

also holds.

The arguments for t_0 and $t_{1\_}$ are similar. In the case of t_0, $\phi^*(m_n JM_z(F))$ and $M(P)$ differ by $\phi^*((m_{n+k})JM_t(F))$, so again by Nakayama's lemma, if $\phi^*(JM_t(F))$ lies in one module it lies in the other. For the case of $t_{1\_}$, the modules $m_1\phi^*(m_n JM(F)_P + I(P)JM_t(F))$ and $m_1\phi^*(m_n JM_z(F) + (t_1, ..., t_k)JM_t(F))$ differ by $m_1(\phi^*(m_n JM_t(F)$, so we can again apply Nakayama's lemma. $\square$

Corollary 2.6. *Suppose X and Y as above, then t_0 holds for the pair (X_0, Y) at the origin, iff W holds for the pair at the origin.*

proof. The argument of the above proposition showed that t_0 holds for the pair (X_0, Y) at the origin iff $JM_t(F)) \subset \overline{m_n JM_z(F)}$. But by 1.11 this is equivalent to W. □

Proposition 2.7. *Suppose X and Y as above, P a linear space which is a direct transversal. Then t_1 holds for P iff P is not contained in a limit of tangent hyperplanes to X along any curve tangent to P.*

proof. We first show that in checking the t_1 condition only curves tangent to P are relevant. Since P is a direct transversal it is the graph of a linear function $f = (f_1, ..., f_k)$. Suppose c is a curve which is not tangent to P at the origin. Then $c^*(m_{n+k}) \subset c^*(t_i - f_i)$ by Proposition 1.7, so $c^*(m_n)JM_t(F) \subset c^*(t_i - f_i)JM_t(F)$. So it suffices to work with curves tangent to P.

This means that $c^*(I(P)) \subset m_1 c^*(m_{n+k})$; since P is a direct transversal the tangent line of c cannot lie in Y, so $c^*(m_{n+k}) = c^*(m_n)$. Translating the statement of the theorem into algebra, we need to show that the pair of inclusions

$$c^*(JM_t(F)) \subset c^*(JM(F)_P)$$
$$c^*(m_n(JM_t(F))) \subset c^*(m_n(JM(F)_P) + (t_i - f_i)(JM_t(F)))$$

are equivalent for curves c tangent to P. The first inclusion directly implies the second. Starting with the second inclusion, since c is tangent to P, we know that

$$c^*(I(P)(JM_t(F))) \subset m_1 c^*(m_n JM_t(F)).$$

The second inclusion then implies

$$\begin{aligned} c^*(m_n JM(F)_P + I(P)JM_t(F)) &= c^*(m_n JM(F)_P) + m_1 c^*(m_n JM_t(F)) \\ &= c^*(m_n JM(F)_P) + m_1 c^*(m_n JM(F)_P + I(P)JM_t(F)). \end{aligned}$$

By Nakayama's lemma, we then have

$$c^*(m_n JM(F)_P + I(P)JM_t(F)) = c^*(m_n JM(F)_P)$$

so we have the inclusion

$$c^*(m_n(JM_t(F))) \subset c^*(m_n(JM(F)_P)).$$

But this clearly implies that

$$c^*(JM_t(F)) \subset c^*(JM(F)_P).$$

However the inclusion $c^*(JM_t(F)) \subset c^*(JM(F)_P)$ holds iff $c^*(JM(F)_P) = c^*(JM(F))$. The techniques which are used to prove proposition 1.5 show that the last equality holds iff P does not lie in a limit of tangent hyperplanes to X along c. □

Corollary 2.8. *Suppose X and Y as above, then t_1 holds for the pair (X_0, Y) at the origin, iff every limit of tangent hyperplanes to X at the origin along curves not tangent to Y contains Y.*

proof. Suppose t_1 holds for (X_0, Y). If H is a limit of tangent hyperplanes at the origin along the curve ϕ which fails to contain Y, then H contains a direct transversal to Y, which contains the tangent line to ϕ at the origin, so t_1 fails. If we assume the geometric condition, and t_1 fails for P, then by the previous proposition the geometric condition fails for P. □

In order to show that our definition agrees with that of Trotman in the real case, and to study the equisingularity of the family of sections of X by direct transversals, we will prove a version of Theorem 2.3 that holds in both the real and complex analytic case using our definition. In the Grassman modification of $\mathbf{k}^{n+k}$ we will always be working in a neighborhood of a direct transversal P. (Recall we regard P as a point of the zero section of $E_{n+k,n}$.) Since we are always working in $E_{n+k,n}$, we denote the projection $\beta_{n+k,n}$ by β. Since all planes near P are also direct transversals, they are also graphs over the z-plane, and they have equations $\{t_1 = \sum_i a_{1,i}z_i, ..., t_k = \sum_i a_{k,i}z_i\}$. This means we have local coordinates on $\widetilde{\mathbf{k}^{n+k}}$ given by $(z_1, ...z_n, a_{1,1}, ..., a_{k,n})$. In these coordinates we have

$$\beta(z_1, ...z_n, a_{1,1}, ..., a_{k,n}) = (z_1, ...z_n, \sum_i a_{1,i}z_i, ..., \sum_i a_{k,i}z_i)$$

If $\phi : \mathbf{k}, 0 \to \tilde{X}, P \times \{0\}$, then $\beta \circ \phi$ is tangent to P at the origin.

Theorem 2.9. *Suppose $X \subset \mathbf{k}^{n+k}$ is the germ of an analytic space at the origin defined by F, Y a submanifold embedded as $\{0\} \times \mathbf{k}^k$, $s \geq 0$. Then X_0 is t_{s+1} regular over Y at 0 iff $\tilde{X}_0$ is t_s regular over $\tilde{Y}$ at every point of $\tilde{Y}$.*

proof for s=0. Since F defines X, $G := F \circ \beta$ defines $\tilde{X}$. From the chain rule we note that

$$\frac{\partial G}{\partial a_{i,j}} = z_j \frac{\partial F}{\partial t_i} \circ \beta$$

$$JM_z(G) = (\frac{\partial F}{\partial z_j} \circ \beta + \sum_i a_{i,j} \frac{\partial F}{\partial t_i} \circ \beta)$$

Let P be in $\tilde{Y}$, with equations $\{t_1 = \sum_i a_{0\,1,i}z_i, ..., t_k = \sum_i a_{0\,k,i}z_i\}$, let$\phi : \mathbf{k}, 0 \to \tilde{X}, P \times \{0\}$. Since $\phi_1 = \beta \circ \phi$ is tangent to P, we know that the inclusions

$$\phi_1{}^*(m_n(JM_t(F))) \subset \phi_1{}^*(m_n(JM(F)_P) + (t_i - f_i)(JM_t(F)))$$

$$\phi_1{}^*(JM_t(F)) \subset \phi_1{}^*(JM(F)_P) \tag{2.10}$$

are equivalent by the argument of 2.7. The inclusion (2.10) is clearly equivalent to

$$\phi^*\beta^*(JM_t(F)) \subset \phi^*((\frac{\partial F}{\partial z_j} \circ \beta + \sum_i a_{0\,i,j} \frac{\partial F}{\partial t_i} \circ \beta)) \tag{2.11}$$

We are also interested in the inclusion

$$\phi^*(JM_a(G)) \subset \phi^*(m_n JM_z(G))$$

This inclusion can be written as

$$\phi^*(z_j \frac{\partial F}{\partial t_i} \circ \beta) \subset \phi^*(m_n(\frac{\partial F}{\partial x_j} \circ \beta + \sum_i a_{i,j} \frac{\partial F}{\partial t_i} \circ \beta)).$$

This last inclusion is clearly equivalent to

$$\phi^*(\frac{\partial F}{\partial t_i} \circ \beta) \subset \phi^*(\frac{\partial F}{\partial z_j} \circ \beta + \sum_i a_{i,j} \frac{\partial F}{\partial t_i} \circ \beta) \tag{2.12}$$

Notice that the modules in (2.11) and (2.12) differ by $m_1(\phi^*\beta^* JM_t(F))$; so if either inclusion holds, by Nakayama's lemma, both must hold, and the two modules are in fact the same. So if we assume t_1, we know that (2.11) holds for any curve ϕ, so (2.12) holds, and this implies t_0 holds along $\tilde{Y}$. If we assume t_0 and we want to check t_1 for a plane P and a curve ψ tangent to P at the origin, we pick a lift ϕ for ψ then use (2.12) to deduce (2.11) which implies t_1 for P. □

The proof of the the above result for the case of s$\neq$ 0, will apear elsewhere.

Corollary 2.13. *Suppose $X \subset \mathbf{C}^{n+k}$ is the germ of an analytic space at the origin defined by F, Y a submanifold embedded as $\{0\} \times \mathbf{C}^k$, $Y = S(X)$. Then t_1 holds at the origin for the pair (X_0, Y) iff the family $\{P \cap X\}$ is Verdier regular.*

proof. The condition on X implies that near any point of $\tilde{Y}$, the singular set of $\tilde{X}$ is exactly $\tilde{Y}$ (β is a submersion off the zero section). So we have that $\tilde{X}_0 = \tilde{X} - \tilde{Y}$. So, we have t_1 iff W holds for $(\tilde{X} - \tilde{Y}, \tilde{Y})$ iff the family $\{P \cap X\}$ is Verdier regular. □

As we mentioned before, in the real analytic case the above result is due to Kuo and Trotman.

Having shown that the two notions of t_1 coincide we give a geometric characterization of the t_{1-} condition, so we can study it's relation to t_1.

Proposition 2.14. *Suppose $X \subset \mathbf{k}^{n+k}$ is the germ of an analytic space at the origin defined by F, Y a submanifold embedded as $\{0\} \times \mathbf{k}^k$. Then t_{1-} holds at the* *origin for the pair (X_0, Y) iff every limit of tangent hyperplanes along any curve not tangent to Y contains Y.*

proof. Suppose first that c is a curve which is not tangent to Y at the origin, so that $c^*(m_n) \supset c^*(m_k)$. An argument using Nakayama's lemma then shows that the inclusions

$$c^*(JM_t(F)) \subset m_1 c^*(JM_z(F))$$

$$c^*(m_n JM_t(F)) \subset m_1 c^*(m_n JM_z(F) + (t_i) JM_t(F))$$

are equivalent for such a curve. By an argument analogous to 1.6, the inclusion

$$c^*(JM_t(F)) \subset m_1 c^*(JM_z(F))$$

is equivalent to every limiting tangent hyperplane along c containing Y, while the second inclusion checks the t_1- condition for the curve c.

Now assume t_1- holds; then the second inclusion holds, so the first does as well.

Now assume the geometric condition; this implies that the t_1- condition checks for all curves not tangent to Y. If c is a curve tangent to Y, then t_1- cannot fail for such a curve, for $c^*(m_n) \subset m_1 c^*(m_k)$ □

By putting together 2.14 and 2.8 we get the following relation between t_1 and t_{1-}.

Corollary 2.15. *Suppose X and Y are as in 2.14. Then t_1 holds at the origin for the pair (X_0, Y) iff t_{1-} holds at the origin.* □

This result shows that it is much stronger for t_1 to hold for the pair (X_0, Y) than it is for t_1 to hold for the individual planes P.

Corollary 2.16. *Suppose X and Y are as in 2.14, but in addition assume Y has dimension 1. Then t_1 holds at the origin for the pair (X_0, Y) iff Whitney A holds at the origin.*

proof. Whitney A implies t_{1-} by 2.14, so it implies t_1. Suppose t_1 holds at the origin, and suppose H is a limiting tangent hyperplane along a curve c which fails to contain Y. Then c cannot be tangent to Y, for H contains the tangent line to c, so H would contain Y. (Here is where we need the dimension of Y is 1.) If c is not tangent to Y then H cannot exist by 2.8. □

If we combine 2.15 and 2.9 we can show that t_r holds at the origin for the pair (X_0, Y) iff t_{r-} holds at the origin for the pair.

§3. The Principle of the Specialization of Integral Dependence and the t_r conditions.

So far we have not used very strongly the algebraic character of the theory of integral closure. However there is a remarkable result which comes from this side of the theory. In many applications we have a family of sets, and a sheaf of modules on each set. These modules often come from specializing a sheaf of modules contained in a locally free sheaf on the total space of the family to the members of the family. Often we know that some element of the free sheaf is in the integral closure of the modules on the members of the family generically. Is it in the integral closure of the member-module for all members? Is it in the integral closure of the sheaf of modules which is defined on the total space? This type of question often arises in mathematics, and the answer is usually no. The theory of integral closure in some cases associates an invariant to the member-modules called their multiplicity. If these multiplicities are constant then the answer to both questions is yes! This is called the Principle of the Specialization of Integral Dependence first stated for ideals by Bernard Teissier.

If one considers the maximal ideal m_x of a local ring $\mathcal{O}_{X,x}$, where X is equidimensional at x, then the multiplicity of m_x is just the multiplicity of X at $x-$ the number of points near x in which a generic linear space of dimension complementary to X will intersect X. If one considers an ideal I of finite colength (the colength of a module $M \subset \mathcal{O}_X{}^p$ is the dimension of the quotient module $\frac{\mathcal{O}_X{}^p}{M}$ as a complex vector space.), one can pick a set of generators of I, and use these to define a map-germ f with image $f(X)$. The multiplicity of I is just the product of the multiplicity of $f(X)$ and the degree of f. The multiplicity of a module is defined for all modules of finite colength. If we are working over a ring which is Cohen-Macaulay then the multiplicity of $M \subset \mathcal{O}^p{}_X$ can be gotten by taking the colength of the submodule generated by $d+p-1$ generic elements of M, where d is the dimension of X. Before stating the Principle of Specialization of Integral Dependence precisely, we give an example to make these ideas concrete.

Example 3.1 Consider the ideal $I = (x^2, y^2, txy)$ in $\mathcal{O}_3$; we think of $\mathbf{C}^3$ as the total space of a family of planes parametrised by $\mathbf{C}$ embedded as the t-axis. The ideal I induces a family of ideals of finite colength I_t. Clearly the term $xy \in I_t$ for $t \neq 0$. The multiplicity of I_t is 4 for all values of t, so we know that $xy \in \overline{I}$. □

If we have a mapping between two spaces X and Y, and the restriction of a sheaf of modules to the fiber over Y is supported at a finite number of points, then we denote by $e(y)$ the sum over the points in the fiber of y of the multiplicities of the stalks of the sheaf.

Theorem 3.2. *(Specialization of integral dependence) Assume that X is equidimensional, and that $F\colon (X, x_0) \to (Y, y_0)$ is a map of germs of analytic spaces. Assume that the fibers $X(y)$ are equidimensional of the same dimension d, d at least 1, and that Y has dimension at least 1. Let M be a coherent submodule of $\mathcal{O}^p{}_X$. Set $S := Supp(\frac{\mathcal{O}_X{}^p}{M})$, and assume that S is finite over Y. Assume $y \mapsto e(y)$ is constant on Y. Let h be a section of $\mathcal{O}^p{}_X$ whose image in $\mathcal{O}^p{}_{X_y}$ is integrally dependent on the image of M for y in a Zariski open set. Then (after X and Y are replaced by neighborhoods of x_0 and y_0 if necessary) h is integrally dependent on M.*

proof. The proof will appear in [G-K]. □

We can combine 3.2 with the theorems and propositions of the last section to obtain many nice examples of how the Principle can be applied to geometry. Recall that $\mu(X)$ denotes the Milnor number of X, while $\mu_1(X)$ denotes the Milnor number of a generic hyperplane slice of X. From now on we will be working only in the complex analytic case.

Theorem 3.3. *Suppose $X^d \subset \mathbf{C}^{n+k}$ is a complete intersection with $Y^k = S(X)$, Y smooth. Then $\mu(X \cap P)$, $\mu_1(X \cap P)$ are constant for all planes P which are direct transversals iff t_1 holds for (X_0, Y) at the origin iff the family $\{X \cap P\}$ is Whitney equisingular.*

proof. Assume that the Milnor numbers are constant. It is shown in [G2] that the constantcy of $\mu(X \cap P)$, $\mu_1(X \cap P)$ is equivalent to the constancy of the multiplicity of the jacobian module of $(X \cap P)$ in $\mathcal{O}^{n+2k-d}{}_{X\cap P}$. Denote the mapgerm defining P by f_P. Now, we have that the modules

$$\frac{\mathcal{O}^{n+2k-d}{}_{X\cap P}}{JM(F, f_P)} \approx \frac{\mathcal{O}^{n+k-d}{}_{X\cap P}}{JM(F)_P}$$

so that the multiplicities of the modules $JM(F)_P$ in $\mathcal{O}^{n+k-d}{}_{X\cap P}$ are independent of P as well. Now return to the notation of the proof of Theorem 2.9. Consider the inclusions

$$(3.4) \qquad \left.\frac{\partial F}{\partial t_i}\right|_P \in \overline{(JM(F)_P)} \subset \mathcal{O}^{n+k-d}{}_{X\cap P} \quad \forall i$$

$$(3.5) \qquad \left.\frac{\partial F}{\partial t_i} \circ \beta\right|_{a_{i,j}=a_{0i,j}} \in \overline{\left.\left(\frac{\partial F}{\partial z_j} \circ \beta + \sum_i a_{0\,i,j} \frac{\partial F}{\partial t_i} \circ \beta\right)\right|_{a_{i,j}=a_{0i,j}}} \quad \forall i$$

$$(3.6) \qquad \left.\frac{\partial G}{\partial a_{i,j}}\right|_{a_{i,j}=a_{0i,j}} \in \overline{(m_n JM_z(G_{a_{0i,j}}))} \quad \forall i, j$$

As the proof of 2.9 shows these inclusions are all equivalent. We know that 3.6 holds generically, because the Whitney conditions hold on a Zariski open set. The constancy of the multiplicity of $JM(F)_P$ implies that the multiplicity of $\left.\left(\frac{\partial F}{\partial z_j} \circ \beta + \sum_i a_{0\,i,j} \frac{\partial F}{\partial t_i} \circ \beta\right)\right|_{a_{i,j}=a_{0i,j}}$ is constant since the induced map of quotient modules is an isomorphism. Now apply 3.2 to the sheaf of modules $(\frac{\partial F}{\partial z_j} \circ \beta + \sum_i a_{i,j} \frac{\partial F}{\partial t_i} \circ \beta)$ and to the sections $\frac{\partial F}{\partial t_i} \circ \beta$. It follows that the inclusion 3.5 holds at the level of the total space, hence so does 3.6, hence the family $\{X \cap P\}$ is Whitney equisingular, and t_1 holds for (X_0, Y) at the origin.

If we assume that $\{X \cap P\}$ is Whitney equisingular, then since the Milnor numbers are invariants of Whitney equisingularity type, we have that they must be constant. □

There is another version of the t_r conditions which asks that the direct transversal be transverse to nearby levels of the equations which define X as well as being transverse to X. As is typical in applications of limiting tangent hyperplanes, this theory is no more difficult than the ground already covered. We say that the pair (X_0, Y) satisfies the ambient t_r condition iff

$$m_n{}^r JM_t(F) \subset \overline{m_n JM(F)_P + I(P)JM_t(F)} \subset \mathcal{O}_{n+k}{}^p.$$

All of the results of section 2 have analogues for the ambient t_r conditions. In the complete intersection case, the ambient t_1 condition not only implies W for the pair $(\tilde{X}_0, \tilde{Y})$ but also implies W_f; namely, the distance between the limiting tangent hyperplanes to the fibers of $F \circ \beta$ at points where $F \circ \beta$ is a submersion go to zero as fast as you approach $\tilde{Y}$. If X is a hypersurface, and $S(X) = Y$ this condition implies the Verdier right triviality of $F \circ \beta$, because $F \circ \beta$ is a submersion off $\tilde{Y}$, so you can define a rugouse vectorfield tangent to the fibers of $F \circ \beta$ lifting the constant fields on $\tilde{Y}$. As usual, numerical conditions which draw on the ambient topology give stronger results than those defined solely on the sets themselves. In the hypersurface case the style of argument used in 3.3 yields:

Theorem 3.7. *Suppose $X^{n+k-1} \subset \mathbf{C}^{n+k}$ is a hypersurface with $Y^k = S(X)$, Y smooth. Then $\mu(X \cap P)$ is constant iff ambient t_1 holds for (X_0, Y) at the origin iff the family of mapgerms $\{F \circ \beta|_{a_{i,j}=a_{0i,j}}\}$ are right Verdier equisingular.*

□

Corollary 3.8. *Suppose X is the germ of a complex analytic hypersurface at a point z, with smooth singular locus Y. Suppose that for every plane P of dimension complementay to Y and transverse to Y at z, the Milnor number of $\{X \cap P\}$ is independent of P. Then the family $\{X \cap P\}$ is Whitney Equisingular.*

□

Note that in the above result there is no dimension restriction on $X \cap P$ as in the theorem of Le-Ramanujam [L-R]. This result shows that the constraint of being a family of sections of a hypersurface is so strong that the topological triviality of the family implies the Whitney equisingularity of the family. This means for example, that the 1-parameter family of surfaces in $\mathbf{C}^3$ that make up the Briancon-Speder example of a family of topologically trivial surfaces which are not Whitney equisingular can never appear in any family of direct transversal sections of a hypersurface for which t_1 holds.

It is possible to get a result similar to Theorem 3.7 for the case of general complete intersections X with smooth singular locus Y. However, you must work with a chain of complete intersections $\mathbf{C}^{n+k} \supset X_1 \supset \ldots \supset X$, each a hypersurface inside the preceding element of the chain; then your hypothesis will involve the constancy of the Milnor numbers of $\{X_i \cap P\}$.

We can combine these results to get a numerical criterion for Whitney A in the case when the dimension of Y is 1.

Corollary 3.9. *Suppose X is the germ of a complex analytic set with smooth singular locus Y of dimension 1.*
1) If X is a hypersurface, then Whitney A holds for the pair (X_0, Y) iff for every plane P of dimension complementary to Y and transverse to Y at z, the Milnor number of $\{X \cap P\}$ is independent of P.
2) If X is a complete intersection, then Whitney A holds for the pair (X_0, Y) iff for every plane P of dimension complementary to Y and transverse to Y at z, the Milnor number of $\{X \cap P\}$ and of its generic hyperplane section is independent of P. □

It is instructive to apply these results to an example of Trotman's.

Example 3.10 ([Tr 3]) Consider the family of sets defined by $F_{a,b,c,d} = y^a - (t^b x^c + x^d) = 0$. In his thesis, Trotman worked out the relationship between the values of a, b, c, d and the regularity conditions satisfied by the pair $(X - Y, Y)$, where Y is the t-axis in $\mathbf{k}^3$, and $X = {F_{a,b,c,d}}^{-1}(0)$. For now, let's consider the cases where $\mathbf{k} = \mathbf{C}$, and $b \geq a > 1, c < d < b + c$ and where $a > 1,\ b + c < d$.

Since the dimension of Y is one, a direct transversal will be a hyperplane. We make a substitution $t = ux + vy$ for t into $F_{a,b,c,d}$ to get the equation of $X \cap P$. The substitution yields $G_{a,b,c,d} = y^a - (ux + vy)^b x^c + x^d$. In the second case, the Milnor number of $G_{a,b,c,d}$ is $(a-1)(d-1)$ when $u = v = 0$, and is $(a-1)(b+c-1)$ when $v = 0$. Since the Milnor number is not constant, Whitney A fails. In the first case, the Milnor number is always $(a-1)(d-1)$ because the terms $(ux+vy)^b x^c$ lie above the Newton diagram of $y^a - x^d$, since $b > a$ and $b + c > d$. So, Whitney A holds in this case. □

As Trotman pointed out to me, this example shows that it is impossible to have a numerical criterion for Whitney A which depends only on the members of the family, and which is necessary and sufficient. If you consider the two families $\{X_t\}$, $\{W_t\}$ defined by $y^2 - (t^3 x^4 + x^6) = 0$ and by $y^2 - (tx^4 + x^6) = 0$ as families X and W have the same members, but in the first case Whitney A holds for the pair $(X - Y, Y)$ and in the second it fails for $(W - Y, Y)$. However as our results show the "next best result" is true, namely if we add in numerical information from all of the sections of the total space coming from sections by linear direct transversals through a fixed point, we do get a necessary and sufficient condition.

REFERENCES

[B] G.W. Brumfiel, Real valuation rings and ideals, Geometrique Algebrique Reele et Formes Quadratiques. Rennes 1981. SLN# 959, pp. 55-97 (1982).

[B-R] D. Buchsbaum and D. Rim, A generalized Koszul complex II. Depth and multiplicity, Trans. Amer. Math. Soc. 111 (1968) p. 197-224.

[G1] T. Gaffney, *Integral closure of modules and Whitney equisingularity*, Inv. math. **107** (1992), 301-322.

[G2] T. Gaffney, Multiplicities and equisingularity of ICIS germs,, to appear in Inventiones.

[G3] T. Gaffney, "Aureoles and integral closure of modules", to appear proceedings of the Hawaii-Luminy conference.

[G-K] T. Gaffney and S. Kleiman, *Specialization of integral dependence for modules*, in preparation.

[Gi] C. G. Gibson et al., *Topological stability of smooth maps*, Springer Lect. Notes **552** (1976).

[H-L] J.P. Henry, D.T. Lê, Limites d'espaces tangents, Sem. Norguet, SLN #482.

[K-T] T. C. Kuo and D. Trotman, *On (w) and (t^s)-regular stratifications*, Inv. math. **92** (1988), 633–643.

[L] D.T. Lê, Calculation of Milnor number of an isolated singularity of a complete intersection, Funk. Anal. i Ego Pril. 8, 45-49 (1974).

[L-R] Lê D. T. and C. P. Ramanujam, *The Invariance of Milnor's Number implies the Invariance of the Topological Type*, Amer. Journ. Math. **98** (1976), 67–78.

[L-S] Lê D. T. and K. Saito, *La constance du nombre de Milnor donne des bonnes stratifications*, C.R. Acad. Sci. **277** (1973), 793–795.

[L-T] D.T. Lê, B. Teissier, Limites d'espace tangents en géométrie analytique, Comm. Helv. 63 (1988).

[R1] D. Rees, Gaffney's problem, notes (1989).

[R2] D. Rees, Reduction of modules, Math. Proc. Camb. Phil. Soc. (1987) 101 p. 431.

[T] B. Teissier, *Varietes polaires II: Multiplicites polaires, sections planes, et conditions de Whitney*, in Algebraic Geometry, Proc., La Rabida 1981, Springer Lect. Notes **961** (1982), 314–491.

[Tr1] D. Trotman, *Comparing regularity conditions on stratifications*, Arcata 1981, Proc. Symp. Pure Math. **40** (1983), 575–586.

[Tr2] D. Trotman, *Transverse transversals and homeomorphic transversals*, Topology **24** (1985), 25–39.

[Tr3] D. Trotman, *Whitney Stratifications: Faults and Detectors*, Warwick University, Thesis (1977).

[T-W] D. Trotman and L. Wilson, *Stratifications and finite determinacy*, in preparation.

J.S. Carter & M. Saito

Knot diagrams and braid theories in dimension 4

Abstract

We explore the diagrammatic aspects of surfaces embedded in 4-dimensional space, the braid theory of such surfaces, and related topics from category theory and mathematical physics.

1 Introduction

In this paper we explore the question : WHAT DIAGRAMMATICAL ASPECTS OF CLASSICAL KNOT THEORY CAN BE CARRIED OVER TO KNOTTED SURFACES IN 4-DIMENSIONS ? First we indicate why the question makes sense.

By *classical knot theory* we mean the theory of smoothly knotted and linked closed curves in 3-dimensional Euclidean space. The embedded curves are projected generically onto an affine 2-plane so that the *projection* consists of immersed loops with transverse double points. The *diagram of a knot or link* is obtained from the projection by breaking the underpassing arc at each crossing. For example, when one looks at a knot, the image is projected onto the retina, and the visual information contained in the shadows indicates which of the two arcs is closer to the observer. Most of the known invariants of classical knots can be defined and computed via diagrams. Examples of diagrammatically definable aspects include the fundamental group, a Seifert surface of the knot (from which the Alexander invariants can be defined) [65], and the Jones-HOMPTFLY polynomials [32].

One reason that classical knot theory can be studied diagrammatically is that there is a short list of moves-to-diagrams (the Reidemeister moves) that can be used to transform one diagram to another. Thus to define an invariant, one can define it via a representative diagram and check that the definition remains unchanged under the Reidemeister moves. Moreover, knots can be put into braid form and invariants are often found by means of representations of the braid group. The braid form is useful in a purely geometric context as well, as shown by Birman and Menasco [6].

Most of the geometric aspects of classical knot theory can be carried over to the case of smoothly embedded surfaces in 4-dimensional space (with the standard smooth structure assumed). The purpose of this paper is to give a summary of the diagrammatic theory of knotted surfaces, thereby establishing the geometric foundations in this subject. We also discuss algebraic and category theoretical aspects.

The theory of knotted surfaces has a long and rich history. Suzuki's article [73] contains a nice survey of the subject up to 1976. The "motion picture method" was

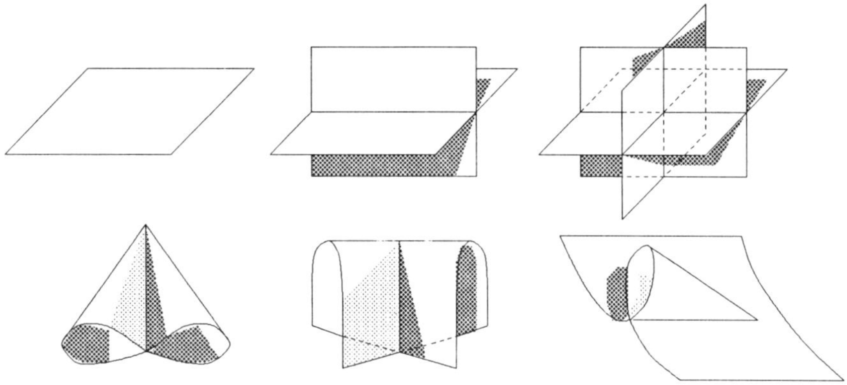

Figure 1: Generic neighborhoods of surfaces in 3-space

popularized in Fox's article [21], and geometric approaches to knotted surfaces can be found in [78, 29, 48, 49, 33, 57, 63] for example. Algebraic aspects and relations to geometric structures of knot theory in dimension 4 are discussed, for example, in [27] and [3]. Applications of Freedman and Donaldson theories are found in [19, 28]. Our focus here, however, is on knot diagrams and the braid theories that were developed recently.

2 Knot Diagrams

2.1 GENERIC PROJECTIONS AND DIAGRAMS. Before we define the diagram of a knotted surface, we develop some notation. Let $K : F \to \mathbf{R}^4$ denote a smooth embedding of a closed surface F into 4-dimensional space (with the standard smooth structure.) Choose a direction $v \in \mathbf{R}^4$ and project $\mathbf{R}^4$ onto a 3-plane perpendicular to the vector v. Choose such an affine hyperplane so that $K : F \to \mathbf{R}^4$ lies in one component of $\mathbf{R}^4 \setminus \mathbf{R}^3$. Denote such a projection by p_v or p.

For almost all choices of the vector v, the composition $p_v \circ K$ will be a *generic surface in 3-space.* In this case, any point of the image surface has a neighborhood U in 3-space such that $p_v(K(F)) \cap U$ looks like one of the pictures indicated in Figure 1. In other words, there is a diffeomorphism of U into $\mathbf{R}^3$ such that the image of F coincides with the intersection of 1, 2, or 3 coordinate planes, or is like the cone on a figure 8. The cone point is called a *branch point;* two other isotopic pictures are depicted in Figure 1.

There is an immersion in $\mathbf{R}^3$ of a compact 1-manifold with boundary (called *the double point manifold*) whose image is the closure of the set

$$\{x \in p(K(F)) | p(K(x_1)) = x = p(K(x_2)) \text{ for some } x_1 \neq x_2, \quad \text{where } x_1, x_2 \in F\}.$$

The closure of the above set is called *the double point set.* The image of the boundary points of the 1-manifold are the branch points of the map $p \circ K$. At a triple point three curves meet transversely, so a triple point is a 6-valent vertex.

Similarly, there is a closed 1-manifold (called *the double decker manifold*) which is immersed into F such that the image (called the *double decker set*) is the closure of the set

$$\{x \in F : \#(p \circ K)^{-1}(p(K(F))) \geq 2\}.$$

The double decker manifold is the 2-fold branched cover of the double point manifold; the branch points of the covering are the preimage of the branch points of the projected surface. There is an involution on the double decker manifold, the fixed points of which are the branch points, and the portion of the double decker set on $K(F)$ that is farthest away from the hyperplane of projection is called the *lower-deck;* the *upper-deck* is the portion of the double decker set that is closest to this hyperplane.

In a *broken surface diagram* of a knotted surface, the portion of the surface that is farthest from the hyperplane of projection is broken. We remove a small open tubular neighborhood of the lower-deck and embed the resulting surface with boundary into 3-space using the restriction of $p \circ K$. In a neighborhood of the branch points, the deleted neighborhood comes to a cone at each branch point. Then we can reconstruct an embedding in 4-space that is isotopic to the given knot $K(F)$, by gluing the neighborhood of the lower-deck to the embedded surface in 3-space: The boundary of this neighborhood lies in 3-space, and the interior protrudes "below" this hyperplane — in other words, the neighborhood lies in the same component of $\mathbf{R}^4 \setminus \mathbf{R}^3$ that $K(F)$ does.

This broken surface diagram is the *diagram of the knotting K(F).* Along open arcs of double points there is a top and bottom sheet, and these are neighborhoods of the upper and lower deck. In a neighborhood of a triple point, there is a top, middle, and bottom sheet, and a branch point is a point along a double curve at which the bottom sheet converges to the top sheet. The notion is easily seen to be equivalent to coloring the generic surface by a palette in which the distance in 4-space from the plane of projection is indicated via some color gradient. Such a coloring convention is found in the video [68].

2.1.1 EXAMPLES. Two well-know examples of surfaces that embed in 4-dimensional space are the projective plane and the Klein bottle. In Figure 2(a) we depict the standard cross-cap embedding, and Figure 2(b) is a depiction of the standard embedding of the Klein bottle. Both figures indicate broken surface diagrams, and the regions at which the diagram is broken corresponds to a piece of the surface that is below the 3-space into which we are projecting. In Figure 2(a), one of the sheets along the intersection curve is cut open.

In Figure 3 one-half of a sphere that is knotted in 4-dimensional space is depicted. The sphere is obtained from the given disk diagram by doubling the diagram along the bounding curve. The represented sphere is knotted in the sense that it cannot be

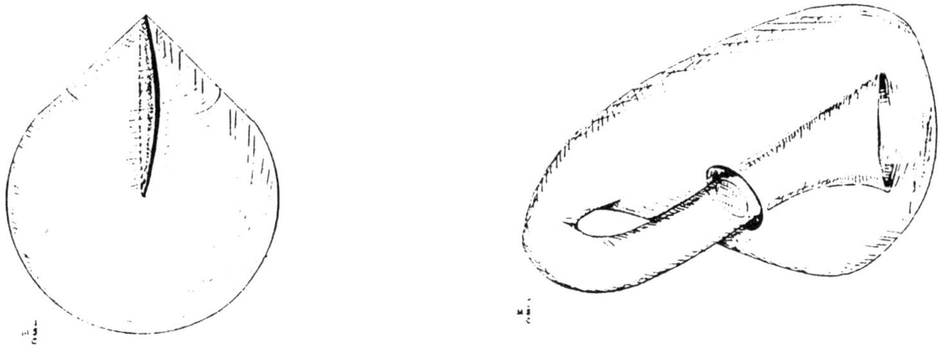

Figure 2: The projective plane and the Klein bottle

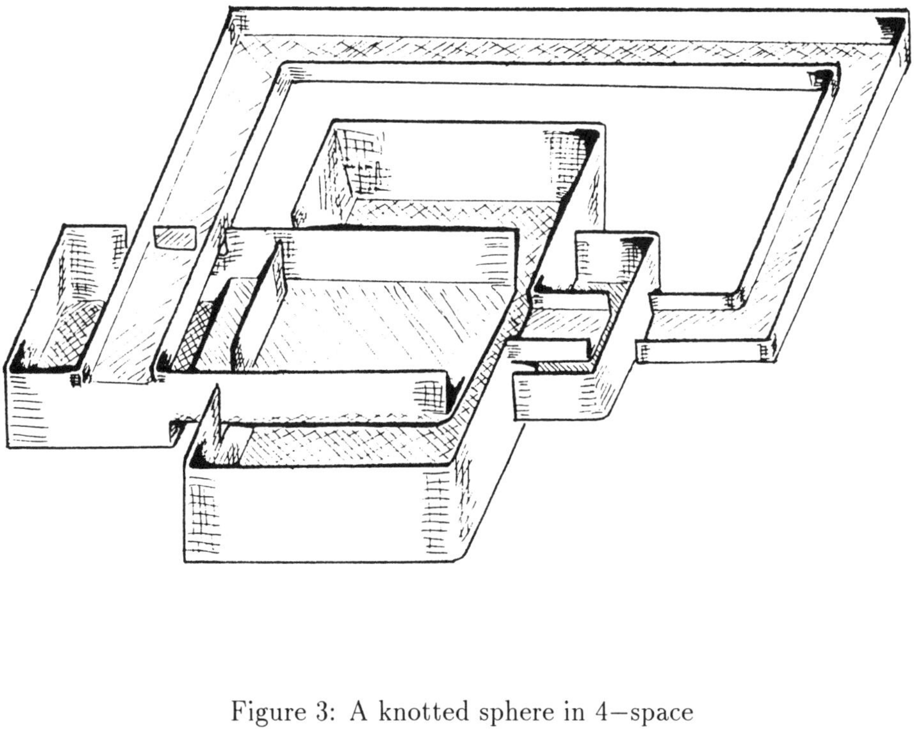

Figure 3: A knotted sphere in 4−space

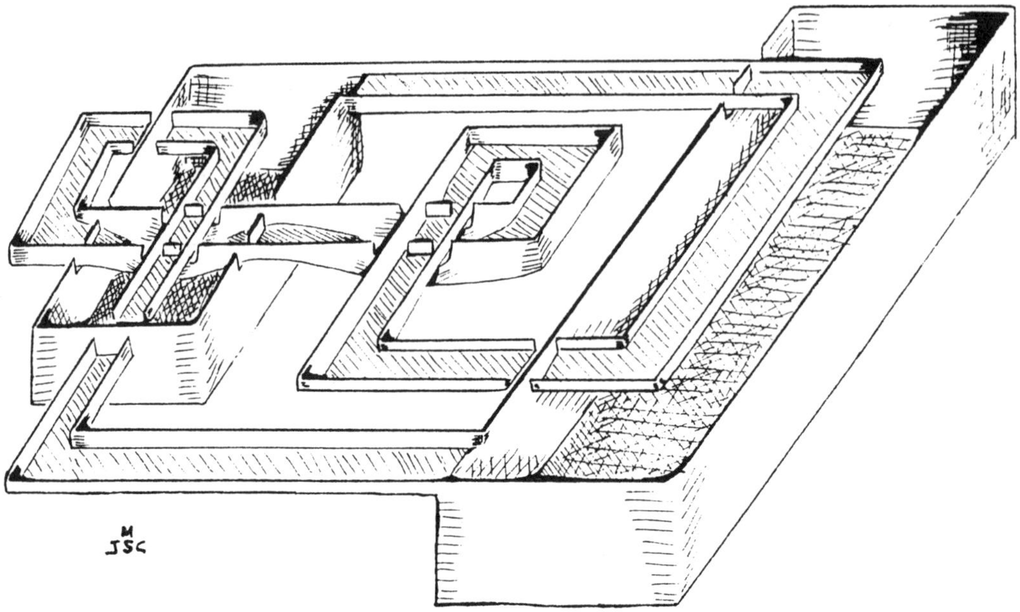

Figure 4: An unknotted sphere in 4− space

ambiently isotoped into a standardly embedded 2-sphere in 3-space. This is knotted because the cross-sectional knot shown in the Figure as the boundary curve is the *Stevedore's* knot and it is known that this knot does not appear as a cross-section of the standard sphere in 4-space (p243, [72]).

In contrast, Figure 4 depicts one-half of an unknotted sphere. The classical knot diagram that is the cross section of Fig. 4 (*i.e.,* the boundary of the disk indicated) is the Kinoshita-Terasaka knot. By performing a band pass move (see [46]) on the portion of the band on the left of the diagram, the unknot can be obtained. The disk indicated can be deformed to a standard embedding of a 2-sphere by using the unknotting disk obtained by the band pass move. We are grateful to Y. Marumoto for telling us of this example [60]. These examples indicate that knotting in 4-dimensions is a subtle phenomenon that is worthy of intense study.

2.2 REIDEMEISTER MOVES AND ROSEMAN MOVES. In classical knot theory two diagrams represent the same knot if and only if one can be moved to another by means of the Reidemeister moves that are depicted in Figure 5. These moves represent local changes in the diagrams of a given knot. A neighborhood of a point in the diagram in which one of the figures appears is replaced by a neighborhood in which the other appears while the diagram remains fixed outside of this neighborhood. These local changes on diagrams clearly do not change knot types. The Reidemeister Theorem states that the converse is true: *If two diagrams represent the same knot then one can*

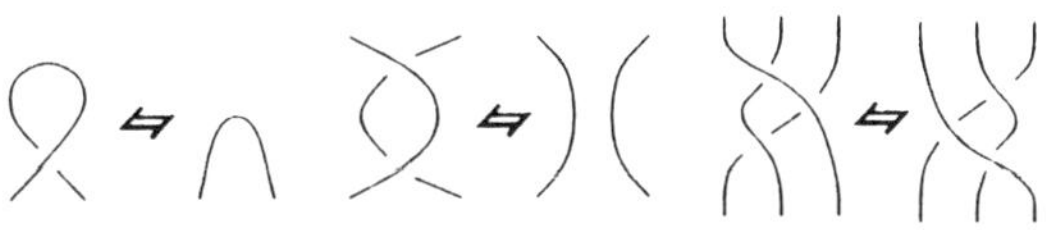

Figure 5: The classical Reidemeister moves

be transformed into the other by means of a sequence of these moves. Dennis Roseman proved an analogous theorem for surfaces embedded in 4-space.

2.2.1 THEOREM (ROSEMAN [66]). Two diagrams represent the same knotting if and only if one can be obtained from the other by a sequence of one of the seven moves depicted in Figure 6.

2.2.2 REMARK. Here we have depicted the moves as moves on projections. When crossing information is included, each move has to respect the crossing information. Thus there might be more than one way to lift each move-to-diagrams. Similarly, in the depiction of the Reidemeister moves we have only indicated crossings in one possible configuration.

2.3 MOVIE MOVES. Roseman's moves do not take into consideration any height function that may be given on the projected surface $p(K(F))$. Similarly, the Reidemeister Theorem does not take into consideration the height of the classical knot diagram. On the other hand, the definitions of a braided monoidal category, a ribbon Hopf algebra, and the topological field theories that result all depend on the auxiliary moves that are needed to change the knot diagram in case a height function is present. These extra moves are shown here in Figure 7.

The movie method is an effective way of visualizing knots in dimension 4. Originally projections were not used, but a direction for the movie was fixed. Combining movies and projections, we can describe knotted surface diagrams as sequences of classical knot diagrams by fixing a height function on the 3-space into which a surface is projected. In [9, 10] we proved that there is a set of moves to projections when a height function is fixed, generalizing Roseman's moves. We described them as moves to movies. An idea put forth to us by Kauffman was that a state sum invariant of knotted surfaces might be achieved in the spirit of the bracket model of the Jones polynomial. This set of moves to movies is a necessary ingredient for such a model.

Let us fix notation. Let $K : F \to \mathbf{R}^4$ be an embedding, let $p : \mathbf{R}^4 \to \mathbf{R}^3$ be a projection such that $p(K(F))$ is a generic surface, and let $h : \mathbf{R}^3 \to \mathbf{R}$ denote a *generic height function* of $p(K(F))$. The map h has the following properties:

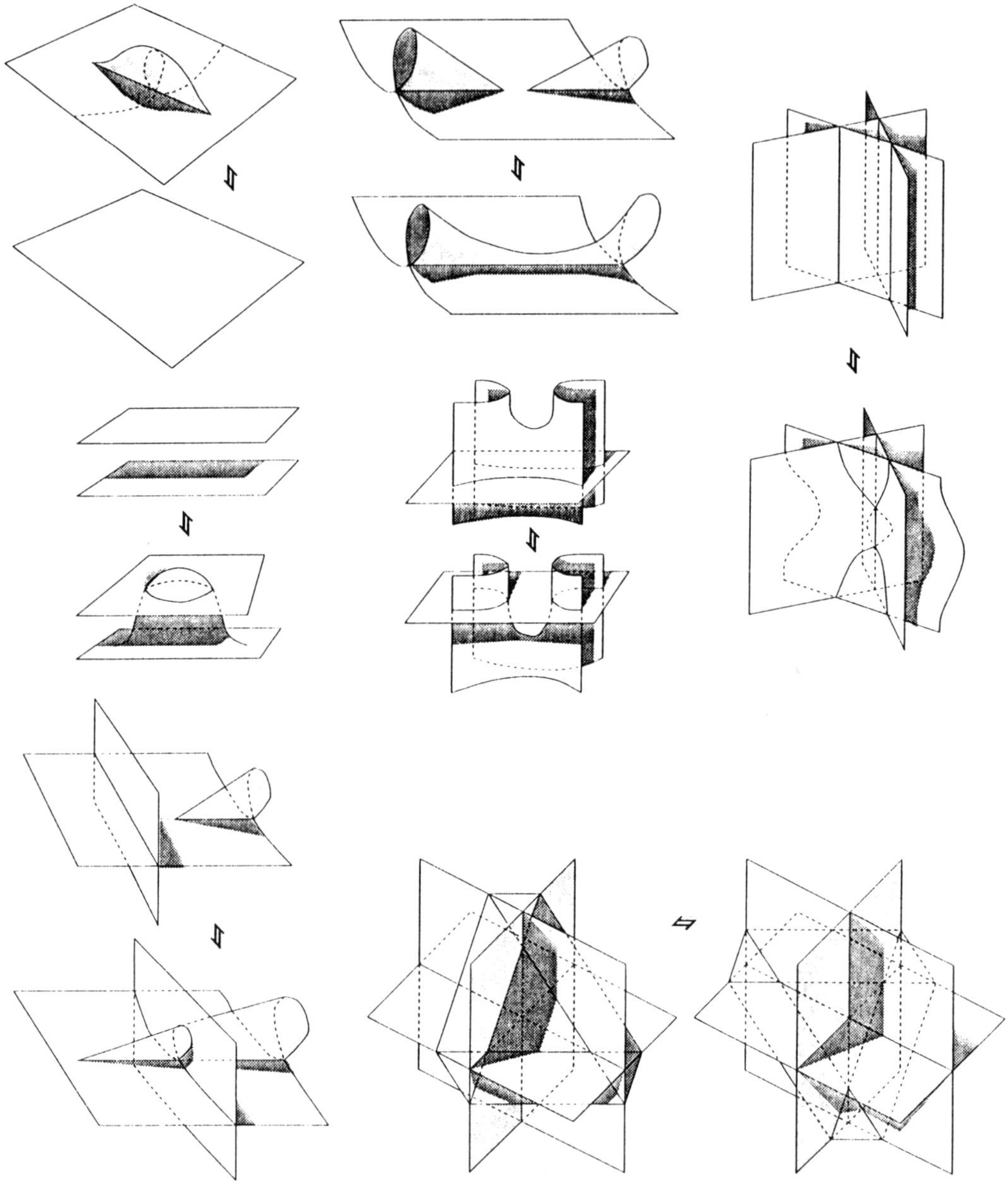

Figure 6: Roseman Moves

Figure 7: Additional moves for classical knot diagrams

1. $h \circ p \circ K$ is a generic height function for F. Thus the critical points of this decomposition are all non-degenerate, and the correspondence between critical points and critical values is a bijection.

2. The function h is a generic height function for the double point manifold of $p \circ K$. So the critical points of the double point manifold are non-degenerate, and distinct critical points have distinct critical values.

3. The critical values of F and those of the double point manifold are distinct, and these values are also distinct from the set of heights achieved by the triple points of $p \circ K$.

Such a height function can be found by using general position arguments.

In this case, if we take slices before/after a critical level, we see one of the following changes, called *elementary string interactions (ESIs)*, on classical knot diagrams:

1. Birth/death of an isolated small circle.

2. Surgery on curves (a saddle point of an embedded surface).

3. Type I classical Reidemeister move (a branch point on the projection).

4. Type II classical Reidemeister move (a min/max point of the double point set of the projection).

5. Type III classical Reidemeister move (a triple point on the projection).

A set of moves among sequences of such changes are depicted in Figures 8, 9, and 10. These moves do not change isotopy type of embedded surfaces. We proved the converse.

2.3.1 THEOREM. [10] *Let h be a generic height function and p a generic projection for isotopic knottings K_1 and K_2. Then the knot diagram of K_1 is obtained from that of K_2 by a sequence of local moves to the fixed-height diagrams depicted in Figures 8, 9, and 10 and isotopy of the diagrams that carry critical points to critical points.*

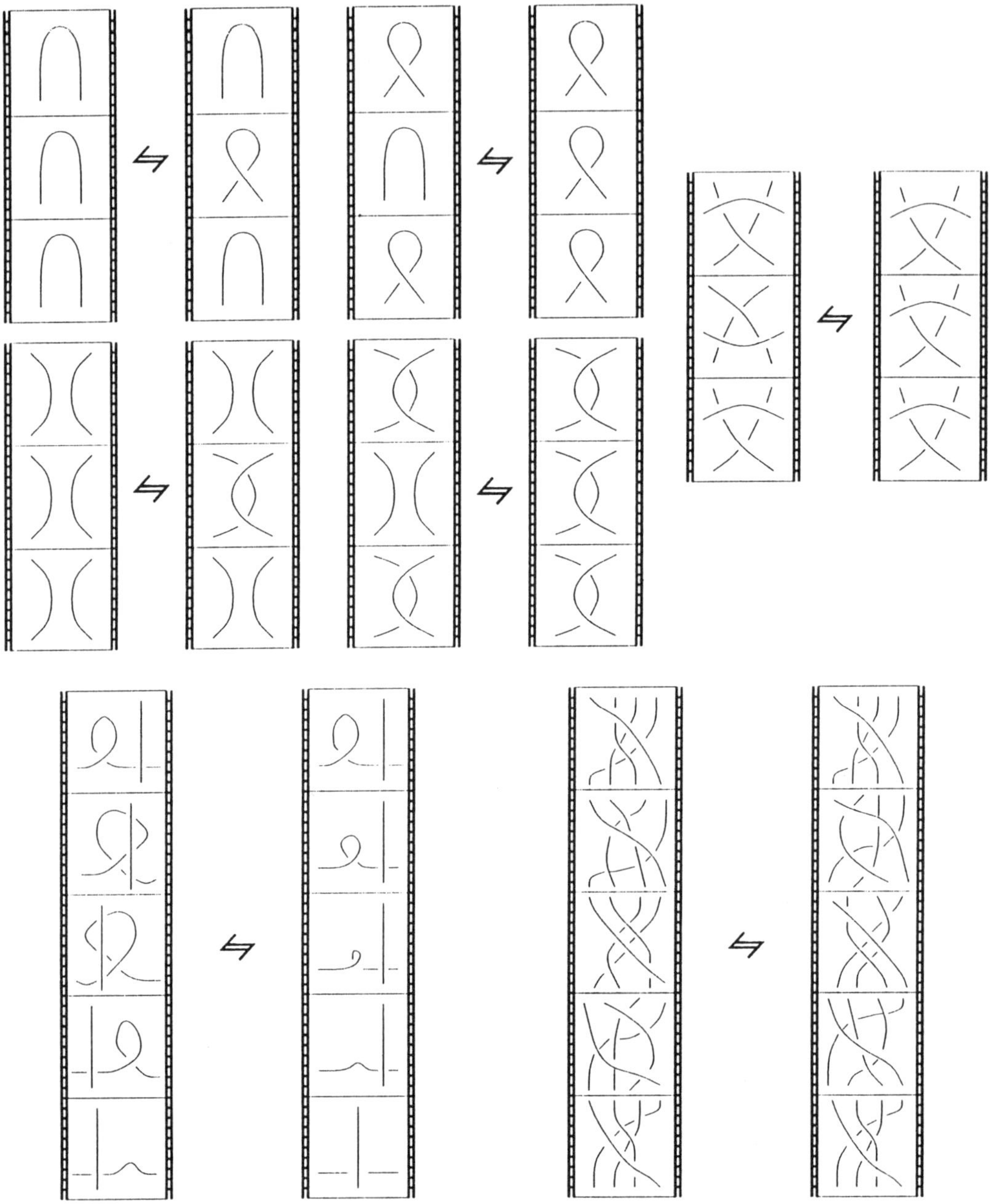

Figure 8: Movie Moves; Part I

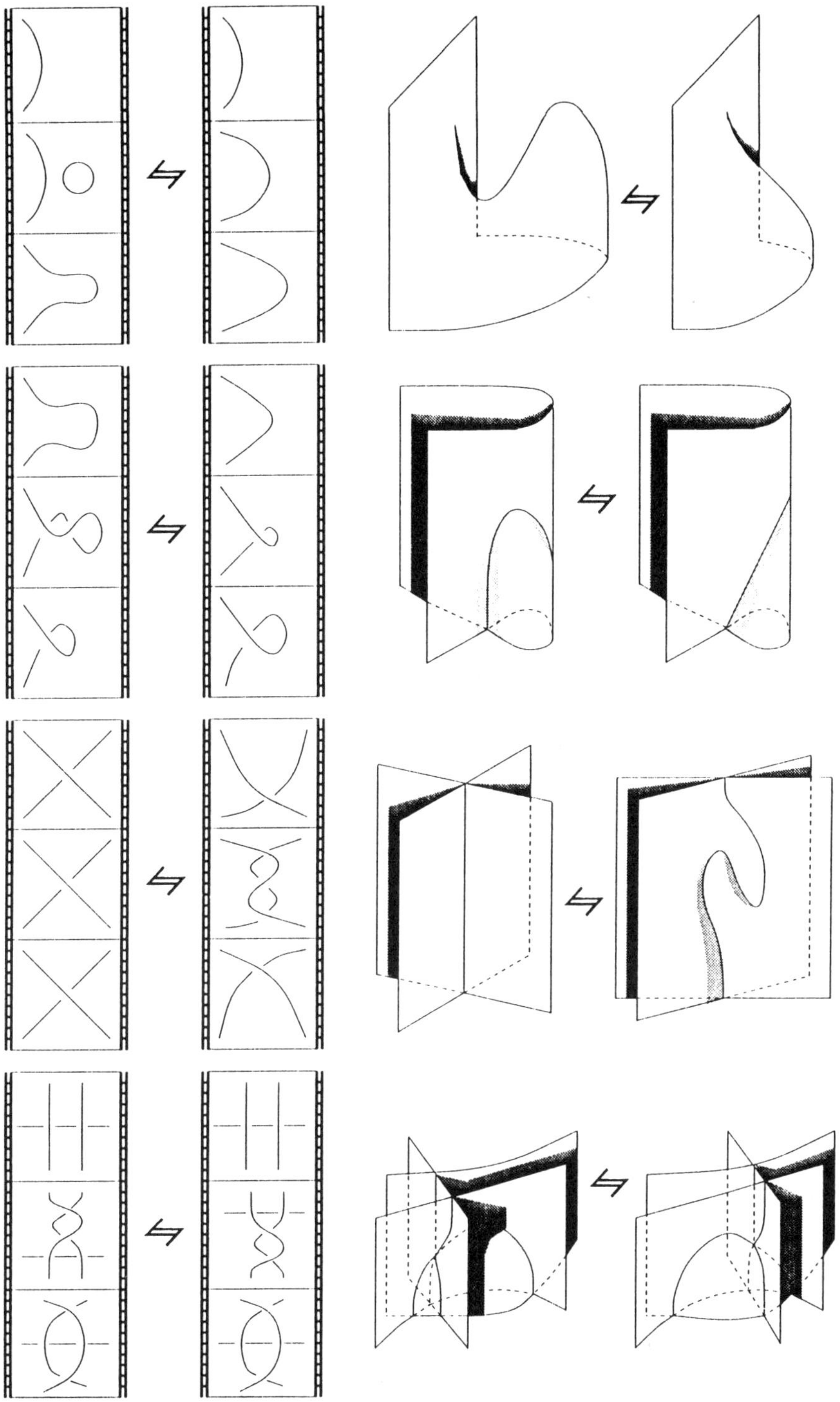

Figure 9: Movie Moves; Part II

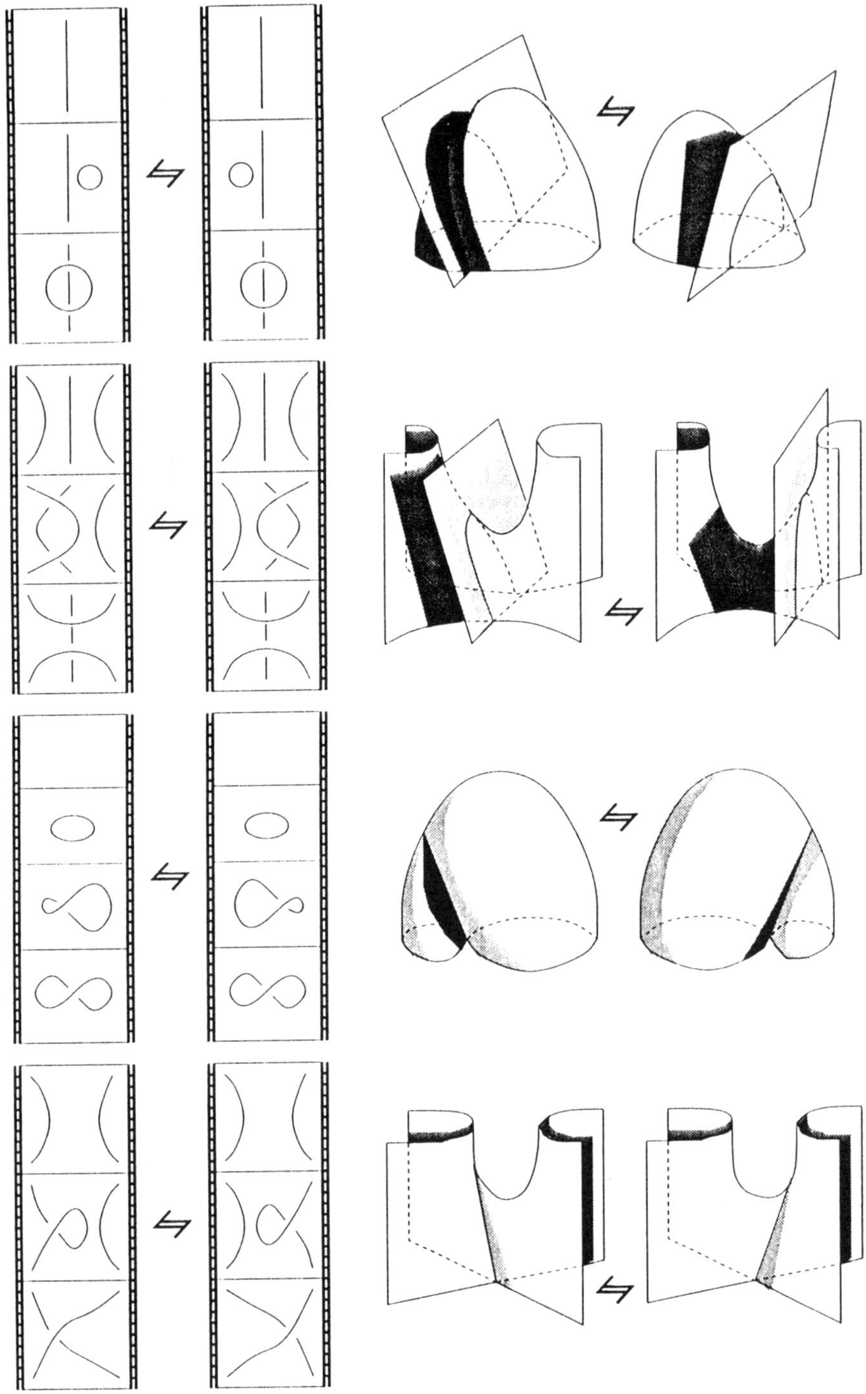

Figure 10: Movie Moves; Part III

Movies, in this case, are used as a convention to locally describe moves in the presence of a height function.

In particular, isotopies on each still are not listed as ESIs. For example, a symmetric diagram of a trefoil knot can be rotated by $2\pi/3$. If such a diagram is found in a still as a part of a movie description, then the surface that results from rotating the diagram can be different from the result of leaving the diagram alone between successive stills. This was pointed to us by Roseman. The movie moves do not deal with such isotopies on stills.

For categorical or computer purposes, one needs a more specific description at each cross sectional diagram to determine the knotted surface. In Section 3, such specificity is achieved using braids.

2.4 IMMERSED PROJECTIONS AND THE SEIFERT ALGORITHM. The projection of a knotted surface is *immersed* if it has no branch points. Thus each arc of double points is a closed curve that may pass through a triple point, and there is no neighborhood of a point in which the image of the surface looks like a cone on a figure 8. Such projections are easier to work with. In this section we discuss when we can get immersed projections, and give two applications of immersed projections: computing fundamental groups and a Seifert algorithm.

If a surface embedded in 4-dimensions is orientable, then it is ambiently isotopic to a surface in which the projection is immersed [24]. More generally, any embedded surface for which the normal Euler class is trivial can be so isotoped [67]. In [8], we showed the following:

2.4.1 THEOREM [8]. *Let F be an embedded surface in $\mathbf{R}^4$ with Euler number e. Then there is an isotopic embedding that projects in $\mathbf{R}^3$ to a map with $|e|$ branch points. This is the minimal number of branch points that can be achieved among all isotopic embeddings.*

2.4.2 LIFTING PROBLEMS. Not every immersed surface in 3-space lifts to an embedding in 4-space. The most famous example of this non-liftability is Boy's surface. This is an immersed projective plane in 3-space. The image cannot be lifted to an embedding in 4-space, for otherwise the embedding would have trivial normal Euler class thereby contradicting a Theorem of Whitney [76, 8] that relates the normal Euler class and the Euler characteristic of the surface.

Giller [24] showed that the orientation double cover of Boy's surface also does not lift to an embedding. This is an immersion of a 2-sphere that can be put into general position. Giller's proof is one in which all possible lifts are considered. One would like to see a proof that depended on an easily computable invariant of immersed surfaces. Furthermore, a classification of non-liftable examples would be pleasant.

2.4.3 COMPUTING FUNDAMENTAL GROUPS. The knot diagram of an orientable surface embedded in 4-space can be used to give a presentation of the fundamental group. This description is well-known; see [78] for example. Since the surface is orientable, we assume that the projection has no branch points. The lower decker set separates the surface into *regions* — connected components of $p(K(F))\setminus$ (a regular neighborhood of the lower decker set). The components are bounded (if at all) by the boundary of the deleted neighborhood of the lower-decker arcs. Loosely speaking, the fundamental group of the complement of $K(F)$ in 4-space is generated by the regions, and subject to Wirtinger relations that occur along the arcs of double points. Details follow.

Orient the surface F. Label the regions in the diagram with elements in the set $\{\alpha_1, \alpha_2, \ldots, \alpha_n\}$. For each region α_j, we consider a small loop, a_j, normal to the region at some point in the region. The loop links the region in 4-space, and we connect this loop to a fixed base point that is in the component of $\mathbf{R}^4 \setminus \mathbf{R}^3$ above the diagram. Furthermore, we orient the loop so that the orientation of the surface followed by the orientation of the loop agrees with the orientation of 4-space. We abuse notation, and let a_j denote this based loop in $\mathbf{R}^4$. Then the set $\{a_1, a_2, \ldots, a_n\}$ generates the fundamental group.

Suppose that a region labeled α_i contains the upper-deck arc of the double point manifold, and suppose that the corresponding lower-deck arc is the boundary of regions with labels α_j and α_k. Then a relation of the form $a_i^{\mp} a_j a_i^{\pm 1} = a_k$ occurs along the double point arc. The sign of the exponent can be fixed by carefully examining the orientation conventions. There is such a relation whenever two regions α_j and α_k are locally separated by an arc in the image of the lower-decker set in F. Finally, the set of all such relations can be used to obtain a presentation of $\pi_1(\mathbf{R}^4 \setminus K(F))$. Compare this presentation with the one given in [21]. Recently, Kamada [36] has given a characterization of the presentations of the fundamental groups of orientable surfaces in $\mathbf{R}^4$. His characterization is based on the braided surface form of the knot.

2.4.4 THE SEIFERT ALGORITHM. A Seifert surface for a classical knot is an important tool for studying the knot. For example, branched covering spaces of the complement can be constructed by means of cutting copies of space along a Seifert surface and regluing them. In this way various invariants involving covering spaces can be computed via such surfaces. Seifert's algorithm for constructing these surfaces is quite simple and beautiful. We review this construction.

Orient a given classical knot diagram. At each crossing of the diagram a pair of intersecting arrows are associated to the orientation. Then the crossing is smoothed in such a way that the orientations match up. Think of the intersection of two one-way streets and the pair of turns that are legal at such an intersection. A collection of oriented simple closed curves results from smoothing all of the crossings. Each such curve bounds a disk in the plane of projection and these disks can be lifted in 3-space so that they are mutually disjoint. To the collection of these disks, we reintroduce crossings by adding twisted bands. An orientable surface embedded in 3-space bounded by the knot results. The example worked in Figure 11 captures the details of the construction.

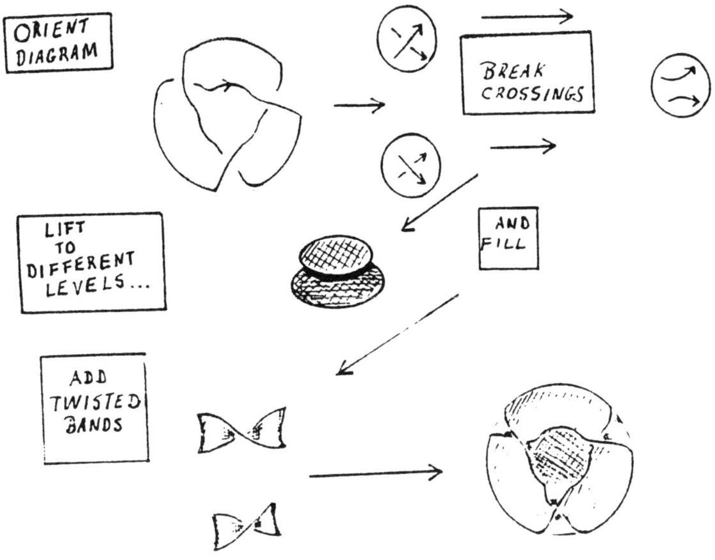

Figure 11: The classical Seifert algorithm

In the rest of this section, we sketch a generalization of this algorithm to surfaces in 4-space. We start with the diagram of knot $K(F)$ where F is an orientable surface, and the projection has no branch points. By Theorem 2, there is such a projection.

We orient the surface, and smooth the surface along the arcs of double points, as indicated in the lower portion of Figure 12. The smoothing is chosen so that the resulting surface is orientable. If there are no triple points, this smoothing can be globally carried out along the double point loops; see [24].

If there are triple points, then we first perform the smoothing indicated in the upper portion of Figure 12; this procedure eliminates a triple point and introduces a pair of branch points. There is a choice of such smoothing so that the result is again an oriented surface. The figure indicates a 1-handle attached between the top and middle sheets. After all triple points have been smoothed, the resulting surface has embedded double point arcs with branch points as end points. The same smoothing of the double point curve in the preceding paragraph can be used at the presence of such end points as well. One can remove the neighborhoods of the double point set and replace it by embedded bands or annuli.

The final result of smoothing is a collection of disjoint embedded surfaces in 3-space; these are called the *Seifert shells*. Each shell bounds an embedded compact 3-manifold in 3-space, and each 3-manifold is, at worst, a cube-with-knotted holes [65]. If the Seifert shells are nested, as they might be, we push them up to different levels in 4-space until they are disjoint. There is partial order on the set of Seifert shells that is induced by the degree of nesting among the various shells. If one shell is

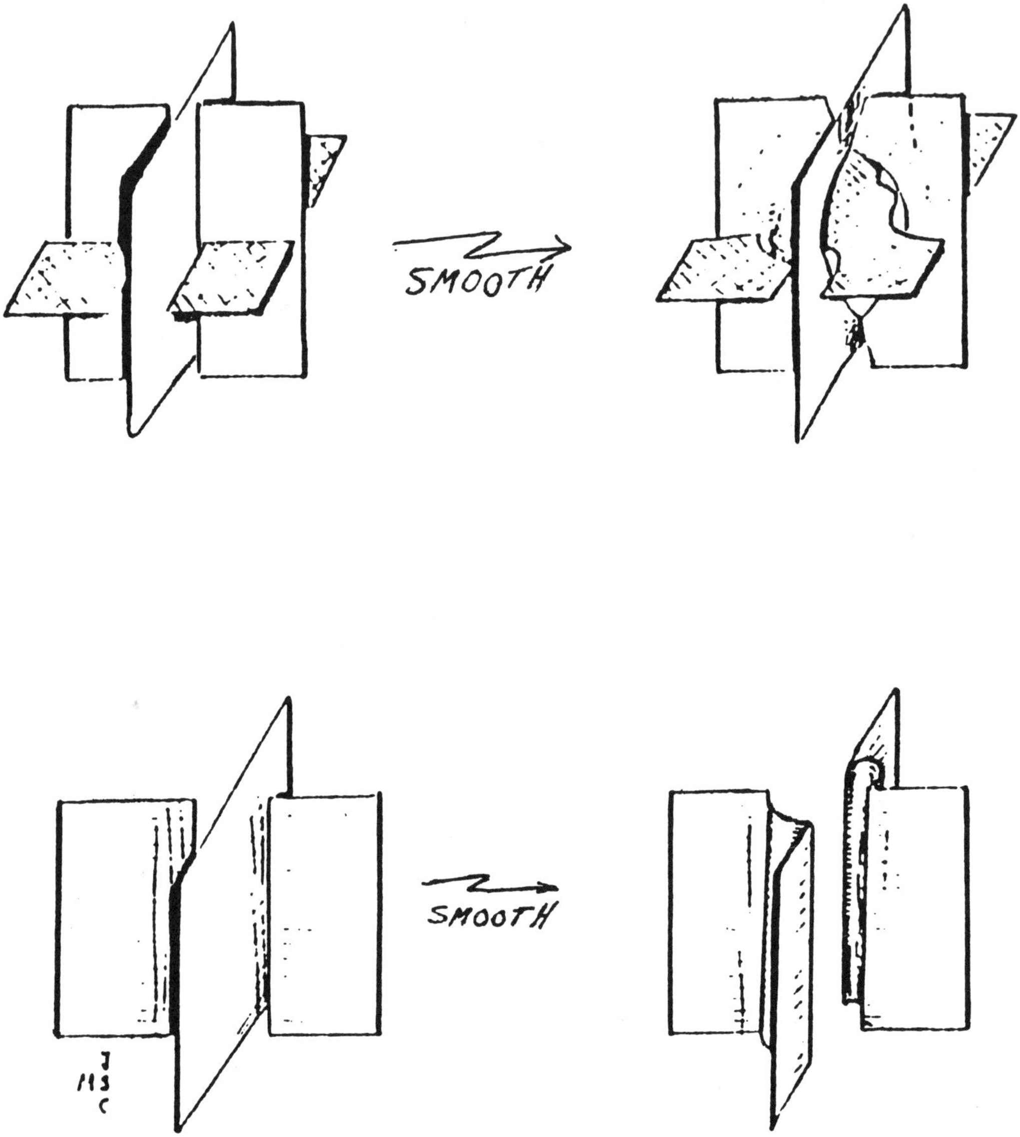

Figure 12: Smoothing double points and triple points

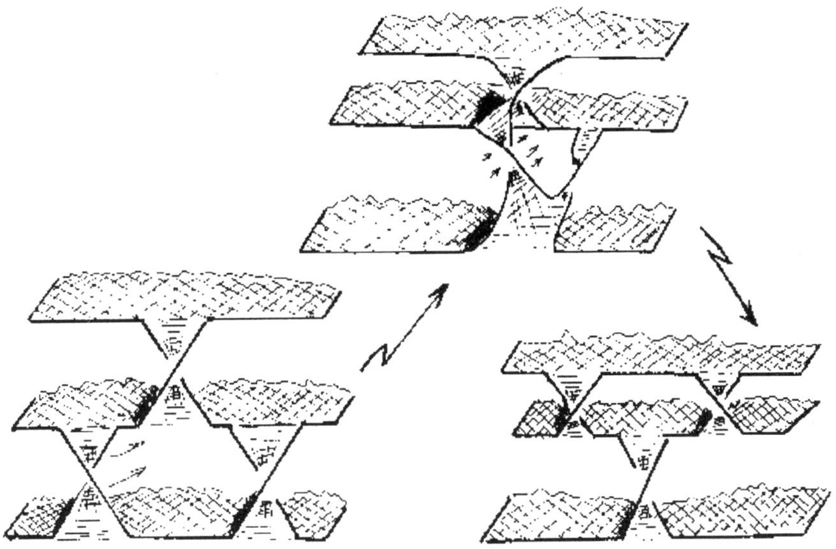

Figure 13: The twisted brick near a triple point

contained in the bounded component of a second shell, then the solid bounded by the first is pushed in 4-space above that of the second with respect to the direction of projection determining the given diagram. Then the crossing arcs are reintroduced by attaching, to the 3-manifolds bounded by the Seifert shells, a collection of *twisted bricks*. In other words we take the twisted bands that are used in the classical algorithm, cross them with intervals, and sew these back into the solids along the double point curves. In this way, the double point arcs can be reintroduced locally.

Now we recover the triple points we smoothed in the first step. This can be done by attaching $D^2 \times [0,1]$ into the "hole" introduced by the smoothing. Along each double point curve, we attach a (twisted band)$\times I$. At a branch point introduced by a triple point smoothing, this attachment ends. Figure 13 shows that in fact such attachment extends through a triple point. Such an extension, in turn, means that we can fill a 3-ball in a hole introduced by the smoothing. A triple point is a type III Reidemeister move in the movie description. Therefore we see the movies of surfaces in Figure 13 that involves a type III move.

This completes the sketch of the algorithm. For two of the examples given in [21], we identified the 3-manifolds obtained by the algorithm. Kamada [37] has given a version of this algorithm for surface braids.

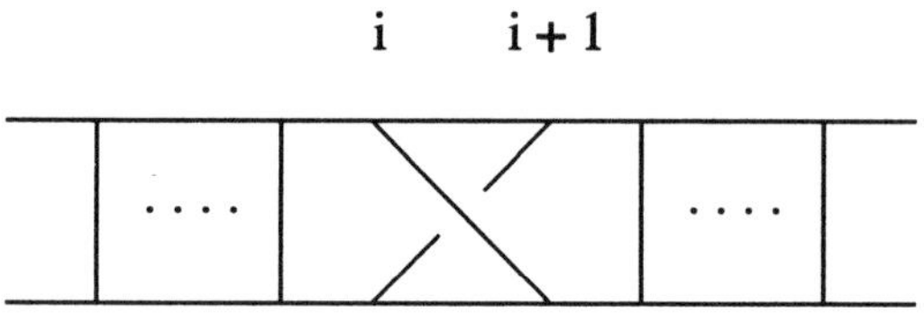

Figure 14: Geometric generators of the braid group

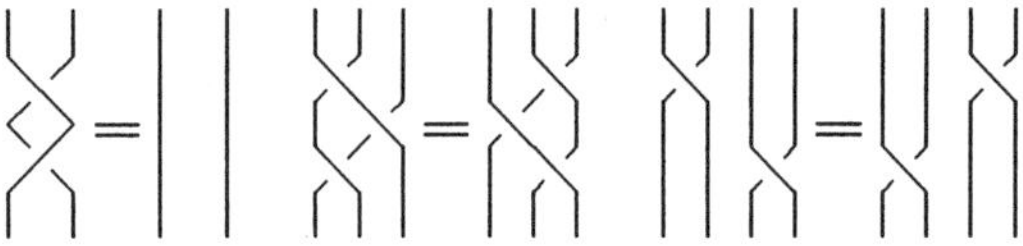

Figure 15: The braid relators

3 Braid Theory

In classical knot theory, the role played by the braid group is quite fundamental. The three theorems that allow braids to play such an important role are the Alexander Theorem [1], the Markov Theorem [59], and the Artin Theorem [2]. Each has an analogue for surfaces in dimension 4. We will state the analogous results in Section 3.2 after we review the classical theory.

3.1 Braid groups and closed braids. A *geometric n-braid* is a set of strings in $\mathbf{R}^2 \times I$ in 3-space connecting n (n a positive integer) fixed points in $\mathbf{R}^2 \times \{0\}$ with the corresponding n points in $\mathbf{R}^2 \times \{1\}$ such that they intersect each plane $\mathbf{R}^2 \times \{t\}$ in exactly n points. An example is depicted on the left of Figure 16.

Two such braids are *equivalent* if they are ambiently isotopic (*i.e.,* one can be continuously deformed to the other through braids) via an isotopy that fixes the boundary points.

The equivalence classes form a group by juxtaposition (and rescaling). It is called the *braid group* of n-strings and denoted by B_n. A standard set of generators of this group is depicted in Figure 14.

The Artin Theorem gives a presentation of the braid group as follows:

$$B_n = \langle s_1, s_2, \ldots, s_{n-1} : s_i s_{i+1} s_i = s_{i+1} s_i s_{i+1} \text{for } i = 1, \ldots n-2; \ \& \ s_i s_j = s_j s_i \text{if } |i-j| > 1 \rangle.$$

These relations are depicted in Figure 15.

The Alexander Theorem states that every knot or link in 3-space can be represented as a closed braid. This can be proven diagrammatically as indicated in Kauffman [47]. A *closed braid* consists of a geometric braid in which the points on the top and bottom

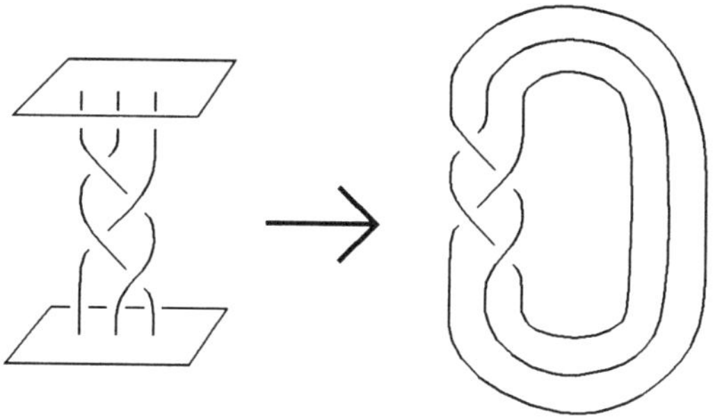

Figure 16: The braid closure

are joined together by a collection of unentangled arcs outside the box containing the braid. Figure 16. indicates the configuration of such arcs in a planar diagram.

The Markov Theorem says that the closures, $\hat{\sigma_1}$ $\hat{\sigma_2}$, of two braids σ_1 and σ_2 (which might involve differing numbers of strings) represent the same knot or link if and only if one can be obtained from the other by a sequence of braid equivalences, conjugations in the braid group, or stabilizations. The *stabilization of a braid* $\sigma \in B_n$ is the braid $\sigma' s_n \in B_{n+1}$ where σ' is the braid in B_{n+1} that is obtained from σ via juxtaposing an extra string to σ on the right.

3.2 SURFACE BRAIDS AND BRAID MOVIES. The braided form of an orientable surface was described to us by Oleg Viro in 1992. Similar notions had been considered much earlier by Lee Rudolph [69]; X. S. Lin [56] developed some portions of this theory independently. This notion has been used in algebraic geometry [61], [62], [55]. Kamada [34, 35, 36, 37, 41, 38, 42] has extensively studied Viro's version of surface braids. The presentation given here follows Kamada quite closely.

3.2.1 DEFINITION. A *surface braid* is a compact oriented surface F which is properly embedded in $D_1^2 \times D_2^2$ (where D_i^2 is a 2-dimensional disk for $i = 1, 2$) such that

1. the projection $p_2 : D_1^2 \times D_2^2 \to D_2^2$ when restricted to F is a branched covering, $p_2|_F : F \to D_2^2$.
2. The boundary of F is embedded in $\partial(D_1^2 \times D_2^2) \simeq S^3$ as the standard unlink: $\partial F = (n \text{ points}) \times S^1 \subset (D_1^2 \times \partial D_2^2) \subset \partial(D_1^2 \times D_2^2)$ where the n points are fixed in the interior of D_1^2. The positive integer n is the *(surface) braid index*.

A branched covering is *simple* if every branch point has branch index 2. A surface braid is *simple* if the branched covering in condition 1 above is simple. We will consider only simple surface braids here except in the discussions of the Markov Theorem.

Two surface braids are *equivalent* if one is the image of the other under a bundle isotopy where we regard the projection $p_2 : D_1^2 \times D_2^2 \to D^2$ as a disk bundle over a disk.

3.2.2 BRAID MOVIES. Closely related to the notion of a surface braid is a *braid movie.* A braid movie is a sequence $(1 = w_0, w_1, \ldots, w_{k-1}, w_k = 1)$ of words, w_i, in the free monoid generated by symbols $s_1^{\pm 1}, \ldots, s_{n-1}^{\pm 1}$ that satisfies the following condition: For $i = 1, \ldots, k$ either $w_i = w_{i-1}$ or w_i differs from w_{i-1} by one of the following *elementary braid changes (EBCs)*:

1. Insertion/deletion of s_j^{ϵ},
2. Insertion/deletion of a pair $s_j^{\epsilon} s_j^{-\epsilon}$,
3. Replacement of $s_j^{\epsilon_1} s_\ell^{\epsilon_2}$ by $s_\ell^{\epsilon_2} s_j^{\epsilon_1}$ where $|j - \ell| > 1$,
4. Replacement of $s_\ell^{\epsilon} s_j^{\epsilon} s_\ell^{\epsilon}$ by $s_j^{\epsilon} s_\ell^{\epsilon} s_j^{\epsilon}$ where $|\ell - j| = 1$.

Here the exponents $\epsilon, \epsilon_1, \epsilon_2 = \pm 1$.

Figure 17 depicts the elementary braid changes. Moreover this illustration indicates that each elementary braid change can be interpreted as an embedding of a surface braid in $D_1^2 \times D_2^2$. Each word in the movie is represented by the corresponding braid diagram in $[0,1] \times [0,1]$; the diagram represents an embedding of arcs in $[0,1] \times [0,1] \times [0,1]$; and the fourth dimension is represented by interpolating a surface between two braid words. Thus every braid movie represents a surface braid. The diagram representing a given braid word in a braid movie is called a *still from the movie.*

On the other hand, every surface braid can be decomposed into a braid movie. Details follow. Each of D_1^2 and D_2^2 is the product of intervals, and we have projections $\pi_i : F \to [0,1]_i$ for $i = 1, 2, 3, 4$. We assume (without loss of generality) that $P = \pi_2 \times \pi_3 \times \pi_4$ restricted to F is generic — thus the image has double point arcs, isolated triple points, and branch points. We further regard the fourth interval factor in $D_1^2 \times D_2^2 = [0,1]_1 \times [0,1]_2 \times [0,1]_3 \times [0,1]_4$ as a time direction. In other words, we consider the family of 2-disks $\{D_t = [0,1]_2 \times [0,1]_3 \times \{t\} : t \in [0,1]_4\}$. Except for a finite number of values of $t \in [0,1]_4$ the intersection $P(F) \cap D_t$ is a collection of properly immersed curves on the disk D_t. Thus we get a braid movie by slicing the braided surface by the disks D_t. Note the following correspondences: a branch point corresponds to the EBC (1), a local maximum/minimum of the double point set corresponds to the EBC (2), and a triple point corresponds to EBC (4). The geometric meaning of the EBC (3) is found in the context of the projection of the surface F to the disk D_2^2 : The arcs of double points of P, when projected, can intersect generically in D_2^2. A double point of these arcs corresponds to the commutant braid relation.

3.2.3 CLOSED SURFACE BRAIDS. Let $F \subset D_1^2 \times D_2^2$ be a surface braid with braid index n. Embed $D_1^2 \times D_2^2$ in 4-space in a standard way. Then the boundary of F is unlink in $S^3 = \partial(B^2 \times D^2)$. This collection of n circles bound standardly embedded 2-disks in 4-space. By capping-off the boundary, we obtain a closed orientable surface, $\hat{F}$, embedded in 4-space. This surface is called the *closure of F*; it is defined up to ambient isotopy. A surface obtained this way is called a *closed surface braid.*

In [36] Kamada presents a proof of an Alexander Theorem for surfaces.

Figure 17: The elementary braid changes

3.2.4 THEOREM [36]. *Any orientable surface embedded in* $\mathbf{R}^4$ *is isotopic to a closed surface braid.*

In the braid movie form, we can form the braid closure of each diagram representing the braid words. At the beginning and end of the movie a system of unknotted and unlinked nested circles results and each of these can be made to bound an embedding disk in the 3-ball represented by the given still.

Kamada [39, 38] has recently announced and proven a Markov Theorem for closed surface braids in the PL category. He defined notions of conjugations and stabilizations for surface braids, and proved that two isotopic closed braids are related by sequence of such operations (and braid equivalences). However, the proof involves nonsimple braids. In [42] he studies nonsimple braids that can not be deformed by braid equivalences to simple braids. Further study in regard to these points is expected.

3.2.5 KAMADA CHARTS. In [34, 35], Kamada expresses the braided surface by means of charts: *A Kamada chart* is an embedded oriented edge-labeled graph in a disk. The labels are integers between 1 and $n-1$. The graph has nodes of valence 1, 4, and 6, and the labels and orientations must satisfy certain conditions at these nodes.

We describe how to obtain such a graph from a surface braid of braid index n, and leave the reader to provide the conditions on the nodes of the graph. The restriction to the double decker set of the projection $P : F \to [0,1]_2 \times [0,1]_3 \times [0,1]_4$ is composed with the projection onto the last two factors to provide the graph. The double point sets and their projections are illustrated in 17.

Figure 17 shows that each edge of the graph thus obtained is a continuous trace of crossings corresponding to a braid generator. The subscript of the braid generator is the label on the edge. We also see from the Figure that the valence 1 vertices correspond to the introduction or deletion of a braid generator, the 6-valent vertices correspond to the type III Reidemeister move, and the 4 valent vertices on the graph correspond to interchanging distant braid generators.

Let us remark further about the Kamada charts.

1. This notion of charts was used by Kamada [40] to generalize *the braid monodromy* that is used in algebraic geometry [61] to this topological setting.

2. Charts are used to obtain moves to braid movies that will be discussed in the next section.

3. Charts are defined purely geometrically as the projection of the double point set. However, these planar graphs are directly related to the maps from disks to the classifying space BG of the braid groups $G = B_n$. Furthermore, the homotopies among such maps are related to the moves to the braid movies [16, 71]. The braid group is the fundamental group of the configuration space; similarly, surface braids are related to homotopies of certain disks in BG.

4. Such planar graphs are related to "identities among relations" in combinatorial group theory [7]. The braid movie moves that are discussed in the next section are also expressible in these terms.

3.3 BRAID RELATIONS AND BRAID MOVIE MOVES Here we present the analogue of Artin's Theorem for surface braids. Recall that the geometric braid group has the presentation stated above. We have represented a surface braid as a braid movie. When do two such movies represent equivalent surface braids? The answer to this question is the braid movie move theorem.

3.3.1 Braid movie moves. The following 14 types of changes among braid movies are called braid movie moves :

(C-I-R1) $(s_i, s_is_js_j^{-1}, s_js_is_j^{-1}) \leftrightarrow (s_i, s_js_j^{-1}s_i, s_js_is_j^{-1})$, where $|i-j|>1$,

(C-I-R1') $(s_i, s_is_i^{-1}s_i, s_i) \leftrightarrow (s_i)$,

(C-I-R2) $(s_is_j, s_js_i, s_is_j) \leftrightarrow (s_is_j)$, where $|i-j|>1$,

(C-I-R3) $(s_is_ks_j, s_ks_is_j, s_ks_js_i, s_js_ks_i) \leftrightarrow (s_is_ks_j, s_is_js_k, s_js_is_k, s_js_ks_i)$, where $|i-j|, |j-k|, |k-i|>1$,

(C-I-R4) $(s_ks_is_js_i, s_is_ks_js_i, s_is_js_ks_i, s_is_js_is_k, s_js_is_js_k)$
$\leftrightarrow (s_ks_is_js_i, s_ks_js_is_j, s_js_ks_is_j, s_js_is_ks_j, s_js_is_js_k)$,
where $|i-j|=1$ and $|i-k|>1, |j-k|>1$,

(C-I-M1) (empty word) $\leftrightarrow$ (empty word, $s_is_i^{-1}$, empty word),

(C-I-M2) $(s_is_i^{-1}$, empty word, $s_is_i^{-1}) \leftrightarrow (s_is_i^{-1})$,

(C-I-M3) $(s_is_js_i, s_js_is_j, s_is_js_i) \leftrightarrow (s_is_js_i)$, where $|i-j|=1$,

(C-I-M4) $(s_is_js_ks_is_js_i, s_is_js_is_ks_js_i, s_js_is_js_ks_js_i, s_js_is_ks_js_ks_i,$
$s_js_ks_is_js_ks_i, s_js_ks_is_js_is_k, s_js_ks_js_is_js_k, s_ks_js_ks_is_js_k)$
$\leftrightarrow (s_is_js_ks_is_js_i, s_is_js_ks_js_is_j, s_is_ks_js_ks_is_j, s_ks_is_js_ks_is_j,$
$s_ks_is_js_is_ks_j, s_ks_js_is_js_ks_j, s_ks_js_is_ks_js_k, s_ks_js_ks_is_js_k)$,
where $k=j+1=i+2$ or $k=j-1=i-2$,

(C-I-M5) $(s_js_i, s_i^{-1}s_is_js_i, s_i^{-1}s_js_is_j) \leftrightarrow (s_js_i, s_js_is_j^{-1}s_j, s_i^{-1}s_js_is_j)$, where $|i-j|=1$,

(C-II) $(s_j, s_js_i, s_is_j) \leftrightarrow (s_j, s_is_j)$, where $|i-j|>1$,

(C-III-1) $(s_is_j, s_is_js_i, s_js_is_j) \leftrightarrow (s_is_j, s_js_is_j)$, where $|i-j|=1$,

(C-III-2) $(s_js_j^{-1}, s_js_is_j^{-1}, s_i^{-1}s_js_i) \leftrightarrow (s_js_j^{-1}$, empty word, $s_i^{-1}s_i, s_i^{-1}s_js_i)$, where $|i-j|=1$,

(C-IV) (empty word, $s_i^{-1}s_i, s_i) \leftrightarrow$ (empty word, s_i).

Furthermore, we include the variants obtained from the above by the following rules:

1. replace s_i with s_i^{-1} for all or some of i appearing in the sequences if possible,

2. if $(w_0, \cdots, w_k) \leftrightarrow (w'_0, \cdots, w'_k)$ is one of the above, add $(w_k, \cdots, w_0) \leftrightarrow (w'_k, \cdots, w'_0)$,

3. if $(w_0, \cdots, w_k) \leftrightarrow (w'_0, \cdots, w'_k)$ is one of the above, add $(v_0, \cdots, v_k) \leftrightarrow (v'_0, \cdots, v'_k)$ where v_j (resp. v'_j) is the palindrone of w_j (resp. v'_j);

4. any combination of the above.

The graphical depiction of the braid movie moves is shown in Figures 18, 19, 20.

Suppose that $(w_0, \cdots, w_k)$ and $(w'_0, \cdots, w'_k)$ are two braid movies that satisfy the following condition. There is an integer i, $1 \leq i \leq k$, such that

1. $w_j = w'_j$ for all $j \neq i$,

2. w_p (resp. w'_p) is written as $w_p = u_p v_p$ (resp. $w'_p = u'_p v'_p$) for $p = i-1, i, i+1$ where u's and v's satisfy

   (a) u_i is obtained from u_{i-1} by an elementary braid change, say η, and $v_i = v_{i-1}$, while $u'_i = u'_{i-1}$, and v'_i is obtained from v'_{i-1} by the EBC ξ, and $u_i = u_{i+1}$.

   (b) v_{i+1} is obtained from v_i by the EBC ξ, and $u_i = u_{i+1}$ while u'_{i+1} is obtained from u'_i by the EBC η and $v'_i = v'_{i+1}$.

Then we say that $(w'_0, \cdots, w'_k)$ is obtained from $(w_1, \cdots, w_k)$ by a locality change (or vice versa). If two braid movies are related by a sequence of locality changes then they are called *equivalent under locality*. Colloquially, equivalence under locality changes means that distant EBCs commute.

3.3.2 THEOREM [15]. *Two braid movies are equivalent if and only if they are related by a sequence of braid movie moves and locality changes.*

3.4 REMARK. We mention other geometric approaches. In [48, 49], certain standard forms of knotted surfaces are defined using Morse functions. Yoshikawa [79] used marked vertex knot diagrams to construct a table of surfaces embedded in 4-space. These diagrams encode the surface in a concise fashion. Similar diagrams have been used in [57].

4 Generalizations of the Yang-Baxter equation and categories

A significant aspect of the Jones polynomial [32] and its generalizations [77, 64, 75, 74] is the close relationship with mathematical physics. In particular, the Yang-Baxter equation (YBE) corresponds to the Reidemeister type III move or the braid relation. A structure that encodes the knot theory, the representation theory, and the physical aspects is the braided monoidal category [23].

The Zamolodchikov tetrahedral equation [80, 4] is a higher dimensional analogue of the YBE and it has enjoyed a number of significant generalizations and solutions [58, 54, 18, 26, 13, 11, 14, 22, 44, 20, 50]. The algebraic form of these generalizations is analogous to the algebraic expression of the YBE. Moreover, the generalizations are expressible as either a movie move or a braid movie move. We have constructed

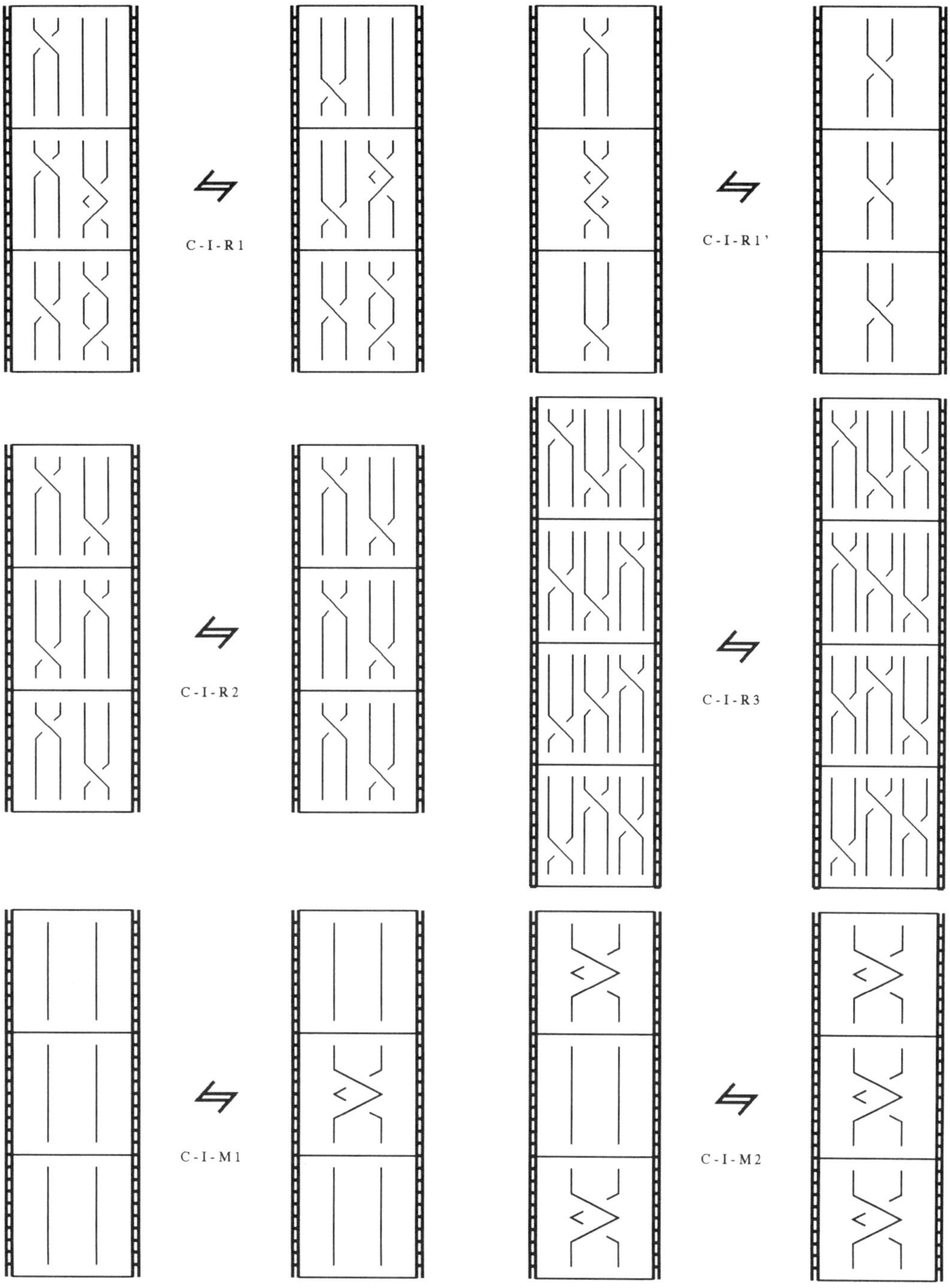

Figure 18: Braid Movie Moves ; Part I

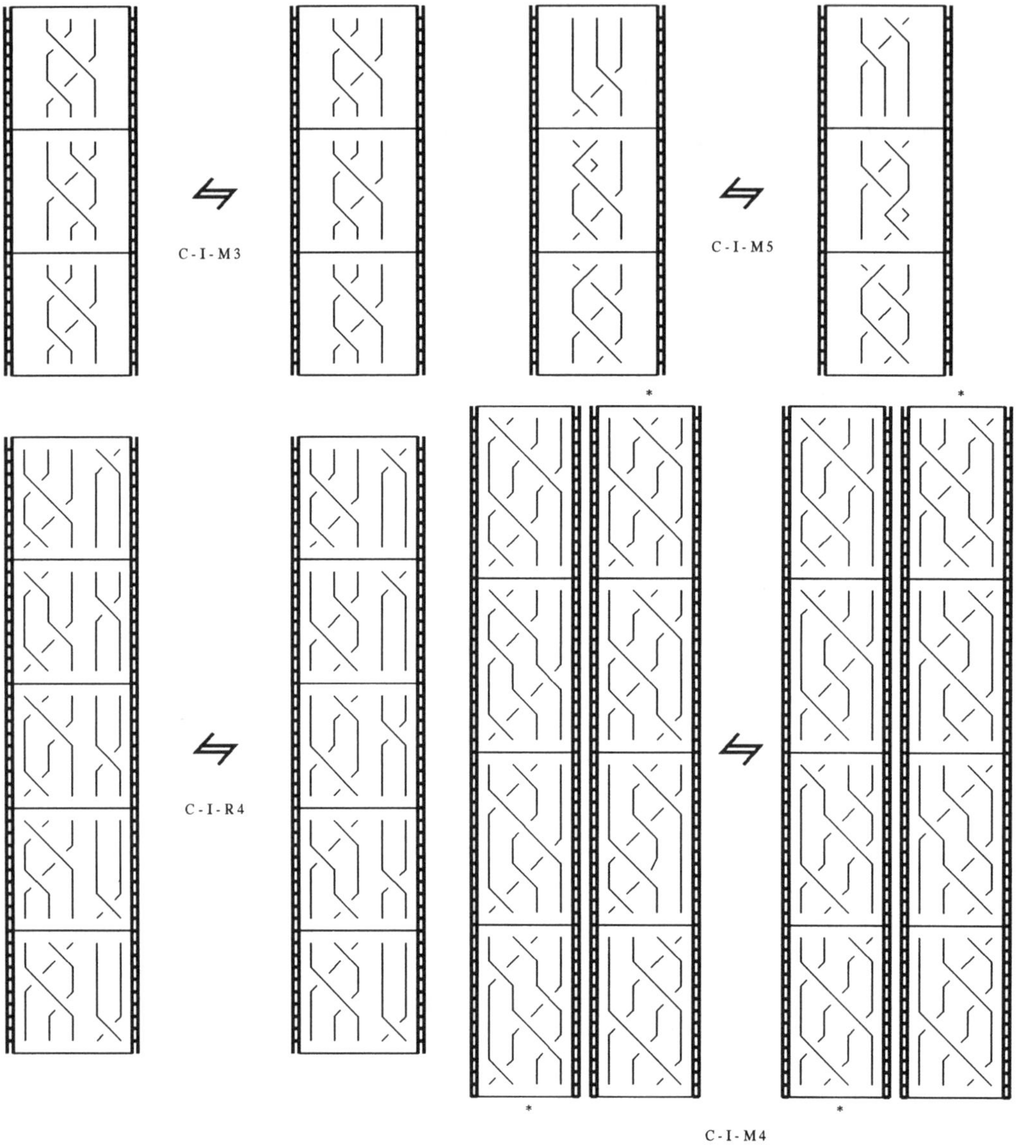

Figure 19: Braid Movie Moves ; Part II

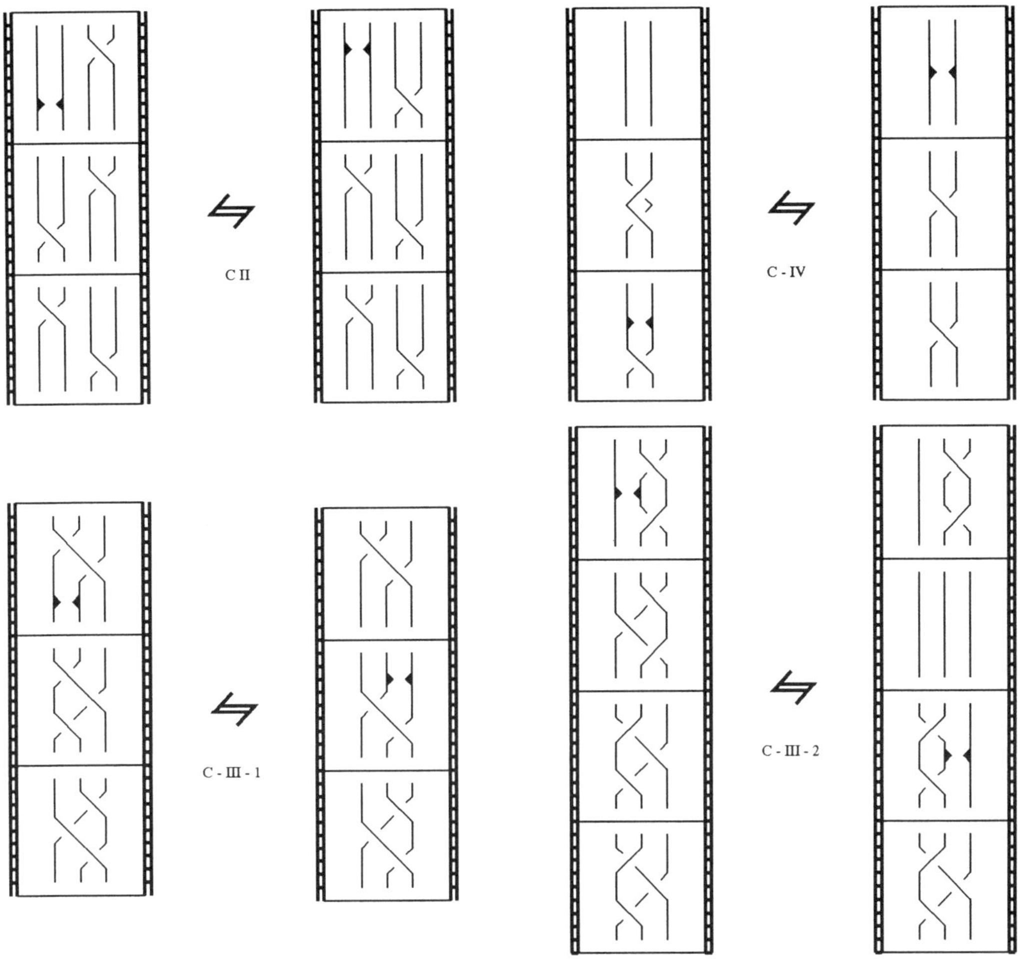

Figure 20: Braid Movie Moves ; Part III

solutions to these equations by using diagrammatic techniques and algebraic interpretations thereof. In this section we review these solutions and a (2-)categorical interpretation of the braid movie moves.

4.1 THE YANG-BAXTER EQUATION AND GENERALIZATIONS. We consider the Yang-Baxter equation in the following two forms:

$$R_{12}R_{13}R_{23} = R_{23}R_{13}R_{12} \tag{1}$$
$$(\hat{R}\otimes 1)(1\otimes \hat{R})(\hat{R}\otimes 1) = (1\otimes \hat{R})(\hat{R}\otimes 1)(1\otimes \hat{R}) \tag{2}$$

One form of the equation can be obtained from the other by composing with a transposition. Here $R : V\otimes V \rightarrow V\otimes V$ where V is some vector space. The subscripts in the first version indicates the two factors of $V \otimes V \otimes V$ on which R acts: R_{12} acts non-trivially on the first two factors of $V \otimes V \otimes V$ and acts as the identity on the third factor. These equations are fundamental in the construction of the quantum groups, in statistical mechanics, and in the construction of the Jones polynomial and its generalization. We refer the reader to Kauffman's book [47] for these applications.

Here we mention three natural generalizations and outline solutions.

4.1.1 THE PLANAR TETRAHEDRAL EQUATION. In [11] we propose the following generalization of the YBE.

$$(R\otimes 1)(1\otimes R)(R\otimes 1)(1\otimes R) = (1\otimes R)(R\otimes 1)(1\otimes R)(R\otimes 1),$$

and we call this the *planar tetrahedral equation.* In the case $R : W \otimes W \otimes W \rightarrow W\otimes W\otimes W$, and the vector space is 2-dimensional, there is a solution R that is found by using the Temperley-Lieb algebra.

Let δ be a complex number. Recall [47, 17] that $TL_n(\delta)$ (the Temperley-Lieb algebra) is an algebra over $\mathbf{C}$ generated by $1, h_1, \ldots, h_{n-1}$.These generators are subject to relations $h_i^2 = \delta h_i$, $h_ih_{i\pm1}h_i = h_i$, and $h_ih_j = h_jh_i$ whenever $|i-j| > 1$. Recall [ibid] also that this algebra is represented by planar diagrams, and each generator is represented by and vertical strings. The solution in TL_3 given in [11] of the planar tetrahedral equation is of the form

$$R = A(h_1h_2 + h_2h_1) + C(h_1 + h_2) + E$$

where $A + E = -\delta C$ and $AE = C^2$. This solution was found by diagrammatic techniques. Further generalizations are formulated and solved in [11]; some of these solutions give representations of certain Coxeter groups.

4.1.2 THE FRENKEL-MOORE TETRAHEDRAL EQUATION AND ITS VARIANTS. In [22], Frenkel and Moore formulate and solve the Zamolodchikov equation in the form

$$S_{123}S_{124}S_{134}S_{234} = S_{234}S_{134}S_{124}S_{123}.$$

An alternative version of this equation is the following:

$$S_{124}S_{135}S_{236}S_{456} = S_{456}S_{236}S_{135}S_{124}.$$

Here S acts on the 3-fold tensor power of a vector space V, and S_{ijk} acts on the 4-fold tensor power of V by acting as the identity on the omitted index.

In [44], a process of going from the Frenkel-Moore version to the alternative version was given (see also [13]). The set of pairs of integers between 1 and 4 is lexically given as $\{12, 13, 14, 23, 24, 34\}$. Under this ordering the pairs correspond to the integers 1 through 6. In the subscript of any tensor S_{ijk} we look at the possible pairs $\{ij, ik, jk\}$ and choose the lexical identification of these. Thus $S_{123} \mapsto S_{124}$ and so forth.

Here we describe solutions to the Zamolodchikov equation by using solutions to the YBE. We start with a vector space over which there is a solution, R, to the YBE $R_{12}R_{13}R_{23} = R_{23}R_{13}R_{12}$. Let $V = W \otimes W$. For given V give a and b as subscripts of W: $V_i = W_{ia} \otimes W_{ib}$ for $i = 1, \cdots 6$. For $S = S_{123}$ acting on $V_1 \otimes V_2 \otimes V_3 = W_{1a} \otimes W_{1b} \otimes W_{2a} \otimes W_{2b} \otimes W_{3a} \otimes W_{3b}$ set

$$S_{123} = R_{1a2a}R_{1b3a}R_{2b3b}.$$

We compute

$$\begin{aligned}
&S_{124}S_{135}S_{236}S_{456} \\
&= R_{1a2a}R_{1b4a}R_{2b4b}\ R_{1a3a}R_{1b5a}R_{3b5b}\ R_{2a3a}R_{2b6a}R_{3b6b}\ R_{4a5a}R_{4b6a}R_{5b6b} \\
&= R_{1a2a}R_{1a3a}R_{2a3a}\ R_{1b4a}R_{1b5a}R_{4a5a}\ R_{2b4b}R_{2b6a}R_{4b6a}\ R_{3b5b}R_{3b6b}R_{5b6b} \\
&= R_{2a3a}R_{1a3a}R_{1a2a}\ R_{4a5a}R_{1b5a}R_{1b4a}\ R_{4b6a}R_{2b6a}R_{2b4b}\ R_{5b6b}R_{3b6b}R_{3b5b} \\
&= R_{4a5a}R_{4b6a}R_{5b6b}\ R_{2a3a}R_{2b6a}R_{3b6b}\ R_{1a3a}R_{1b5a}R_{3b5b}\ R_{1a2a}R_{1b4a}R_{2b4b} \\
&= S_{456}S_{236}S_{135}S_{124}
\end{aligned}$$

Here the second and the fourth equalities follow from the fact that linear maps acting on disjoint tensor factors commute.

We found this solution entirely by diagrammatic means [13]. The idea is that the double decker set on each sheet of four planes involved in the Roseman tetrahedral move under-goes a Reidemeister type III move. An algebraic interpretation of this geometric fact yields the above solutions.

This observation and some generalizations allowed us to construct families of solutions to related higher dimensional analogues. We mention that other solutions were found by using the quantum group representations [44, 50].

4.1.3 THE PERMUTAHEDRON EQUATION. This equation is the following:

$$S_{123}S_{145}R_{25}R_{16}S_{246}S_{356}R_{34} = R_{34}S_{356}S_{246}R_{25}R_{16}S_{145}S_{123}.$$

The equation seems to combine aspects of both the Zamolodchikov equation and the YBE; in [43] a relation between this expression and the Zamolodchikov equation was

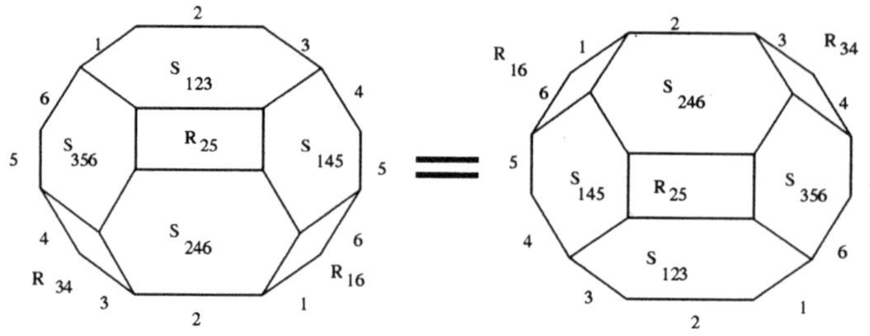

Figure 21: The permutahedron equation

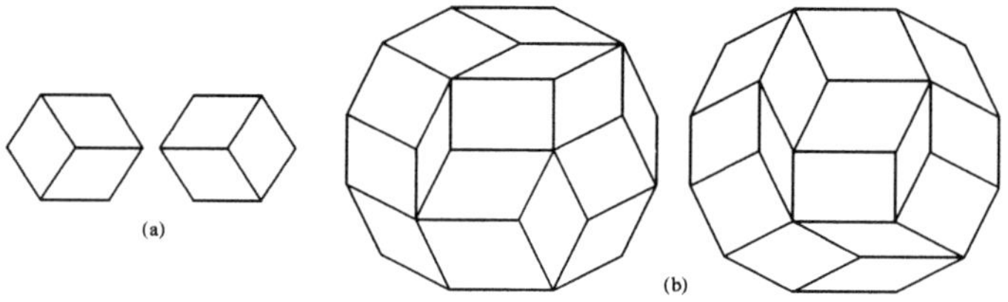

Figure 22: The solution to the permutahedron equation

given. The permutahedron equation can be expressed graphically via the picture in Figure 21.

The hexagons represent the tensors S and the parallelograms represent the tensors R. The solution to the permutahedron equation that we gave in [14] is too long to present. Here is the diagrammatic idea. Tile each hexagon with parallelograms as in Figure 22. There are two possibilities for doing this as indicated in Figure 22 (a) but choose your favorite. One such filling is depicted in Figure 22 (b).

QUIZ: (1) Pick any hexagon in the left of Figure 22 (b) and change the tiling by parallelograms in the hexagon you picked (for example you can change the choice you made in the above). Change this tiling to the one on the right of (b) by a sequence of such changes.

(2) Use your answer to the quiz (1) to give solutions to the permutohedron equation. (Hint: the change of tiling is dual to Reidemeister type III move.)

Also in [14] we formulated and solved certain versions using Burau representations. It was interesting to note that the solutions found using the Burau representation did not hold in the Temperley-Lieb representation, confirming the unfaithfulness of the Burau representation.

4.2 CATEGORIES AND 2-CATEGORIES. Category theoretical descriptions of knot diagrams in the classical case played a significant role in quantum topology [23, 64]. The categorical structure that lies behind the classical knot diagrams and quantum group representations is the notion of a braided monoidal category. The Yang-Baxter relation is a theorem in any such category. See [23, 20, 43, 45] for a definition. In [43, 45], Kapranov and Voedvoski formulate the axioms for a braided monoidal 2-category. Thus it is expected that such a 2-category will play an important role in the study of knotted surfaces.

Such categorical aspects of movie moves were developed in detail in John Fischer's Yale Ph.D. thesis [20]. His result virtually ensures that *any* other 2-category of the sort that he defined would yield an invariant of knotted surfaces. Here we describe such a 2-category using braid movies instead.

In a category, there are axioms or relations satisfied by compositions of morphisms. Morphisms are represented by arrows, and such relations are represented by planar polygons as commutative diagrams. In a 2-category, 1-morphisms are represented by arrows, 2-morphisms are represented by polygons. There are axioms and relations satisfied by compositions of 2-morphisms; they are now represented by polytopes in 3-space.

Rather than going into the intricate details of this definition — it occupies 7 pages in [43] — we prefer to assert that the braid movie moves form a braided monoidal 2-category. To justify that assertion, we will describe the 2-category, BrMov, of braid movies by giving a set of generating objects, morphisms, and 2-morphisms.

The objects in the category are the non-negative integers. The integer n is represented by n dots arranged along a horizontal line. Given integers n and m we have $\hom(m,n)$ is the empty set unless $m = n$. If $m = n$, $\hom(n,n)$ consists of words in the free monoid generated by $\{s_1^{\pm 1}, s_2^{\pm 1}, \ldots, s_{n-1}^{\pm 1}\}$. Such a word is represented by a braid *diagram*. Words (or braid diagrams) are related by a 2-morphisms if one can be obtained from another by a sequence of EBCs.

There is a tensor product in this 2-category that is given on objects by addition: $[n] \otimes [m] := n + m$. A given element of $\hom(m,m)$ or $\hom(n,n)$ can be extended to $\hom(n+m, n+m)$ by adjoining straight strings. The tensor product satisfies certain conditions that ensure that a monoidal 2-category structure is present; see [44] for details of these conditions.

An essential datum in a braided 2-category is the braiding. In BrMov the braiding is an element in $\hom(m+n, n+m)$ that is diagrammatically represented by the braid diagram in which m strings on the left cross over n strings on the right.

The axioms among 2-morphisms and other data in the braided monoidal 2-category can be verified for BrMov by using braid movie moves. We have additional data in BrMov such as the inverse braiding, and some braid movie moves enforce additional relations among 2-morphisms. A careful analysis of this 2-category in the spirit of Fischer's thesis is sure to bring new insight to the study of knotted surfaces.

5 Conclusion

The diagrammatic theory of knotted surfaces is rich. By closely examining the diagrams, a great deal of intuition about 4-space can be deemed. The diagrams themselves contain visual and tactile information. Good computer graphics packages such as GEOMVIEW or a virtual reality box could allow us to manipulate these diagrams just as strings can be manipulated to provide models for classical knot theory.

We have described the basic geometric and diagrammatic study of knot theory in dimension 4. Basic theorems that offer new tools have been presented. Furthermore, we used such diagrams to solve algebraic problems. Knotted surfaces should have a broad influence on the areas mentioned here: categorical aspects, classifying space of groups, and real algebraic geometry are some of the areas touched upon. The story of knotted surfaces is far from complete; we hope this brief introduction will serve as an impetus to further studies by others.

Acknowledgements

We are grateful for valuable conversations with J. Baez, D. Flath, C. Gordon, S. Kamada, M. Kapranov, L. Kauffman, Y. Marumoto, K. Murasugi, L. Rudolph, O. Viro, D. Yetter, and K. Yoshikawa. This paper was delivered at the 3rd International Workshop on Real and Complex Singularities in São Carlos, Brazil August 1994 by the first named author; he is grateful for that support. Alabama EPSCoR provided additional financial support. Finally, Cameron Gordon's financial support has been crucial to our continuing work. We are also grateful to the referee for valuable comments.

BIBLIOGRAPHIC NOTE Items [32, 52, 74, 77] can be found in [51]. Items [4, 5, 80] can be found in [31].

References

[1] Alexander, J.W., *A lemma on systems of knotted curves,* Proc. Nat. Acad. Science USA, 9 (1923), 93-95.

[2] Artin, E., *Theory of braids,* Ann. of Math., 48 (1947), 101-126.

[3] Apanasov, Boris, *Representations of hyperbolic groups and two-dimensional knots,* Preprint.

[4] Baxter, R.J. *On Zamolodchikov's solution of the tetrahedron equations*, Comm. Math. Phys. 88 (1983), 185-205;

[5] Bazhanov, V.V. and Stoganov, Yu. G. *Chiral Potts models as a descendant of the six-vertex model.*

[6] Birman, J. and Menasco, W., *Studying links via closed braids,* I. Pac. J. Math. 154 (1992) 71-82, II. Top. appl. 40 (1991) 71-82, III. Pac. J. Math. 161 (1993) 25-113, IV. Invent. math. 102 (1990) 115-139, V. Trans. AMS. 329 (1992) 585-606, VI. Pac. J. Math. 156 (1992) 265-285.

[7] Brown, R., and Huebchumann, J., *Identities among relations,* in Low-dimensional Topology, eds R. Brown and T. L. Thickstun, London Math. Soc. Lecture Note Series 48, pp153-202, (1982).

[8] Carter, J. Scott and Saito, Masahico, *Canceling branch points on projections of surfaces in 4-space*, Proc. of the AMS. 116, No 1. (Sept 1992), 229-237.

[9] Carter, J. Scott and Saito, Masahico, *Syzygies among elementary string interactions in dimension 2+1*, Letters in Mathematical Physics 23 (1991), 287-300.

[10] Carter, J. Scott, and Saito, Masahico, *Reidemeister moves for surface isotopies and their interpretation as moves to movies*, J. of Knot Theory and its Ram., vol. 2, no. 3, (1993), 251-284.

[11] Carter, J. Scott, and Saito, Masahico, *Planar generalizations of the Yang-Baxter equation and their skeins*, Journal of Knot Theory and its Ram. Vol 1, No. 2 (1992), 207-217.

[12] Carter, J. Scott, and Saito, Masahico, *A Seifert algorithm for knotted surfaces*, Preprint.

[13] Carter, J. Scott, and Saito, Masahico, *On formulations and solutions of simplex equations*, Preprint.

[14] Carter, J. Scott, and Saito, Masahico, *Some new solutions to the permutohedron equation,* to appear in the conference proceedings on quantum topology (eds. L. Crane and D. Yetter).

[15] Carter, J. Scott, and Saito, Masahico, *Braids and Movies*, Preprint.

[16] Carter, J. Scott, and Saito, Masahico, *Knotted surfaces, braid movies and beyond,* in Baez, J., "Knots and Quantum Gravity," Oxford Science Publishing (Oxford 1994), 191-229.

[17] Carter, J. Scott, Flath, Dan, and Saito, Masahico, *Classical and quantum 6j symbols,* Preprint.

[18] Ewen, H. and Ogievetsky, O., *Jordanian solutions of simplex equations*, Preprint.

[19] Finashin, S. M., Kreck, M., Viro, O. Ya., *Non-diffeomorphic but homeomorphic surfaces in the 4-sphere,* in Viro, Oleg Ya., "Topology and Geometry – Rohlin Seminar," LNM #1346, Springer (1988), 157-198.

[20] Fischer, John, *2-categories and 2-knots,* Duke Journal of Mathematics 75, No. 2, (August 1994), 493-526.

[21] Fox, R.H., *A Quick trip through knot theory,* in Cantrell and Edwards, "Topology of Manifolds," Prentice Hall (1962).

[22] Frenkel, Igor, and Moore, Gregory, *Simplex equations and their solutions,* Comm. Math. Phys. 138 (1991), 259-271.

[23] Freyd, P.J., and Yetter, David, *Braided compact closed categories with applications to low dimensional topology,* Adv. in Math., 77 (1989), 156-182.

[24] Giller, Cole, *Towards a classical knot theory for surfaces in* $\mathbf{R}^4$, Illinois Journal of Mathematics 26, No. 4, (Winter 1982), 591-631.

[25] Golubitsky, Martin, and Guillemin, Victor, "Stable Mappings and Their Singularities," GTM # 14, Springer Verlag, (1973).

[26] Hietarinta, J., *Some constant solutions to Zamolodchikov's tetrahedron equations,* Preprint.

[27] Hillman, Jonathan,"2-knots and Their Groups," Australian Mathematical Society Lecture Series No. 5, Cambridge University Press (1989).

[28] Hillman, Jonathan, and Kawauchi, Akio, *Unknotting orientable surfaces in the 4-sphere,* Preprint.

[29] Hosokawa, Fujitsugu, and Kawauchi, Akio, *Proposals for unknotted surfaces in four-space,* Osaka J. Math. 16 (1979), 233-248.

[30] Homma, T. and Nagase, T., *On elementary deformations of the maps of surfaces into 3-manifolds I,* Yokohama Mathematical Journal 33 (1985), 103-119.

[31] Jimbo, M., "Yang-Baxter Equation in Integrable Systems," World Scientific Publishing Co., (Singapore 1989).

[32] Jones, V. F. R., *A polynomial invariant for knots and links via von Neumann Algebras,* Bull. AMS 12 (1985), 103-111.

[33] Kamada, Seiichi, *Non-orientable surfaces in 4-space,* Osaka J. Math. 26 (1989), 367-385.

[34] Kamada, Seiichi, *Surfaces in* R^4 *of braid index three are ribbon,* Journal of Knot Theory and its Ramifications 1 (1992), 137-160.

[35] Kamada, Seiichi, *2-dimensional braids and chart descriptions,* "Topics in Knot Theory", Proceedings of the NATO Advanced Study Institute on Topics in Knot Theory, Turkey, (1992), 277-287.

[36] Kamada, Seiichi, *A characterization of groups of closed orientable surfaces in 4-space*, Topology 33 (1994), 113-122.

[37] Kamada, Seiichi, *Seifert circles for surface braids*, in Surikaiseki-Kenkyujo-Kokyuroku, (Seminar note at RIMS, Kyoto) vol. 813, (1992), 144-154 (in Japanese).

[38] Kamada, Seiichi, *Alexander's and Markov's theorems in dimension four*, Bull. Amer. Math. Soc., 31 (1) 1994, 64-67.

[39] Kamada, Seiichi, *Generalized Alexander's and Markov's theorems in dimension four*, Preprint.

[40] Kamada, Seiichi, *Braid monodromies of nonsimple braided surfaces*, Preprint.

[41] Kamada, Seiichi, *On 2-dimensional braids and 2-links*, to appear in the Proceedings of the 3rd Korea-Japan school of knots and links, Taejon 1994.

[42] Kamada, Seiichi, *Survey on 2-dimensional braids,* Preprint.

[43] Kapranov, M., and Voevodsky, V., *Braided monoidal 2-categories, 2-vector spaces and Zamolodchikov's tetrahedra equations,* (first draft) Preprint.

[44] Kapranov, M., and Voevodsky, V., *2-Categories and Zamolodchikov tetrahedra equations,* Preprint.

[45] Kapranov, M., and Voevodsky, V., *Braided monoidal 2-categories and Manin-Schechtman higher braid groups,* J. Pure and Applied Alg., 92 (1994), 241-267.

[46] Kauffman, Louis, " On Knots", Ann. of Math. Studies, Princeton U. Press, 1987.

[47] Kauffman, Louis, "Knots and Physics," World Science Publishing (1991).

[48] Kawauchi, A. Shibuya, T., and Suzuki, S., *Descriptions of Surfaces in four-space, I*, Math. Sem. Notes Kobe Univ. 10 (1982), 75-125.

[49] Kawauchi, A. Shibuya, T., and Suzuki, S., *Descriptions of surfaces in four-space, II*, Math. Sem. Notes Kobe Univ. 11 (1983), 31-69.

[50] Kazhdan, D., and Soibelman, Y., *Representations of quantized function algebras, 2-categories and Zamolodchikov tetrahedra equation,* Preprint.

[51] Kohno, T., "New Developments in the Theory of Knots," World Scientific Publishing (Singapore 1989).

[52] Kirillov, A. N. and Reshetikhin, N. Yu, *Representations of the Algebra* $U_q(sl(2))$, *q-Orthogonal Polynomials and Invariants of Links.*

[53] Korepin, V.E., *Completely integrable models in quasicrystals*, Commun. Math. Phy. 110, (1987), 157-171.

[54] Lawrence, Ruth, *On algebras and triangle relations,* in Topological and Geometric Methods in Field Theory, eds. J. Mickelsson and O. Pekonen, World Scientific (1992), 429-447.

[55] Libgober, A., *Invariants of plane algebraic curves via representations of the braid groups*, Invent. math. 95 (1989), 25-30.

[56] Lin, X.-S., *Alexander-Artin-Markov theory for 2-links in* $\mathbf{R}^4$, Preprint circa 1990.

[57] Lomonaco, Sam, *The homotopy groups of knots I. How to compute the algebraic 2-type,* Pacific J. Math. 95, (2), (1981), 349-390.

[58] Maillet, J.M., Nijhoff, F.W. *Multidimensional integrable lattices, quantum groups, and the D-simplex equations*, Instit. for Non-Linear Studies Preprint 131 (1989).

[59] Markov, A. A., *Über die freie Äquivalenz der geschlossner Zopfe,* Rec. Soc., Math. Moscou 1 (1935), 73-78.

[60] Marumoto, Yoshihiko, *Univ. of South Alabama topology seminar,* (1992)

[61] Moishezon, B., *Stable branch curves and braid monodromies*, Lect. Notes in Math. 862, Springer (1981), 107-192.

[62] Moishezon, B. and Teicher, M., *Braid group technique in complex geometry, I : Line arrangements in* $\mathbf{C}P^2$, in "Braids" eds J. Birman and A. Libgober, Contemp. Math. 78 (1986), 425-555.

[63] Nakanishi, Yasutaka, and Teragaito, Masakazu, *2-knots from a view of moving picture*, Kobe J. Math., 8 (1991), 161-172.

[64] Reshetikhin, N. and Turaev, V, *Invariants of 3-manifolds via link polynomials*, Inventiones Math. 103 (1991), 547- 597.

[65] Rolfsen, Dale, "Knots and Links," Publish or Perish Press, (Berkley 1976).

[66] Roseman, Dennis, *Reidemeister-type moves for surfaces in four dimensional space,* Preprint.

[67] Roseman, Dennis, *Projections of codimension two embeddings,* Preprint.

[68] Roseman, Dennis, *Twisting and turning in four dimensions* , A video made at Geometry Center, (1993), presented at University of South Alabama conference on Knot Theory, Low Dimensional Topology, and Quantum Groups.

[69] Rudolph, Lee, *Braided surfaces and seifert ribbons for closed braids,* Comm. Math. Helv. 58, (1983), 1-37.

[70] Rudolph, Lee, *Algebraic functions and closed braids*, Topology 22 (2), (1983), 191-202.

[71] Rudolph, Lee, Private conversations.

[72] Sumners, De Witt, *Inverible Knot Cobordisms*, Commentarii Mathematici Helvetici, Vol. 46 fasc. 2 (1971), p. 240-256.

[73] Suzuki, Shin'ichi, *Knotting problems of 2-spheres in 4-sphere,* Math. Sem. Notes Kobe Univ., vol.4, No.3 (1976), 241-371.

[74] Tureav, V., *The Yang-Baxter equation and invariants of links*, Invent. Math. 92 (1988), 527-553.

[75] Turaev, V. and Viro, O, *State Sum Invariants of 3-Manifolds and Quantum 6J-Symbols*, Topology **31**, No 4 (1992), 865-902.

[76] Whitney, H., *On the topology of differentiable manifolds*, in Lectures in Topology, (Wilder and Ayres, eds.), University of Michigan Press (1941), 101-141.

[77] Witten, Edward, *Quantum field theory and the Jones polynomial,* Comm. Math. Phys., 121 (1989), 351-399.

[78] Yajima, T., *On simply knotted spheres in* $\mathbf{R}^4$, Osaka J. of Math 1 (1964), 133-152.

[79] Yoshikawa, Kazuyuki, *An enumeration of surfaces in four space,* Preprint.

[80] Zamolodchikov, A. B., *Tetrahedron equations and the relativistic S-matrix of straight-strings in* $2+1$*-dimensions*, Comm. Math. Phys. 79 (1981), 489-505.

J.W. Bruce P.J. Giblin F. Tari[1]

Families of surfaces: height functions, Gauss maps and duals

1 Introduction

Let M be a smooth surface in euclidean space $\mathbb{R}^3$ and p a point of M. The intersection of M with its tangent plane at p contains useful geometrical information on the geometry of the surface at p. More precisely, we can associate to p a *contact singularity* between the tangent plane and the surface, containing this geometrical information. To define this singularity we use the *height function* on M in the direction of the normal at p. The level sets of this function are the planes in $\mathbb{R}^3$ parallel to the tangent plane at p, and the germ of the height function at p defines the contact singularity. If p is elliptic or hyperbolic, then the height function has a singularity that is of a stable 'Morse' type (also known as A_1), and the intersection is a point (elliptic case) or a pair of intersecting smooth curves (hyperbolic case). The remaining case, when p is parabolic, leads to a contact singularity of 'worse' type (the height function has type A_2 or 'worse'); in the simplest case (A_2) the tangent plane meets the surface in a curve with an ordinary cusp at p. Height functions can also be used to study the set of planes which are tangent to M at more than one point. These correspond to multi-local singularities of the height function (singularities at all of which the height function has the same value). These 'bitangent planes' have arisen several times in applications of differential geometry. See for example [19], where the reader can also find intuitive descriptions of some of the other phenomema we describe in this article. Recently bitangent planes have arisen again in the study of shadows on surfaces thrown by a light source which is itself a (convex) surface [20].

The connexion between the extrinsic geometry of a surface and the singularities of functions (such as height functions) defined on the surface was pointed out by R.Thom in the 1960s, and this point of view has proved extremely fruitful in understanding and extending the differential geometry of surfaces in $\mathbb{R}^3$.

The collection of all tangent planes to M constitutes the *dual* of M, and this object, which exists in the space dual to $\mathbb{R}^3$, carries a good deal of information too. The dual of M is a surface which is singular at points corresponding to parabolic points of M, and the dual has correspondingly 'worse' singularities when the tangent plane has a higher contact with M. The dual clearly also contains all the information present in the Gauss map of M, and more besides.

[1]Supported by the Esprit grant VIVA.

In this paper we apply the techniques of singularity theory to study height functions on surfaces which are evolving in a generic 1-parameter family in $\mathbb{R}^3$. We begin in §2 by recalling the results for a single surface and introducing the concepts needed later. In §2 we also make a geometrical study of the curves of contact of bitangent planes to a smooth surface M. Our study of families is partly *local* and partly *multilocal.* In the local case (§3), we find the changes which occur on parabolic sets and on the images of these sets under the Gauss map. The generic changes on the dual have already been described by Arnold, and are drawn in [1, Figure 36 in the 3rd edition]. In the multilocal case (§4), we concentrate on the image under the Gauss map of the curves of contact of bitangent planes. In fact we show that all transitions which are theoretically possible in this context (as enumerated in [11]), with one exception, do actually occur. This is in sharp contrast to the example considered in [11], namely the evolution of symmetry sets in the plane, where many of the transitions are ruled out for geometrical reasons.

The definitions of height functions and Gauss map both depend on the use of inner products. It is therefore rather surprising that the family of height functions is *invariant under affine transformations.* We do not make direct use of this fact here, but in Appendix A we prove a general result which shows affine independence for any family of linear projections. This applies to projections to planes as well as height functions, which are projections to lines. One consequence of the result is that, when studying the family of height functions on M we can make an initial affine transformation. So, for example, suppose we are working at the origin O, where M has (locally) the equation $z = h(x, y)$ with no linear or quadratic terms (so O is a 'flat umbilic'). Then we can by an affine transformation turn M into the surface M_1 with equation $z = x(x^2 \pm y^2) + h_1(x, y)$, where h_1 has order 4 and higher. The surface M_1 has an *equivalent* family of height functions.

For a different approach to some of the material of this paper, see [21].

The present paper is one of a series which treats the differential geometry of 1-parameter families of surfaces by means of singularity theory. In [13] there is a detailed exploration of the very striking duality which exists between the family of height functions and the family of projections to planes, which was introduced for a single generic surface in [14]. In [12] the family of distance-squared functions is treated; these measure the contact between a surface M and *spheres.* The points of M where there is high (A_3) contact make up the *ridges* of M and the evolution of parabolic curves is intimately associated with that of ridges. This is because cusps of Gauss occur at intersections of the parabolic set with the ridge set. See §3.1 and Figure 1. In §2.1 we give a geometrical criterion for measuring the singularity of a height function in terms of *contact* between the parabolic curve and ridge. The structure of asymptotic curves on M, at the moments of transition occurring in the present paper, is analysed in [15].

As a general reference for background results and terminology in singularity theory we recommend [28].

2 Generic geometry of one surface via the family of height functions

Let M be a smooth surface (which is also compact and without boundary) in $\mathbb{R}^3$, and let S^2 be the unit sphere in $\mathbb{R}^3$, which is here regarded as the set of unit vectors, that is directions, in $\mathbb{R}^3$. The *family of height functions* on M is the family

$$h : M \times S^2 \to \mathbb{R},$$

defined by

$$h(p,u) = \langle p, u \rangle, \tag{1}$$

the Euclidean inner product of p with u. We write h_u for the function h with u fixed. The function h_u is the height function on M in the direction u, and it measures the contact between M and planes perpendicular to u.

We also use the family

$$H : M \times S^2 \to \mathbb{R} \times S^2, \tag{2}$$

defined by $H(p,u) = (h(p,u), u)$.

We begin by recalling some facts about height functions, for which it is convenient to take M in Monge form.

2.1 Height functions on a surface: type and versal unfolding by the family of height functions

Let M be given locally by $z = f(x,y)$, where f and the partial derivatives f_x and f_y all vanish at $x = y = 0$. Let $p = (0,0,0)$ and $u = (0,0,1)$. Then f itself is the height function on M in the direction u. We may parametrize the directions near to (0,0,1) by $(a,b,1)$ for a, b close to 0, so that the family of height functions takes the local form

$$h(x,y,a,b) = f(x,y) + ax + by.$$

This family versally unfolds the height function f if and only if $\partial h/\partial a = x$ and $\partial h/\partial b = y$ span $\mathcal{M}(x,y)/J(f)$, where $\mathcal{M}$ is the maximal ideal of functions vanishing at (0,0) and $J(f)$ is the Jacobian ideal of f, generated by $\partial f/\partial x$ and $\partial f/\partial y$. (See [2, 4, 9, 22] for information on versal unfoldings.) We use the now standard notation of Arnold for singularities of functions (A_k, D_k). A function is of type A_k^+ (resp. A_k^-) if it is $\mathcal{A}$-equivalent to $x^2 + y^{k-1}$ (resp. $x^2 - y^{k-1}$). The germs A_{2k}^+ and A_{2k}^- are equivalent.

Type A_1 This requires that the quadratic terms of f do not form a perfect square, which is equivalent to saying that p is not a parabolic point of M. The two types, A_1^+ and A_1^-, correspond to elliptic and hyperbolic points, respectively. Versal unfolding in this case is automatic (virtually by definition of an A_1 singularity).

Type A_2 This requires that the quadratic terms do form a perfect square, so that rotating about the z-axis, we can assume that the quadratic term is of the form a_0x^2. Thus p is parabolic and the asymptotic direction at p is along the y-axis. We can write

$$f(x,y) = a_0x^2 + b_0x^3 + b_1x^2y + b_2xy^2 + b_3y^3 + \phi_4,$$

where ϕ_4 is of order 4 or higher. Provided $a_0 \neq 0$ and $b_3 \neq 0$ this function has type A_2 at the origin, and versal unfolding is easily checked to be automatic.

Type A_3 Here, $b_3 = 0$. Write f as

$$f(x,y) = a_0x^2 + b_0x^3 + b_1x^2y + b_2xy^2 + c_0x^4 + c_1x^3y + c_2x^2y^2 + c_3xy^3 + c_4y^4 + \phi_5 \quad (3)$$

where ϕ_5 is of order 5 or higher. When $a_0 \neq 0$ we complete squares to kill off the terms beyond x^2 which involve an x, arriving at a 4-jet of the form $a_0x^2 + (c_4 - b_2^2/4a_0)y^4$. So if $b_2^2 - 4a_0c_4 \neq 0$ we have a 4-$\mathcal{R}$-determined jet which is $\mathcal{R}$-equivalent to $\pm x^2 \pm y^4$, an A_3 singularity. The versal unfolding condition in this case is easily checked to be $b_2 \neq 0$. (The contrary case, $b_2 = 0$, is a 'non versally unfolded A_3' which we shall meet in §3 as a transitional phenomenon in a family of surfaces.) Geometrically, we have a 'cusp of Gauss', which occurs when the parabolic curve is *nonsingular* at p and is tangent to the asymptotic direction there (with ordinary contact). (Compare [3].) There are two types: *elliptic* cusps, corresponding to A_3^+ (where $b_2^2 < 4a_0c_4$) and *hyperbolic* cusps (A_3^-, where $b_2^2 > 4a_0c_4$).

Type A_1^2 We are also concerned in this article with *multi-local* singularities. In the case of A_1^2, the height function h_u has two A_1 singularities, at $p = p_1, p = p_2$ say, for which $h(p_1, u) = h(p_2, u)$; this amounts to saying that there is a single plane perpendicular to u and tangent to M at p_1, p_2. Neither point is parabolic. It turns out that the family of height functions always versally unfolds such a singularity; see Prop. 5.1.

Type A_1A_2 Here, a single plane is tangent to M at a non-parabolic point p_1 and a parabolic point p_2. This time there is a condition for h to versally unfold the singularity: the asymptotic direction at p_2 must not pass through p_1; see Prop. 5.1. The contrary case provides one of our transitions in a 1-parameter family of surfaces; see §4.

Type A_1^3 Here, a single plane is tangent to M in three non-parabolic points. The condition for a versal unfolding is that the three points are not collinear (see Prop. 5.1); the contrary case is a transition in a 1-parameter family of surfaces; see §4.

We pause here to point out geometrical criterion for detecting A_k singularities of the height function on a surface M. Recall that the *ridges* of M consist of points where one of the spheres of curvature has A_3 contact with M. That is, the ridges are the 'A_3 sets of the family of distance squared functions'. (See [24].)

Theorem 2.1 *The height function has an A_k singularity ($k \geq 3$) at a smooth point on the parabolic curve if and only if the parabolic curve and the ridge have $(k-2)$-point contact at this point.*

Proof The height function has an A_3-singularity at p (i.e., p is a cusp of Gauss, see [3]) if and only if the parabolic curve crosses the ridge transversally at p. Taking M in Monge form as above we may use the 'modified Gauss mapping' [3, p.11] $F(x,y) = (\partial f/\partial x, \partial f/\partial y)$. The order of contact of parabolic curve and ridge is measured by the *number of cusps* of F at $(0,0)$ (compare [17, 25]).

The function f has an A_k singularity if and only if F is $\mathcal{K}$-equivalent to (x, y^k). Suppose that this is the case, so that, by [16, Prop.4.8], F can be written in a suitable choice of co-ordinates in the form

$$F = (x, y^k + P_{k-2}(x)y^{k-2} + ... + P_1(x)y)$$

with $P_i \in \mathcal{M}_1$. Since the parabolic curve (which is the critical set of the Gauss map F) is smooth P_1 is of order 1 and a further change of coordinates yields

$$F = (x, y^k + P_{k-2}(x)y^{k-2} + ... + xy) = (x, g(x,y))$$

say.

Write $\mathcal{O}$ for the ring of germs at the origin of functions of x, y. Working over the complex numbers the number of cusps of F (see [25]) is given by

$$c(F) = dim_C \frac{\mathcal{O}}{\mathcal{O} < g_y, g_{yy} >}.$$

Clearly we have $c(F) = k - 2$.

Conversely, if $c(F) = k - 2$ and the critical set of F is smooth then $F = (x, y^k + P_{k-2}(x)y^{k-2} + ... + xy)$, that is f has an A_k-singularity.

The most important geometric invariant of a versal unfolding is its bifurcation set and we recall the definition now.

Definition 2.2 *The* bifurcation set *of h or H consists of the set of parameter values u for which the map h_u is* unstable, *that is* not *of type A_1, for some point p. This is precisely the set of normals at parabolic points of M, as in 2.1. We say that the height function has 'type A_2 or worse'.*

The full bifurcation set *of h or H is the union of the bifurcation set and the set of parameter values u for which there exist at least two distinct points p_1, p_2 at which h_u is singular, and for all of which points the value of h is the same. The latter is precisely the set of normals to planes which are tangent to M at two or more distinct points. We say that the height function has 'type A_1^2 or worse'.*

Note that, since h_u and h_{-u} have the same singularities, the bifurcation set and full bifurcation set contain $-u$ whenever they contain u. For this reason we can regard them both as lying in the real projective plane $\mathbb{R}P^2$ as well as in the sphere S^2.

Next we recall briefly some other standard definitions concerning a family of functions and note their application to the present situation. We take M to be oriented, that is with a well-defined unit normal vector field.

Consider the following maps:

$$M \xrightarrow{i} M \times S^2 \xrightarrow{H} \mathbb{R} \times S^2 \xrightarrow{\pi} S^2.$$

Here, $i(p) = (p, n(p)), n$ being the unit normal to M at p, and π is projection to the second factor. Note that the composite $\pi \circ H \circ i$ is simply the Gauss map of M. For some purposes below, the sign of n will be unimportant, so we write also $i^-(p) = (p, -n(p))$. We have the following, all of which are easy to check.

Proposition 2.3 *1. The critical set Σ of H is the union of the images of i and i^-, namely, the set of points $(p, \pm u)$ where u is the normal to M at p.*

2. *$H(\Sigma)$, otherwise known as the* discriminant *of H, can be identified with the* dual *of M, that is, the set of tangent planes to M, by identifying a plane $\langle x, u\rangle = d$ in $\mathbb{R}^3$ with the pair $(d, u) \in \mathbb{R} \times S^2$. (See the note below on duals.)*

3. *Assume that the origin of $\mathbb{R}^3$ does not lie on M. Mapping $\mathbb{R} \times S^2$ to $\mathbb{R}^3$ by $(d, u) \to du$ turns the discriminant of H into the classical* pedal surface *of M.*

4. *The dual $H(\Sigma)$ has a local singularity at $H(p, u)$ precisely when $(p, u) \in \Sigma$ and p is a parabolic point of M. It has a self-intersection singularity at $H(p_1, u) = H(p_2, u)$ precisely when u is the normal at two distinct points p_1, p_2.*

5. *The projection by π of the full singular set of $H(\Sigma)$ (local and self-intersections as above) is the set of normals at parabolic points and bitangent contact points of M. This is precisely the full bifurcation set of h in S^2.*

Note on duals If we embed $\mathbb{R}^3$ in $\mathbb{R}P^3$ by $(x, y, z) \mapsto (x : y : z : 1)$, then our surface M, regarded as a surface in $\mathbb{R}P^3$, has a dual M^* in the dual space $(\mathbb{R}P^3)^*$. This dual is the image of $H(\Sigma)$ under the map

$$j : \mathbb{R} \times S^2 \to (\mathbb{R}P^3)^*$$

given by $j(d, u) = (u : -d)$. The map j is a local diffeomorphism but a double cover of its image, since $j(d, u) = j(-d, -u)$. We can project this dual into $\mathbb{R}P^2$ by $(d, u) \mapsto u$ (note that $u \neq 0$) and then the image of the full singular set of the dual is the full bifurcation set of h, here regarded as lying in $\mathbb{R}P^2$.

For each of the singularity types in §2.1 above there is a geometrical interpretation in terms of the dual and the bifurcation set, as follows.

2.2 The dual of M and the bifurcation set of h

A_1 singularities The dual is smooth at $H(p,u)$. The local intersection of M with its tangent plane at p is a point or a pair of transversally intersecting smooth curves.

A_2 singularities The dual has a cuspidal edge at $H(p,u)$ (Figure 1) . The point u now belongs to the bifurcation set of h (or H), the Gauss map having a fold singularity at p, so that the bifurcation set is smooth at u. The local intersection of M with its tangent plane at p is an ordinary cusp. (There is a beautiful picture of this in [18, p.199].)

A_3 singularities, versally unfolded The parabolic curve is smooth at p and is tangent to the asymptotic direction at p. The dual has a swallowtail point at $H(p,u)$ (Figure 1). The Gauss map has a cusp singularity at p, and the bifurcation set of h has an ordinary cusp at u. The *full* bifucation set has an extra branch ending in the cusp: this is the projection to S^2 of the self-intersection line on the swallowtail. Note that a cusp of Gauss thereby has a family of nearby bitangent planes; compare [3, Theorem 3.1(c)].

The local intersection of M with its tangent plane at p is a pair of smooth tangential curves (hyperbolic cusp of Gauss) or a point (elliptic cusp).

A_1^2 singularities, versally unfolded The dual has a transverse crossing of two smooth sheets at $H(p_1,u) = H(p_2,u)$ (Figure 1). The point u belongs to the full bifurcation set of h (or H), which is smooth at u.

A_1A_2 singularities, versally unfolded The dual has a crossing of a smooth sheet and a cusp edge at $H(p_1,u) = H(p_2,u)$ (Figure 1) and the full bifurcation set has a cusp at u.

A_1^3 singularities The dual has a pairwise transverse crossing of three sheets at $H(p_i,u), i = 1,2,3$, where p_i are the points of contact of the tritangent plane. The full bifurcation set has a pairwise transverse crossing of three smooth curves at u (Figure 1).

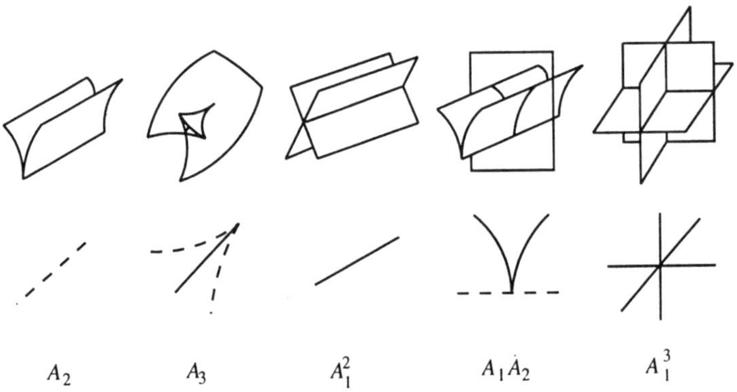

Figure 1

2.3 Geometry of bitangent planes

To conclude this section we consider the geometry of bitangent planes. Specifically we consider the *curves of contact of bitangent planes.* Of course, we are strictly studying *multigerms* of height functions on the surface M.

If a bitangent plane π has contact points p_1, p_2 and neither is parabolic, then the singularity is of type A_1^2. In general we can certainly expect that there will be other bitangent planes whose points of contact are close to p_1 and p_2: the bitangent plane will 'roll' on the surface. Thus on M we have a set (or stratum) which can also be labelled A_1^2: the set consisting of points of contact of such planes. If the plane π happened to be tangent at a third point p_3, then we would label the three points of contact A_1^3. In that case, through each of the p_i we expect *two* curves of A_1^2 points on M, corresponding to planes tangent to M near p_i and one of the other two points.

The (full) bifurcation set in the Gauss sphere S^2 also has strata which are labelled A_1^2, A_1A_2, A_1^3, as in §2.2 above.

Given a bitangent plane to M, let l be the line joining the points of contact p_1, p_2. Passing to the dual M^*, the bitangent plane corresponds to a point q on the self-intersection curve of M^*, and the points p_i correspond to the distinct tangent planes to the dual at q. The line l corresponds to the intersection of these two tangent planes, which is clearly tangent to the curve of self-intersection of the dual.

Projecting from the dual in $\mathbb{R}P^3$ to the bifurcation set in $\mathbb{R}P^2$ (compare Prop. 2.3 and the note following it), we obtain the following.

Proposition 2.4 *Let p_1, p_2 be points of contact of a bitangent plane to M. Then the tangent to the A_1^2 set (in the bifurcation set) at the point corresponding to the pair p_1, p_2 is determined by the line l joining p_1 and p_2.*

Note that the proposition holds in limiting cases: in the A_1A_2 case, for example, where the A_1^2 stratum in the bifurcation set acquires a cusp at the A_1A_2 point, the line l, under duality and projection, becomes the cuspidal tangent.

Corollary 2.5 *For a tritangent plane π, with p_3 a further point of contact, the branches of the A_1^2 stratum in the bifurcation set, corresponding to p_1, p_2 and p_1, p_3, are tangent if and only if the points p_1, p_2, p_3 are collinear.*

The 'if' part of the corollary follows from Proposition 2.4. It is not hard to check that the collinearity condition is necessary as well as sufficient. Of course, if two branches are tangent, then all three must be, in view of the symmetry of the collinearily condition.

Let C_i be a curve on M through p_i, containing the points of contact of bitangent planes close to some bitangent plane π. Then it is easy to check that C_1 will be smooth close to p_1 provided the following hold:
(a) p_2 is not a parabolic point of M and,

(b) if p_1 *is* a parabolic point of M then the asymptotic direction there does not pass through p_2.
(Similarly with points 1 and 2 reversed.)

Thus both C_1 and C_2 will be smooth provided neither p_1 nor p_2 is parabolic; this latter condition also ensures the bifurcation set is smooth at the point corresponding to this bitangent plane.

Proposition 2.6 *Suppose that C_i is smooth at p_i. Then the tangent to C_i is the conjugate, at p_i, of the tangent vector in the direction of the line l joining p_1 and p_2. As a special case, if say p_1 is parabolic, (the A_1A_2 case; here the singularities are at p_2, p_1 respectively) then the tangent to C_1 is necessarily in the asymptotic direction at p_1, unless l is itself in the asymptotic direction (in which case C_1 will be singular at p_1).*

Proof The normal to M at p_i is instantaneously turning about l, since it is also the normal to the bitangent plane. Thus the derivative of the normal along C_i is perpendicular to l, and this implies that the conjugate direction to the tangent to C_i is *along* l.

It is interesting to pursue a little further the nature of the curves C_i on M. Assume that p_1 is parabolic, and that p_2 *does* lie along the asymptotic direction at p_1. In that case the height functions do not versally unfold the A_1A_2 singularity (see §2.1 and Prop. 5.1). But we can still say something about the local structure of C_1 and C_2.

To be specific, let M have local equation, near $p_1 = (0, 0, 0)$,

$$z = f_1(x, y) = a_{01}x^2 + b_{01}x^3 + b_{11}x^2y + b_{21}xy^2 + b_{31}y^3 + \dots,$$

where $a_{01} \neq 0$ and $b_{31} \neq 0$. Let M have local equation, near $p_2 = (0, \lambda, 0)$,

$$z = f_2(x, y) = a_{02}x^2 + a_{12}x(y - \lambda) + a_{22}(y - \lambda)^2 + \dots,$$

where $\lambda \neq 0$ and the Gauss curvature at p_2, namely $K = 4a_{02}a_{22} - a_{12}^2$, is nonzero. We can write down the conditions for a plane to be tangent to M at $(x_1, y_1, f_1(x_1, y_1))$ close to $(0, 0, 0)$ and at $(x_2, y_2, f_2(x_2, y_2))$ close to $(0, \lambda, 0)$. These amount to three conditions on the four variables x_1, y_1, x_2, y_2, and we find that x_2, y_2 are functions of x_1, y_1. The three conditions then reduce to a single condition on x_1, y_1, the linear terms of which vanish, and the quadratic terms of which distinguish between the two expected 'Morse' singularities. We find the following.

Proposition 2.7 *The curve C_1 is an isolated point or a crossing of two smooth branches at p_1 according as*

$$\lambda(\lambda(b_{21}^2 - 6b_{11}b_{31}) + 6a_{01}b_{31}(1 + \frac{4a_{01}a_{22}}{K}))$$

is < 0 or > 0.

If this expression is equal to 0, then the situation is more degenerate than A_1A_2. Note that there is a nonzero value of λ which this brings in.

It can be shown that, in a generic family of surfaces, passing through this 'non-versal A_1A_2' situation, the curve C_1 will undergo a standard Morse transition, modelled by the curves $x^2 - y^2 = t$ for a crossing at $t = 0$ and $x^2 + y^2 = t$ for an isolated point at $t = 0$. The curve C_2 remains smooth throughout.

3 Generic geometry of a 1-parameter family of surfaces via height functions: local singularities

Let M be a given surface. We will be interested in families of embeddings $f : M \times I \to \mathbb{R}^3$ where I is some open, connected and finite interval. So for each $t \in I$ the set $f_t(M)$ is an embedded surface in $\mathbb{R}^3$, and will be denoted by M_t.[2] We first consider the contact singularities occurring generically in such families, beyond those in §2. To establish a list, and some other useful facts concerning unfoldings, we need a transversality theorem. See [27] for background information.

The important consequences are contained in the corollary.

Theorem 3.1 *Let M be a compact surface, I an open interval. Let k be a positive integer and $\mathcal{S}$ an $\mathcal{A}$-invariant Whitney regular stratification of the multijet-space ${}_rJ^k(M,\mathbb{R})$. Let $Emb^\infty(M,I,\mathbb{R}^3)$ denote the open subset of the space of smooth mappings $f : M \times \mathbb{R} \to \mathbb{R}^3$ with $f_t : M \to \mathbb{R}^3$ an embedding for each t. Then the set of $f \in Emb^\infty(M,I,\mathbb{R}^3)$ with the jet-extension*

$$ {}_rj_1^k h \circ f : M^{(r)} \times I \times S \to {}_rJ^k(M,\mathbb{R}) $$

given by

$$ {}_rj_1^k h \circ f(p,t,u) = {}_rj^k(h_u \circ f_t)(p) $$

transverse to $\mathcal{S}$ is residual. (We can replace residual by open and dense if we ask for transversality over a compact subinterval J of I.)

Proof This is a consequence of Theorem 1.1 in [8], which is in turn a straightforward generalization of a theorem of Montaldi [23]. The key fact required to apply this result to the present situation is that the family of height functions on the ambient space $\mathbb{R}^3$ is $\mathcal{A}$-versal. This is a trivial consequence of the definition.

The key consequence that we require is the following result.

[2] If we wish to think of our surface as a level set, that is determined by some equation $g = 0$ where g is a function on $\mathbb{R}^3$, then it is natural to consider the family given by setting g equal to a constant t. One would then expect that for certain values of t the surfaces $g = t$ would be singular. In particular they might acquire new components. This raises some interesting problems, but we shall concentrate in this paper on embeddings.

Corollary 3.2 *Let M, I, J be as before. Then for an open dense set of mappings f in $Emb^\infty(M, I, \mathbb{R}^3)$ there are finitely many points $\{t(1), \ldots, t(r)\}$ in the interval J such that*

(i) If $t \notin \{t(1), \ldots, t(r)\}$ then the only singularities of the height functions h_u for the surface M_t are of type A_1, A_2 and A_3. Moreover these singularities are versally unfolded by the family $h : M_t \times S \to \mathbb{R}$.

(ii) If t is one of the $t(j)$ then the only additional singularities are of type A_4 and $D_4^{\pm}$. All singularities (including these) are versally unfolded by the family $h \circ f : M \times I \times S \to \mathbb{R}$, defined by $(h \circ f)(p, t, u) = h_u(f_t(p)) = \langle f_t(u), u \rangle$.

Proof The basic idea is fairly simple. If f is a generic family of embeddings then the flat geometry of the family of surfaces $f_t(M) = M_t$ for $t \in J$ has finitely many catastrophic events. These are of two types: (i) some point of the surface $M_{t(j)}$ may have contact with its tangent plane which is more degenerate than an A_3 singularity (an A_4 or $D_4^{\pm}$) or (ii) the singularity may be of type A_1, A_2 or A_3 but not versally unfolded by the family h of height functions alone. In case (ii) this means that the contact is one of the prescribed types, but the geometry associated to that singularity (the Gauss map and dual) does not follow the standard model. We have already remarked in §2.1 that versal unfolding is *automatic* for A_1 and A_2 singularities.

The latter means that the various possible changes corresponding to A_1 and A_2 contact are ruled out. Another, more geometric, reason why they cannot occur is furnished by looking at the corresponding changes in the critical sets of the families of height functions. If M is a surface embedded in $\mathbb{R}^3$ then the family of height functions has as its critical set the pairs $(x, n(p))$ where $p \in M$ and $n(p)$ is one of the two unit normal at p. In other words the critical set is simply two copies of the surface. Now the dual of the surface is the image of this critical set (Prop. 2.3), and our models for changes in the dual also give (simpler) changes in the critical set as the evolution of the surface takes place. For example, in the case of the A_2 catastrophes these changes are standard Morse transitions or surgeries, given by either the emergence of a new component (a sphere) from nothing, or the transition locally modelled by that from a hyperboloid of two sheets to a hyperboloid with one sheet. In both cases there is a change in the topology. However our family of surfaces M_t consists of embeddings of a single surface M, so the topology of M_t, and hence that of the critical sets, remains unchanged. Thus these transitions cannot occur.

We now turn to the transitions which *can* occur. The following result is proved by 'Monge-Taylor' arguments; see §5 for details.

Theorem 3.3 *In a generic 1-parameter family of surfaces the parabolic curve can undergo changes as follows.*

(1) Morse transitions at a non-versal A_3. This occurs precisely when the Monge-Taylor map is not transverse to the A_2 stratum. Figure 2; §3.1.

(2) The birth/annihilation of a pair of cusps of Gauss (A_3 points) on a smooth parabolic curve occurring at an A_4 point. Figure 3; §3.2.

(3) Cone sections at a D_4 point (flat umbilic). Figures 4, 5; §3.3.

We now give examples of families realising each of the transitions stated in the theorem, and make various additional comments on the transitions. For all of our models we will take families of surfaces in the 'Monge form'

$$z = f(x, y, t) \tag{4}$$

where $f(x, y, 0)$ is a function with 1-jet 0.

3.1 Non-versal A_3 transitions

Suppose that $f(x, y, 0)$ has a non-versal A_3 singularity. In generic 1-parameter family of surfaces the family $F(x, y, t, a, b) = f(x, y, t) + ax + by$ is a versal unfolding of the A_3 singularity, so the bifurcation set of F is a cuspidal edge. We need to check that the t-sections of this cuspidaledge are generic.

Proposition 3.4 *If the family F above is versal, then the projection along the t-parameter yields a generic Morse function on the bifurcation set of this family.*

Proof Since we have a versal unfolding the set

$$\Sigma F = \{(x, y, a, b, t) : F_x = F_y = F_{xy}^2 - F_{xx}F_{yy} = 0\}$$

is the germ of a smooth manifold at the origin and the projection to the (a, b, t) variables $\Sigma F \to \mathbb{R}^3, 0$ is $\mathcal{A}$-equivalent to the map $\phi : \mathbb{R}^2, 0 \to \mathbb{R}^3, 0$ given by $(X, Y) \mapsto (X^2, X^3, Y)$. Now, we claim that a function $f : R^3, 0 \to \mathbb{R}, 0$ is of Morse type on the bifurcaton set if and only if $f \circ \phi$ has a Morse singularity. For if $f(u, v, w) = \alpha u + \beta v + \gamma w + \eta u^2 + ... + \delta w^2 + ...$ then $f \circ \phi$ is Morse precisely when $\gamma = 0$ and $\alpha\delta \neq 0$, which are exactly the conditions for f to be of Morse type on the bifurcation set [5].

We deduce that the projection from the bifurcation set in (a, b, t)-space to the t variable is of Morse type if and only if the map $\Sigma F, 0 \to \mathbb{R}, 0$ given by the projection to the t variable is. But this is equivalent to

$$g(x, y, a, b) = (F_x, F_y, F_{xy}^2 - F_{xx}F_{yy})(x, y, a, b, 0)$$

being $\mathcal{K}$-equivalent to a submersion with folds [5]. Now $g = (a + f_x, b + f_y, f_{xy}^2 - f_{xx}f_{yy})$ and since $f_{xy}^2 - f_{xx}f_{yy}$ has a Morse singularity the result cleary follows.

From the point of view of the singularities of the Gauss map the transitions correspond to the more classical 'lips' and 'beaks' transitions on the image of the parabolic curve under the Gauss map, and the birth/death of a pair of Gauss cusp points. It can be shown that both Gauss cusp points will be of the same type, *i.e.* both elliptic or both hyperbolic, depending on whether the contact between surface and tangent plane at the transition moment $t = 0$ is A_3^+ or A_3^-. This choice is independent of the choice between lips and beaks, and the examples below cover all the possible combinations.

Let

$$f(x, y, t) = x^2 + \epsilon_1 x^2 y^2 + \epsilon_2 y^4 + ty^2,$$

where $\epsilon_1 = \pm 1, \epsilon_2 = \pm 1$. We first note that (when $t = 0$) the function $f_0 = f(x, y, 0)$ does have a singularity of type A_3 (complete squares to kill off the x^2y^2 term and use the fact that the germ is 4-$\mathcal{R}$-determined). We get A_3^+ for $\epsilon_2 = 1$ and A_3^- for $\epsilon_2 = -1$. It is a simple exercise to check that the family

$$F(x, y, t, a, b) = x^2 + \epsilon_1 x^2 y^2 + \epsilon_2 y^4 + ty^2 + ax + by$$

is versal. So the bifurcation set is a cuspidaledge and the t-sections are generic. It is also easy to determine the behaviour of the parabolic curve in our examples; this is summed up in Figure 2.

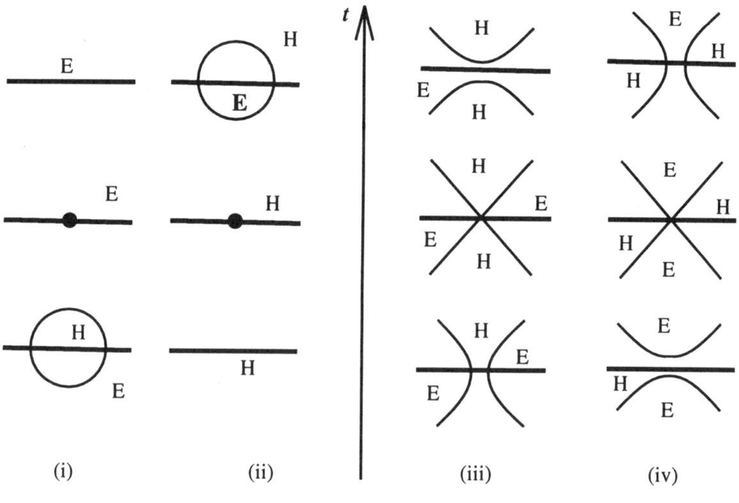

Figure 2

3.2 A_4 transition

Using the same construction as above the functions giving our family of surfaces $z = f(x, y, t)$ induce a family $F(x, y, t, a, b) = f(x, y, t) + ax + by$. The bifurcation set of F is a swallowtail when F is a versal unfolding of the A_4 singularity $f(x, y, 0)$. There is only one generic section of the swallowtail. We seek to determine whether the projection along the t parameter provides such a section.

Proposition 3.5 *If F is a versal unfolding of the A_4 singularity, then the projection along the t-parameter is a generic section of the bifurcation set of F.*

Proof The set

$$\Sigma F = \{(x, y, a, b, t) : F_x = F_y = F_{xy}^2 - F_{xx}F_{yy} = 0\}$$

is smooth again and the projection $\pi : \Sigma F, 0 \to \mathbb{R}^3, 0$ is $\mathcal{A}$-equivalent to $(X, Y) \mapsto (X, -10X^3 - 3YX, 15X^4 + 3YX^2)$. It is now easy to check that the projection g to the t-coordinate of the bifurcation set $\pi(\Sigma F)$ is generic in Arnolds sense if and only if the composite $g \circ \pi : \Sigma F, 0 \to \mathbb{R}, 0$ is a submersion. This is so if and only if the map $(F_x, F_y, F_{xy}^2 - F_{xx}F_{yy})$ restricted to the $t = 0$ set is a submersion, which it clearly is.

Let

$$f(x, y, t) = x^2 + y^5 + 2xy^2 + y^4 + ty^3.$$

We note that (when $t = 0$) the function $f_0 = f(x, y, 0) = (x + y^2)^2 + y^5$ does have an A_4-singularity, and the intersection of the surface with its tangent plane is now a *rhamphoid* cusp. If we set $x = X - y^2$ the germ F becomes

$$X^2 + y^5 + ty^3 + a(X - y^2) + by \tag{5}$$

and this clearly is a versal unfolding.

From Theorem 3.3 we know that two cusps of Gauss are born or annihilated on the parabolic set. See Figure 3. The parabolic set is given by $2x + 2y^2 + 3ty + 10y^3 = 0$, remaining smooth throughout the family. It is not hard to show that of the cusps of Gauss born, one is elliptic and one hyperbolic.

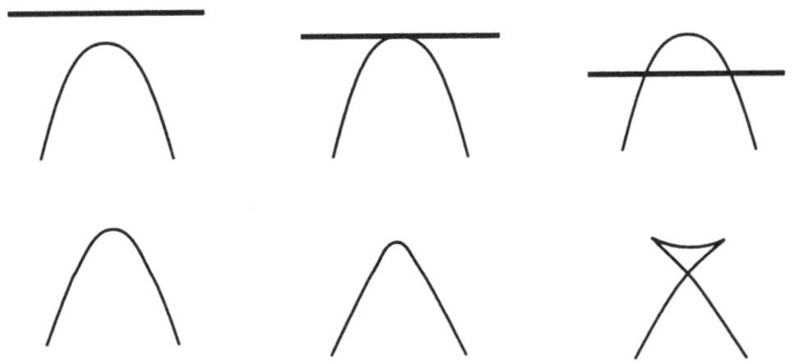

Figure 3

3.3 D_4 singularities

At a D_4 singularity of the height function, the surface in Monge form (4) has no quadratic part. The origin is called an *elliptic* or *hyperbolic* umbilic according as the cubic terms have three real factors (D_4^-) or one (D_4^+).

We shall take the family of surfaces to be given by $f(x, y, t) = x^3 \pm xy^2 + ty^2$. The family $F(x, y, t, a, b) = f(x, y, t) + ax + by$ is then a versal unfolding of the $D_4^\pm$

singularities $x^3 \pm xy^2$ at $t = 0$, and the dual undergoes a standard D_4 transition as in [1, Figure 36 (3 and 4)].

However when we are considering the changes in the Gauss map we run into the problem of a smooth modulus in the classification of generic functions on the corresponding bifurcation sets, and hence in principle there are infinitely many distinct types. However this is not too troublesome. By using a natural stratified equivalence, rather than local diffeomorphism, there are *two* different types of generic function on the bifurcation set of the hyperbolic umbilic, the differences being intrinsic, indeed topological, and not a consequence of using too strong an equivalence relation. In §5 we give some details of the calculations which are used to distinguish the two types of D_4^+ transition, illustrated in Figure 5.

We now consider the changes in the parabolic curves. The parabolic curve has equation

$$3x^2 \pm 3tx \mp y^2 = 0.$$

For both choices of sign the equation in the variables (x, y, t) determines a quadric cone, as in Theorem 3.3. The two cases are distinguished by taking sections by the planes t = constant. In the D_4^- case we have elliptic sections of the cone. In the D_4^+ case we have sections which are hyperbolas. (Luckily the elliptic and hyperbolic umbilics are well named!) See Figure 5.

On the parabolic sets there are various cusps of Gauss, and we now determine their nature. First we note that in the (a, b, t) unfolding space there are 1 (resp. 3) A_3 strata through a hyperbolic (resp. elliptic) D_4.

Consider the elliptic umbilic case D_4^-, where the intersection of the surface with the tangent plane is three smooth transverse curves. It is clear that the intersection deforms to two tangential branches and another branch (Figure 4). Since we have two tangential branches and not an isolated point, this means that the A_3 is of type A_3^- (hyperbolic cusp of Gauss).

Similar remarks hold for the hyperbolic umbilic case D_4^+, sketched in the right-hand part of Figure 4. Consequently there are 1 or 3 *hyperbolic* cusps of Gauss which merge and are then reborn. Figure 5 shows the evolution of bifurcation sets.

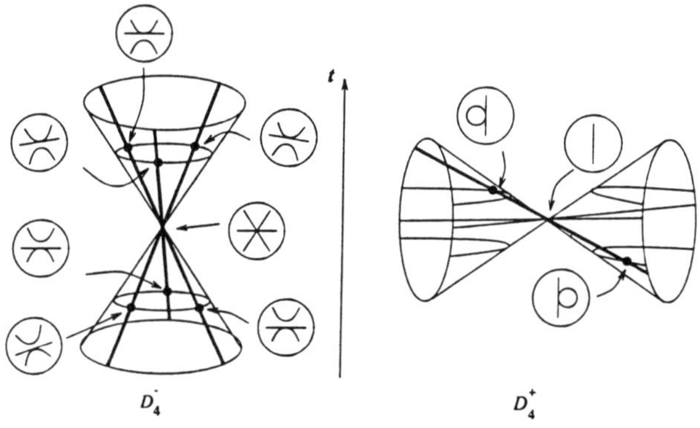

Figure 4

4 Generic geometry of a 1-parameter family of surfaces via height functions: multilocal singularities

In [11] there is a study of the general situation of a 1-parameter family of functions having bifurcation sets in the plane $\mathbb{R}^2$, and all the generic transitions occurring on these bifurcation sets are found. In fact the list contains a number of cases where there are two quite distinct transitions possible for a given singularity type. We reproduce the diagram of transitions in Figure 5. In this figure, the dashed lines represent the 'local' part of the bifurcation set and the solid lines the multilocal part.

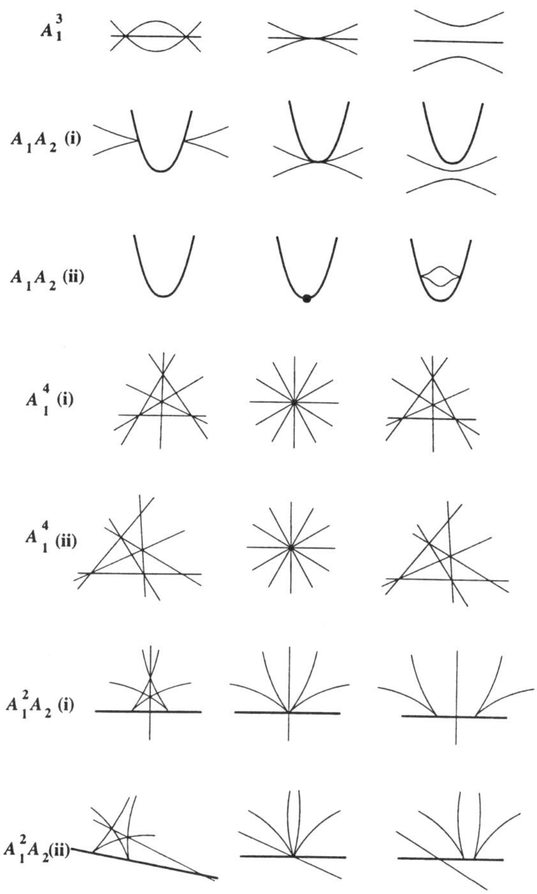

Figure 5

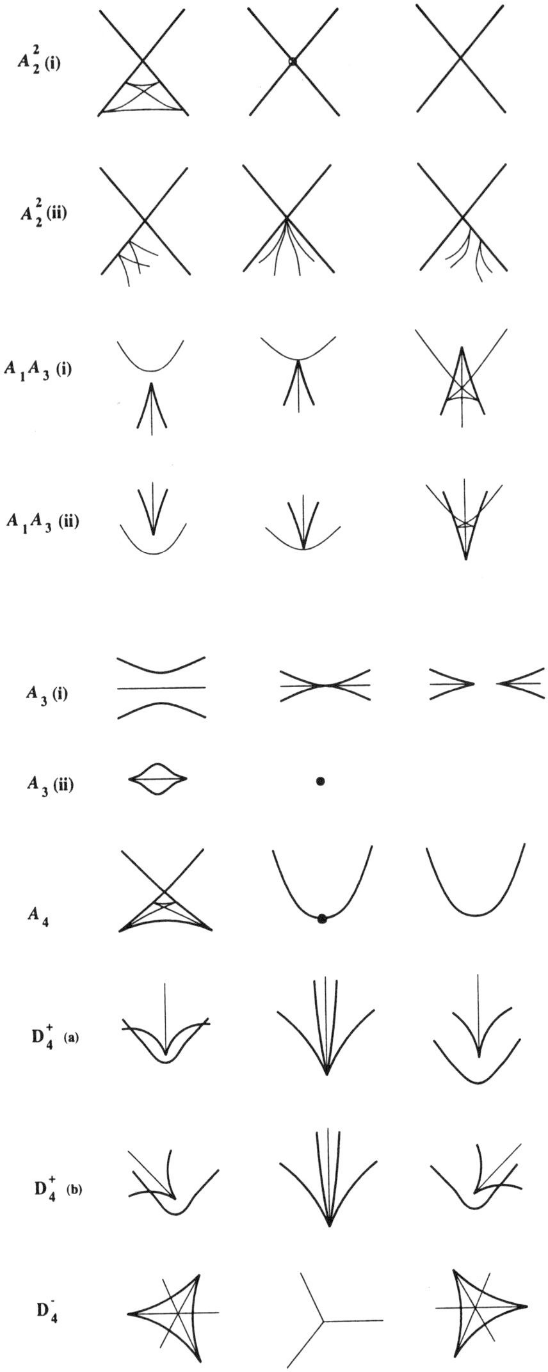

Figure 5 (cont.)

In what follows, we determine which of these possible cases can occur in the present context of height functions on surfaces. It turns out that, in contrast to the case of evolving symmetry sets of plane curves studied in [11], all possible cases with the exception of A_1^2 do occur.

To begin, we state the geometrical meaning of the various transitional cases which we need to consider.

(i) A_1^3: a tritangent plane, where the points of contact are collinear.
(ii) A_1A_2: a plane tangent at an ordinary point and a parabolic point, where the asymptotic direction at the parabolic point passes through the other point. (See Prop. 2.7 for some geometrical information on this case.)
(iii) A_1^4: quadritangent plane.
(iv) $A_1^2A_2$: plane tangent at two ordinary points and one parabolic point.
(v) A_1A_3: plane tangent at one ordinary point and one cusp of Gauss.
(vi) A_2^2: plane tangent at two parabolic points.

From Figure 5 it can be seen that there are two distinct transition pictures for each of the singularities A_1A_2, A_1^4, $A_1^2A_2$, A_2^2 and A_1A_3. In all except the second of these cases, the difference between the two pictures is apparent even at the *transitional moment.* For example, the two central A_2^2 pictures are distinguished by the difference between an isolated point and a whole curve, in fact two cusps with a common tangent. On the other hand the central pictures for the two A_1^4 singularities are indistinguishable, consisting of six smooth branches through a point. (6 is the number of choices of 2 from 4, so is the number of A_1^2 curves through the A_1^4 point.) In the situation of [11] (namely, symmetry sets of families of curves in the plane) it was found that only certain of the transitions were geometrically possible; for example only *one* of the A_1^4 transitions actually occurred. In fact this came down to the geometrical observation that, given 4 points lying on a circle, each of them is outside the triangle formed by the other 3.

In the present situation of families of height functions on one-parameter families of surfaces, we need to examine each multilocal singularity and if necessary each different transition on the bifurcation set arising from that singularity and to decide whether it is geometrically possible for that transition to occur. If it is, then we seek the corresponding geometrical conditions.

It is remarkable that, in the present context of height functions on generic families of surfaces, *all* the possible multilocal transitions actually occur, with the exception of A_1^2. This is always versally unfolded by the height functions alone and so does not give a transition (§2.1, Prop. 5.1).

4.1 Methods for analysing the transitions

The calculations can be done in (at least) two ways, for the majority of transitions. The first way is to compare the unfolding provided by the geometrical situation of a family of surfaces with the standard unfolding of a normal form of the singularity. This has already been mentioned in connexion with the case D_4 in Example 3.3, with further details given in §5 of [10]. The same method was used in §4 of [11]. In Example 3.3 we were of course considering only germs, whereas here, as in [11], we need to consider multigerms.

In most cases a second method, based on studying the transitional moment alone, can be used. This is because the bifurcation set at the transitional moment takes a different form for the different possible transitions. The diagrams of Figure 5 make it clear that, except for the two A_1^4 transitions and possibly the two A_1A_3 transitions, all these multilocal transitions can be distinguished by their bifurcation sets at the transitional moment. In fact the two cases for A_1A_3 also fall into this category. For it is not hard to show that there is no local diffeomorphism of the plane taking the triple consisting of an ordinary cusp $x^2+y^3=0$, its cuspidal tangent $\{x=0, y\leq 0\}$ and the singular 'swallowtail' curve $x^4=y^3$ to the triple consisting of $x^2=y^3, \{x=0, y\geq 0\}, x^4=y^3$. On the other hand the two pictures for the transitional moment on the bifurcation set of A_1^4 are completely indistinguishable, being simply six smooth mutually transverse curves through a point.

To use the second method, we set up a general situation of two (or more) pieces of surface which give rise to the singularity type under consideration, and calculate directly from there. We give some brief indication of one case where this works, namely A_2^2.

4.2 Direct calculation in case (vi), A_2^2

Take two general pieces of surface, each with a parabolic point, say at (0,0,0) and (0,1,0) respectively. The common tangent plane can be taken as $z=0$. Let the Monge forms up to degree 3 of the two surface pieces M_1, M_2 be

$$z=\lambda(a_{11}x+y)^2+b_{01}x^3+b_{11}x^2y+b_{21}xy^2+b_{31}y^3,$$

and

$$z=\mu(a_{12}x+(y-1))^2+b_{02}x^3+b_{12}x^2(y-1)+b_{22}x(y-1)^2+b_{32}(y-1)^3.$$

The idea is to calculate, on say M_1, the set of points which are points of contact of bitangent planes, close to $p_1=(0,0,0)$ on M_1 and $p_2=(0,1,0)$ on M_2. For the 'moth' transition (see Figure 5) this will be (near to the origin) a single point, while for the 'nib' transition it will be two transverse intersecting curves. In fact writing (x,y) for points of contact of bitangent planes on M_1 and (u,v) for points of contact on M_2, we find that the conditions for a plane with normal $(a,b,1)$ to be bitangent are

$$a=-z_x(x,y)=-w_u(u,v);$$

$$b=-z_y(x,y)=-w_v(u,v);$$

$$ax+by+z(x,y)=au+bv+w(u,v).$$

Now the second equalities in the first two lines here can be solved for y, v as functions of x, u *provided the asymptotic directions at* p_1 *and* p_2 *are not parallel*, i.e. provided $a_{11} \neq a_{12}$. Substituting for a and b as functions of x and u in the last equation, we obtain a single equation in x and u which turns out to have no linear terms, and quadratic terms equal to

$$2\gamma\mu x^2 + 2\delta\lambda u^2,$$

where

$$4\gamma\mu(a_{11} - a_{12}) = -6(a_{11}^2 b_{21} + b_{01} - a_{11}b_{11} - a_{11}^3 b_{31}) = -S_1, \text{say}$$

and

$$4\delta\lambda(a_{11} - a_{12}) = 6(a_{12}^2 b_{22} + b_{02} - a_{12}b_{12} - a_{12}^3 b_{32}) = S_2, \text{say.}$$

The condition for an isolated point on M_1 and M_2 is therefore $S_1 S_2 < 0$, which has a simple geometrical interpretation. In fact consider the cusp in which the surface M_1 say meets its tangent plane $z = 0$. It is straightforward to calculate the condition for this cusp to be on one side or the other of the y-axis, which is the line joining the two points p_1 and p_2. The decisive sign is that of λS_1 in fact. We deduce that the cusps are on the *same* side of the y-axis if and only if $\lambda\mu S_1 S_2 > 0$.

Finally, we can interpret λ and μ geometrically, too. At a parabolic point, all the planes containing the surface normal *except* that which also contains the asymptotic direction meet the surface in a curve which is concave in the same direction at the parabolic point. In particular the parabolic curve on the surface is on the side of the tangent plane given by this direction. For example, the surface piece M_1 has normal sections which are concave upwards (in the direction $z > 0$) if $\lambda > 0$ and downwards if $\lambda < 0$. In particular the parabolic curve on M_1 is in the region $z > 0$ close to the origin if and only if $\lambda > 0$. Thus the concavity is in the *same* direction for the two surface pieces M_1 and M_2 if and only if $\lambda\mu > 0$. See Figure 6, which shows two pieces of surface with the same concavity (downwards).

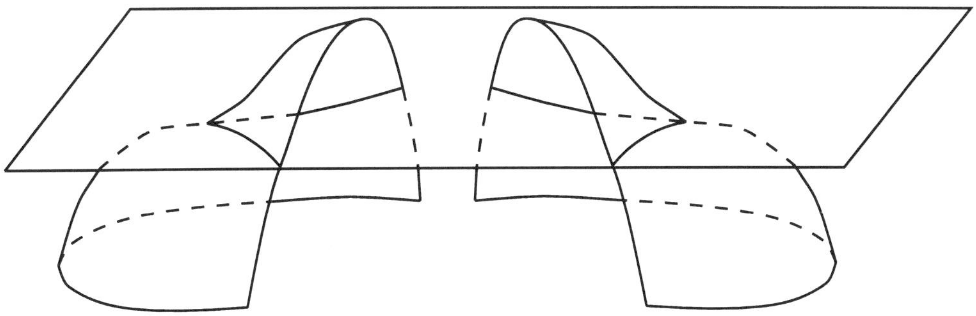

Figure 6

Hence we have:

Suppose that an A_2^2 singularity of the height functions occurs in a generic family of surfaces. We assume that neither of the asymptotic directions at the two parabolic points is along the line segment l joining the two parabolic points, and that the asymptotic directions are furthermore not parallel.

Suppose the concavities as above are in the same direction for the two surface pieces. (Of course these are determined, for example, by the direction of concavity given by the particular plane containing the normal and l.) Then the transition is a moth if and only if the two cusps in which the surface pieces meet the bitangent plane are on opposite sides of l.

If the concavities are opposite, then the transition is a moth if and only if the two cusps are on the same side of l.

In both cases, the alternative transition, the 'nib', occurs if and only if the moth fails to occur.

Thus, in Figure 6, the transition will be a nib, assuming (as in intended in the diagram!) that the cusps of intersection of the two surface pieces with the common tangent plane are on the same side of the line joining the points of contact.

The same result follows, of course, by using the alternative method, where we compare the versal unfolding provided by the geometrical situation with a standard versal unfolding, as in Example 3.3, and §4 of [11].

4.3 (i) A_1^3: a tritangent plane

The only transition here is when the three points of contact are collinear; see §2.1, Prop. 5.1. The transition is shown in Figure 5. Note that, in agreement with Cor. 2.5, all three branches in the bifurcation set are tangent.

4.4 (ii) A_1A_2, a plane tangent at an ordinary point and a parabolic point

The only transition occurs when the (unique) asymptotic direction at the parabolic point of contact passes through the other point of contact. (See §2.1, Prop. 5.1.) There are then two alternative transitions on the bifurcation set, as in Figure 5. These transitions are essentially beaks and lips on the A_1^2 stratum, and the distinction can be detected on the piece of surface where the parabolic point lies. In the former case, the set of points of contact of bitangent planes will be a crossing and in the latter case it will be an isolated point. In fact the distinction between these two cases has already been found in Proposition 2.7 above.

Explicit families of surfaces giving this transition are, using the conditions of Proposition 2.7 above, as follows. Let M_1 be given by $z = (x + ty)^2 + y^3$, where t is the parameter in the family, so passes through 0, and let M_2 be given by $z = x^2 + (y \pm 1)^2$.

Then we get the upper (resp. lower) transition picture of Figure 5 for the upper (resp. lower) sign of 1.

4.5 (iii) A_1^4, quadritangent plane

The calculation is very similar to that given in [11] for the case of a quadri-tangent circle to a plane curve. It turns out that the distinction between the two cases (upper and lower A_1^4 pictures in Figure 5) can be stated at the transitional moment after all: at this moment there are 4 points of the surface M with the same tangent plane, and therefore 4 points of contact $p_i, i = 1, 2, 3, 4$, all lying in this plane. The result is this: *The distinction between the two transitions is between sets of 4 points where one is inside the triangle formed by the other 3 (upper picture of Figure 5), and sets where this does not occur (lower picture).*

Note that only *one* of the four points can be inside such a triangle! Of course we expect that generically no 3 of the 4 points will be collinear, and indeed this would result in a more degenerate singularity. One case passes into the other when one of the 4 points moves through the line joining two others.

4.6 (iv) $A_1^2 A_2$, plane tangent at two ordinary points and one parabolic point

In this transition, at the transitional moment there is a plane tangent to M at 3 points, say p_1, p_2, p_3, one of which, p_1 say, is parabolic. The bifurcation set has 3 branches, corresponding to bitangent planes having contact near each of the 3 pairs of points which can be chosen from p_1, p_2, p_3. The branches where contact is close to p_1, p_2 or p_1, p_3 give cusps on the bifurcation set, while the smooth branch comes from bitangent planes with contact close to p_2, p_3. There is a visible distinction between the two possible cases, looking only at the transitional moment. The two cuspidal tangents, which have a *direction* (away from the cusp, say) are either separated or not separated by the tangent to the smooth branch. This is something that can be detected quite easily in an example, and calculation shows the following.

Consider the transitional moment and the asymptotic direction at p_1 (the parabolic point). In general neither of the other points lies along this direction, and *we obtain the upper (resp. lower) picture of Figure 5 when the points p_2, p_3 lie on opposite sides (resp. the same side) of this asymptotic direction.*

4.7 (v) $A_1 A_3$, plane tangent at one ordinary point and one cusp of Gauss

Here, at the transitional moment, we have a plane tangent to M at a cusp of Gauss (p_1, say) and a non-parabolic point. The distinction between the two cases (the two diagrams of Figure 5) is more subtle this time. Consider a surface piece M_1 with Monge form

$$z = a_{01}x^2 + b_{01}x^3 + b_{11}x^2y + b_{21}xy^2 + \ldots,$$

where ... represents terms of order at least 4, and the coefficient c_{41} of y^4 satisfies $c_{41} \neq b_{21}/a_{01}$ and $b_{21} \neq 0$. This has a cusp of Gauss at $(0, 0, 0)$ and the asymptotic direction there is along the y-axis. The other piece of surface M_2 will be tangent to the same plane $z = 0$ at some point $(\lambda, \mu, 0)$. Then it is the sign of $\lambda a_{01} b_{21}$ which determines the type of transition. In fact:

We have the upper (resp. lower) transition of Figure 5 if and only if $\lambda a_{01} b_{21} < 0$ (resp. > 0).

Of course when λ changes sign that is simply a matter of the second piece of surface moving from one side to the other of the asymptotic direction on the first piece of surface. At the cusp of Gauss $(0, 0, 0)$ the parabolic curve has its tangent along the y-axis, so locally points on one side or the other of the plane $x = 0$ (say, points $(x, 0, z)$ for small x), will be elliptic or hyperbolic. Changing the sign of $a_{01}b_{21}$ amounts to changing this from one case to the other. Thus the surface piece M_1 'shows its hyperbolic or elliptic side' towards the surface piece M_2, and which side determines the sign of $a_{01}b_{21}$. In fact, if points with small $x > 0$ are elliptic, then $a_{01}b_{21} > 0$.

Thus the overall result is the following

We obtain the upper (resp. lower) transition of Figure 5 if and only if M_1 shows its hyperbolic (resp. elliptic) side to M_2.

5 Proofs of three results above

5.1 Proof of the transversality result §3.3

As in [6] let p be a point on a surface M. We choose a smooth unit normal vector field, and a non-zero unit tangent vector field, in a neighbourhood U of p. This allows us to determine, at each point near p, an orthonormal set of co-ordinates, and at any point q we write our surface locally as the graph of a function of two variables x and y in the given normal z-direction. Let V_4 denote the set of triples (f_1, f_2, f_3), where Q_1 is a quadratic form, f_2 a cubic and f_3 a quartic form in the variables x and y. Then taking the 4-jet of the function defining M locally, we obtain a smooth map $\theta : U \to V_4$.

The set V_4 has a natural $SO(2)$-action given by a change of (x, y) co-ordinates (a different choice of initial tangent vector field). A subset Z of V_4 which is of any geometric significance will be $SO(2)$-invariant. Moreover if Z is furnished with an $\mathcal{A}$-regular Whitney stratification then for any generic surface M the map germ $M, p \to V_4$ will be transverse to (the strata of) Z. Since we wish to describe the parabolic curve on M we will need the corresponding subset Z of V_4.

Suppose M is given in a Monge form $(x, y, f(x, y))$. We consider in V_4 the set of expansions $f = f_2 + f_3 + f_4$. We can set $f_2 = a_0x^2 + a_2y^2$ by an $SO(2)$-change of coordinates, and write $f_3 = \sum_{i=0}^{3} b_i x^{3-i} y^i$ and $f_4 = \sum_{i=0}^{4} c_i x^{4-i} y^i$.

We start by analysing the situation where the origin is not a flat umbilic (*i.e.* the height function does not have a D_4 singularity), so that a_0 or a_2 are non-zero. A transversal to the $SO(2)$-orbit of f in V_4 consists of points of the form $((a_0 + \overline{a_0})x^2 + (a_2 + \overline{a_2})y^2, f_3 + \overline{f_3}, f_4 + \overline{f_4})$, where $\overline{f_3}$ (resp. $\overline{f_4}$) is a general cubic (resp. quartic) in x and y. The closure of the A_2 stratum is then the union of the hypersurfaces given by $a_0 + \overline{a_0} = 0$ and $a_2 + \overline{a_2} = 0$.

Following Proposition 2.2 in [6] the the image of $d\theta(p)$ is generated by u_1 and u_2 with

$$\begin{aligned} u_1 = \; & j^4\{-f_{xx}(0,0)x - f_{xy}(0,0)y + f_x(x,y) - f_{xx}(0,0)f_x(x,y)f(x,y)- \\ & f_{xy}(0,0)f_y(x,y)f(x,y)\}, \\ u_2 = \; & j^4\{-f_{xy}(0,0)x - f_{yy}(0,0)y + f_y(x,y) - f_{yy}(0,0)f_y(x,y)f(x,y)- \\ & f_{xy}(0,0)f_x(x,y)f(x,y)\}. \end{aligned}$$

In general the Monge-Taylor map intersects the A_2 stratum transversally in a smooth curve. The condition for transversality to fail is easily calculated. In the case when we are dealing with the component $a_2 + \overline{a_2} = 0$ it turns out to be $b_3 = b_2 = 0$. (In this case we can work in V_2. The condition that the tangent vectors to the stratum in the transversal, the tangent space to the orbit, and u_1 and u_2 fail to span V_2 is easily found.) The condition $b_3 = 0$ means that we are at a cusp of Gauss and $b_2 = 0$ is the condition for the family of height functions to fail to be versal at this point.

In a generic 1-parameter family the Monge-Taylor map becomes transverse to the $A_{\geq 2}$ set and intersects it in a smooth surface. The transitions on the parabolic set are obtained by taking generic Morse sections of this surface, as in Figure 2.

We shall now examine the situation at an A_3-singularity, *i.e.* when $a_2 = b_3 = 0$. The A_3 set in V_4 is given by $\overline{a_2} = \overline{b_3} = 0$, where $\overline{b_3}$ is the coefficient of y^3 in the cubic form $\overline{f_3}$. This is a smooth set of codimension 2 in V_4. If we use the coefficients of the monomials as a set of co-ordinates on V_k a calculation shows that the tangent space to this stratum is given by the kernel of the 1-forms $\xi_1 = da_2$ and $\xi_2 = -b_2 da_1 + 2a_0 db_3$.

The Monge-Taylor map fails to be transverse to the A_3 stratum if there exists a non zero vector $\lambda u_1 + \mu u_2$ such that

$$\xi_1(\lambda u_1 + \mu u_2) = \xi_2(\lambda u_1 + \mu u_2) = 0$$

This is equivalent to

$$b_2 = 0 \quad \text{or} \quad b_2^2 - 4a_0c_4 = 0.$$

The case $b_2 = 0$ is dealt with above. When $b_2^2 - 4a_0c_4 = 0$, the height function has an A_4 singularity. This is expected in generic 1-parameter family of surfaces. The image of the Monge-Taylor map intersects the A_2 startum transversally at an A_4-point. So the parabolic curve is smooth in this case and remains smooth in the family. We have generically a birth/annihilation of two A_3-points of this curve.

At a flat umbilic (D_4 singularity, $a_0 = a_2 = 0$) a transversal has the form $(\overline{a_0}x^2 + 2\overline{a_1}xy+\overline{a_2}y^2, f_3+\overline{f_3}, f_4+\overline{f_4})$, where $\overline{f_4}$ is a general quartic and $\overline{f_3}$ can be chosen to span a 2-dimensional subspace of the cubics. The A_2 stratum is the cone $\overline{a_1}^2 - \overline{a_0}\overline{a_2} = 0$, which, in a generic 1-parameter family, intersects the image of the Monge-Taylor map in a cone. There are two generic sections of the cone as shown in Figure 4.

5.2 The two D_4^+ transitions; see §3.3

We discuss the method used to distinguish the two types of D_4^+ transition, as shown in Figure 5.

The Monge form of a surface with a hyperbolic umbilic will have no quadratic terms, and cubic terms which have only one real factor. By an affine change of coordinates (which leaves the family of height functions invariant up to the natural equivalence; see Appendix A) we can make the cubic terms take the form $x^3 + y^3$ and we therefore need to consider a family of height functions $F(x, y, a, b, t)$ of the form

$$ax + by + txy + x^3 + y^3 + c_1x^4 + c_2x^3y + c_3x^2y^2 + c_4xy^3 + c_5y^4 + \dots \qquad (6)$$

This family versally unfolds the singularity D_4^+, so it is R-equivalent, as an unfolding, to

$$G(u, v, \tau, \alpha, \beta) = u^3 + v^3 + \tau uv + \alpha u + \beta v.$$

Hence we have an equivalence of the form

$$G(u, v, \tau, \alpha, \beta) = F(A(u, v, \tau, \alpha, \beta), B(\tau, \alpha, \beta)), \qquad (7)$$

for some $A : \mathbb{R}^2 \times \mathbb{R}^3, 0 \to \mathbb{R}^2, 0$, with $A(u, v, 0, 0, 0) = (u, v)$ and some germ of diffeomorphism $B : \mathbb{R}^3, 0 \to \mathbb{R}^3, 0$. We need to have some precise information about the isomorphism of unfoldings in order to discover which projection from Bif(F) to the t-axis we actually have in our geometrical example.

The bifurcation set of G is mapped by B to the bifurcation set of F, and B_1, the first component of B, is the function on Bif(G) corresponding to the projection of Bif(F) to the t-axis, as in the following commutative diagram. In the top line the projection π_3 is standard (projection on to the third (t) factor) and in the bottom line the unfolding is standard.

$$\begin{array}{ccccccc}
\mathbb{R}^2 \times \mathbb{R}^3 & \xrightarrow{F} & \mathbb{R}^2 \times \mathbb{R}^3 & \xrightarrow{\pi} & \mathbb{R}^3 & \xrightarrow{\pi_3} & \mathbb{R} \\
\uparrow & A \times B & \uparrow & I \times B & \uparrow B & \nearrow B_1 & \\
\mathbb{R}^2 \times \mathbb{R}^3 & \xrightarrow{G} & \mathbb{R}^2 \times \mathbb{R}^3 & \xrightarrow{\pi} & \mathbb{R}^3 & &
\end{array}$$

What we actually want to know is the way in which the bifurcation set of F in $\mathbb{R}^3$ is sliced up by the fibres $\pi_3^{-1}\{t\}$. We use the diffeomorphism B to study instead the slicing up of the bifurcation set of G by the fibres $B_1^{-1}\{t\}$, so we need some information on B_1, and in fact the 1-jet is enough. (In §5 of [10] it is shown how to obtain *any* jet of the diffeomorphism B by an effective method.) To obtain this 1-jet we apply the chain rule to the formula (7), to find

$$\begin{pmatrix} 1 & 0 & 0 \\ 0 & 1 & 0 \\ 0 & 0 & 1 \end{pmatrix} = \begin{pmatrix} ? & uv \text{ coeff.of } \frac{\partial F}{\partial a} & uv \text{ coeff.of } \frac{\partial F}{\partial b} \\ ? & 1 & 0 \\ ? & 0 & 1 \end{pmatrix} \begin{pmatrix} \text{Jacobian} \\ \text{of } B \\ \text{at } 0 \end{pmatrix}.$$

Here, $\partial F/\partial a$ and $\partial F/\partial b$ are evaluated at $A(u,v,0,0,0)$ and the identity matrix is the 'jet matrix' for G. Calculating the uv-coefficients of $\partial F/\partial a$ and $\partial F/\partial b$ amounts to calculating these coefficients in $A_1(u,v,0,0,0)$ and $A_2(u,v,0,0,0)$, where $A=(A_1,A_2)$. These are obtained from (7) at $\tau=\alpha=\beta=0$, and come to $-c_2/3$ and $-c_4/3$ respectively.

The above matrix equation implies that the 1-jet of B_1 is given by the cofactors of the first column of the middle matrix, that is

$$B_1(\tau,\alpha,\beta)=\tau-\frac{c_2}{3}\alpha-\frac{c_4}{3}\beta+\text{higher order terms}=0.$$

We use the results of [7], §2.10, which gives the classification of generic functions (such as B_1) on the bifurcation set of the standard D_4^+ singularity (such as Bif(G)). The kernel vector of B_1 is $(1,-c_2/3,-c_4/3)$. Regarding this as a point in the real projective plane (coordinates $(\lambda_1,\lambda_2,\lambda_3)$ as in [7]), we divide the plane into four regions by the three lines $\lambda_i=0$. The regions containing the points $(1,-1,-1),(1,1,1)$ give one generic function and the regions containing the points $(1,1,-1),(1,-1,1)$ give the other. Clearly $c_2c_4>0$ corresponds to the first of these and $c_2c_4<0$ to the second. Hence *we distinguish between the two D_4^+ transitions precisely by the sign of c_2c_4 in* (6).

Examples of the two families (which give respectively the cases (a) and (b) of Figure 5) are

$$z=x^3+y^3+x^3y+\epsilon xy^3+txy,$$

for $\epsilon=1,-1$.

5.3 Proof of the versality results

Proposition 5.1 *The following singularity types are always versally unfolded by the family of height functions $F(x,y,a,b)$ alone:*
(a) A_1^2,
(b) A_1A_2 (provided the asymptotic line at the A_2 does not pass through the A_1),
(c) A_1^3 (provided the three A_1-points are not collinear).

Proof (a) The setup we need is two pieces of surface, defined without loss of generality by

$$z=a_{01}x^2+a_{21}y^2+\ldots,\text{ and}$$

$$z = a_{02}(x-\lambda)^2 + a_{12}(x-\lambda)(y-\mu) + a_{22}(y-\mu)^2 + \dots,$$

where there is a common tangent plane $z = 0$ at the points $(0,0,0)$ and $(\lambda,\mu,0)$ respectively.

We need to investigate the family of height functions on this pair, which is given by adding $ax+by$ to each one. We must keep separate track of the coordinates on the two pieces of surface; we use $(x_1,y_1) = (x,y)$ on the first piece and $(x_2,y_2) = (x-\lambda, y-\mu)$ on the second piece, so that x_1, y_1, x_2, y_2 are all close to 0. The family of height functions is then given by the pair $F = (F_1, F_2)$ where

$$F_1(x_1,y_1,a,b) = ax_1 + by_1 + a_{01}x_1^2 + a_{21}y_1^2 + \dots,$$

$$F_2(x_2,y_2,a,b) = a(x_2+\lambda) + b(y_2+\mu) + a_{02}x_2^2 + a_{12}x_2y_2 + a_{22}y_2^2 + \dots.$$

We may work in the space of 2-jets, since A_1^2 is 2-determined. To prove that F is versal, we need to check that $\partial F/\partial a, \partial F/\partial b$ and $(1,1)$ span this jet space, modulo the jacobian ideals of F_1 and F_2, and for the latter we need only use $\partial F_i/\partial x_i, \partial F_i/\partial y_i$ for $i = 1,2$, both at $a = b = 0$, since we are confined to degree ≤ 2. Thus we seek vectors $(1,0), (x_1,0), (y_1,0), (0,1), (0,x_2), (0,y_2)$. One way to set out such calculations is to put everything in a matrix, as follows.

	$(1,0)$	$(x_1,0)$	$(y_1,0)$	$(0,1)$	$(0,x_2)$	$(0,y_2)$
$\partial F_1/\partial x_1$	0	$2a_{01}$	0	0	0	0
$\partial F_1/\partial y_1$	0	0	$2a_{21}$	0	0	0
$\partial F_2/\partial x_2$	0	0	0	0	$2a_{02}$	a_{12}
$\partial F_2/\partial y_2$	0	0	0	0	a_{12}	$2a_{22}$
$\partial F/\partial a$	0	1	0	λ	1	0
$\partial F/\partial b$	0	0	1	μ	0	1
$(1,1)$	1	0	0	1	0	0

The top row gives the vectors we seek. The left entry in each row below that is equal to the linear combination of the top row vectors with the coefficients directly below them. It is easy to check that the 7×6 matrix of coefficients has the required rank of 6 provided a_{01}, a_{21} and $4a_{02}a_{12} - a_{12}^2$ are all nonzero, and λ or μ is nonzero. This says precisely that the singularity is of type A_1^2 and that the two points on the surface are not coincident.

(b) This is similar, except that we must work in the 3-jet space for the height function at the A_2 point, which we take as the one associated with F_1. Let

$$F_1(x_1,y_1,a,b) = ax_1 + by_1 + a_{01}x_1^2 + \dots + b_{31}y_1^3 + \dots.$$

In addition to the vectors used above we use $x_1\partial F_1/\partial x_1$ and $y_1\partial F_1/\partial x_1$. The conditions for versality become $a_{01}, 4a_{02}a_{22} - a_{12}^2, b_{31}, \lambda$ all nonzero. These say precisely that the singularity is A_2A_1 and that the second point $(\lambda,\mu,0)$ is not on the asymptotic direction $x = 0$ at the first point.

(c) This is again similar.

A Affine Invariance of the Family of Projections to Linear Subspaces

In this appendix we show that the dual of a surface, and the critical values of the Gauss map are, in a natural sense, affine invariants. The same is also true when we project to planes and consider the 'viewsphere' and its associated stratification. These results both follow from a quite general theorem which we now formulate.

Let $\mathbb{R}^n$ denote n-dimensional Euclidean space, and let G_k denote the space of k-planes in $\mathbb{R}^n$. There is a natural bundle Γ_k over G_k constructed as following. Consider the set

$$\{(y, L) \in \mathbb{R}^n \times G_k : y \text{ is orthogonal to all vectors in } L\}.$$

Clearly the projection of Γ_k to G_k is a submersion with fibre over L a linear space of dimension $n - k$ which we denote by $L^\perp$. There is a natural map

$$P_k : \mathbb{R}^n \times G_k \to \Gamma_k$$

defined by $P_k(x, L) = (P_L(x), L)$, where $P_L(x)$ is orthogonal projection of x to $L^\perp$. Clearly for any fixed L we can choose a neighbourhood U of L in G_k and straighten out the restriction of the above family $P : \mathbb{R}^n \times U \to \Gamma_k$ into one $\mathbb{R}^n \times U \to \mathbb{R}^{n-k}$.

Now given a submanifold X of $\mathbb{R}^n$ we can restrict the family P to $X \times G_k$. This will give information on the contact between X and linear subspaces of dimension k, indeed we also keep track of simultaneous contact with parallel sets of planes. We define a couple of sets of interest. First note that the set Γ_k can be thought of as the set of all affine k-planes in $\mathbb{R}^n$, for given $(y, L) \in \Gamma_k$ there is a unique k-plane through y parallel with L.

Definition A.1 *The set of critical values of $P_k : X \times G_k \to \Gamma_k$ is defined to be the k-dual of X. Identifying Γ_k with the affine k-planes this consists of all k-planes which fail to be transverse to X. If k+dim $X < n$ then this is all planes of course, and of no interest. If k+dim $X \geq n$ then this is the set of k-planes which are tangent to X. If dim $X = r$ and we consider the map $P_r : X \times G_r \to \Gamma_r$. Then the set of points $(x, L) \in X \times G_r$ where P has rank 0, denoted $\Sigma_0 P$, is diffeomorphic to X and the corresponding set of critical values is the classical dual of X, i.e. the set of tangent planes to X. The natural map from X to $\Sigma_0 P$ followed by P_r and then projection from Γ_r to G_r is the Gauss map of X.*

A.1 Examples

Let M be a surface in $\mathbb{R}^3$ and let $k = 1$. Then the family P_1 is essentially the family of height functions h we have been discussing in the body of the paper. When $k = 2$ we have essentially the family of orthogonal projections of M to planes in $\mathbb{R}^3$.

Now given an affine motion α of $\mathbb{R}^n$, say $x \mapsto Ax + a$ for some invertible matrix A and constant vector a we can also consider the restriction of P to $\alpha(X) \times G_k$, or better the map $P \circ (\alpha, id) : X \times G_k \to \Gamma_k$. To state the required result we need a couple of further definitions.

Definition A.2 *(i)Let $\pi : Y \to Z$ be a smooth locally trivial bundle with fibres diffeomorphic to W and let F_1, $F_2 : X \times Z \to Y$ be smooth mappings with $\pi \circ F_j(x, z) = z$ for all $x \in X$. The F_j can be thought of as Z-parameter families of mappings to W. We say that F_1 and F_2 are $\mathcal{A}_Z$-equivalent if the following is true. For some smooth mappings $\Phi : X \times Z \to X \times Z$, $\Psi : Y \to Y$, $\theta : Z \to Z$ with $\Phi(x, z) = (\phi(x, z), \theta(z))$ and $\pi \circ \Psi = \theta \circ \pi$ we have $F_1(\phi(x, z)) = \psi(F_2(x, z))$. (ii) If $F : X \times Z \to Y$ is as above then we can define a subset* Bif*(F) of Z by*

$$\mathrm{Bif}(F) = \{z \in Z : F_z : X \to \pi^{-1}(z) \text{ not stable}\}.$$

The next result follows immediately from the definition.

Proposition A.3 *If F_1 and F_2 are $\mathcal{A}_Z$-equivalent then the corresponding map $\theta : Z \to Z$ maps* $\mathrm{Bif}(F_1)$ *diffeomorphically to* $\mathrm{Bif}(F_2)$.

The following theorem is the key result.

Proposition A.4 *The families $P : X \times G_k \to \Gamma_k$ and $P \circ (\alpha, id) : X \times G_k \to \Gamma_k$ are $\mathcal{A}_Z$-equivalent, where Z here is G_k.*

Proof. We first deal with the case of a pure translation, $x \mapsto x + a$. But $P(x + a, L) = P_L(x + a) = P_L(x) + P_L(a)$, so set $\Phi(x, L) = (x, L)$, $\Psi(y, L) = (y + P_L(a), L)$ and $\theta(L) = L$.

We now consider the case of a linear transformation $x \mapsto Ax$. For the map θ we choose $L \mapsto A^{-1}(L)$, for Φ choose $(x, L) \mapsto (x, AL)$, and for Ψ the map $(y, L) \mapsto (P_{AL} \circ A(y), AL)$. It is easy to see that $P_{AL} \circ A$ yields an isomorphism from $L^\perp \to AL^\perp$ and a little diagram chasing proves the result.

Corollary A.5 *Any two affine equivalent submanifolds of $\mathbb{R}^n$ have affinely-equivalent k-duals, and affinely-equivalent classical duals. Their Gauss maps are also affinely-equivalent.*

Remark A.6 *By affinely equivalent of course we mean, in the duals case, via a change of co-ordinates of the the form*

$$(y, L) \mapsto (P_{AL} \circ A(y), AL)$$

and for the critical values of the Gauss map a change of co-ordinates of the form $L \mapsto AL$.

Corollary A.7 *For a surface X in $\mathbb{R}^3$ the bifurcation set of X in the viewsphere, and that of an affinely-equivalent surface AX are projectively equivalent. (In point of fact the viewsphere is really a projective plane and then the change of co-ordinates taking the one bifurcation set to the other is a projective equivalence.)*

References

[1] V.I.Arnold, *Catastrophe Theory*, Springer-Verlag, New York-Heidelberg-Berlin, 1984, 1986; third edition 1992.

[2] V.I.Arnold, A.N.Varchenko and S.M.Gussein-Zade, *Singularities of Differentiable Maps, Volume I,* Birkhauser, Monographs in Mathematics, Vol. 82, Birkhauser, Boston, 1985.

[3] T.Banchoff, T.J.Gaffney and C.McCrory, *Cusps of Gauss Mappings,* Pitman, New York, 1982.

[4] J.W.Bruce, 'The duals of generic hypersurfaces', *Math. Scand.* 49 (1981),36-69.

[5] J.W. Bruce, 'Wavefronts and parallels in Euclidean space', *Math. Proc. Camb. Phil. Soc.* (1983), **93**, 323-333.

[6] J.W.Bruce, 'Generic reflections and projections', *Math. Scand.* 54 (1984),262-278.

[7] J.W.Bruce, 'Generic functions on semialgebraic sets', *Quart.J.Math.Oxford* (2), 37 (1986), 137-165.

[8] J.W.Bruce, 'Generic geometry, transversality and projections', *J.London Math. Soc.* 49 (1994),183-194.

[9] J.W.Bruce and P.J.Giblin, *Curves and Singularities*, Cambridge University Press, Second Edition, 1992.

[10] J.W.Bruce and P.J.Giblin, *Outlines and their duals*, Proc. London Math. Soc. (3) 50 (1985), 552-570.

[11] J.W.Bruce and P.J.Giblin, 'Growth, motion and 1-parameter families of symmetry sets', *Proc. Royal Soc. Edinburgh*, 104A (1986), 179-204.

[12] J.W.Bruce, P.J.Giblin and F.Tari, 'Families of surfaces: focal sets, ridges and umbilics', Preprint, University of Liverpool, 1994.

[13] J.W.Bruce, P.J.Giblin and F.Tari, 'Families of surfaces: height functions and projections to planes', Preprint, University of Liverpool, 1994.

[14] J.W.Bruce and M.C.Romero-Fuster, 'Duality and projections of curves and surfaces in 3-space', *Quart. J. Math. Oxford* Ser. (2), 42 (1991), 433-441.

[15] J.W.Bruce and F.Tari, 'On binary differential equations'. To appear in *Nonlinearity.*

[16] A.du Plessis, 'On the determinacy of smooth map germs', *Invent. Math.* 58 (1980), 107-160.

[17] G.-M. Greuel and Lê Dung Tràng, 'Spitzen, Doppelpunkte unt vertikale Tangenten in der Diskriminante verseller Deformationen von vollständigen Durchschnitten', *Math. Ann.* 222 (1976), 71-88.

[18] D.Hilbert and S.Cohn-Vossen, *Geometry and the Imagination*, Chelsea Publishing Company 1952.

[19] J.J.Koenderink, *Solid Shape*, The M.I.T. Press, Cambridge Massachusetts, 1990.

[20] D.C.Knill, P.Mamassian, D Kersten, 'The geometry of shadows', Preprint, Dept. of Psychology, University of Minnesota, May 1993.

[21] E.E.Landis, 'Tangential singularities', *Funct. Anal. Appl.* 15 (1981), 103-114.

[22] J.Martinet, *Singularities of Smooth Functions and Maps*, London Mathematical Society Lecture Notes Series 58, Cambridge University Press, 1982.

[23] J.Montaldi, 'On generic composites of maps', *Bull. London Math. Soc.* 23 (1991), 81-85.

[24] I.R.Porteous, *Geometric Differentiation*, Cambridge University Press, 1994.

[25] J.H.Rieger, 'Families of maps from the plane to the plane', *J. London Math. Soc.* (2) 36 (1987), 351-369.

[26] O.P.Shcherbak, 'Projectively dual space curves and Legendre singularities', *Trudy Tbiliss. Univ.* 232-233(1982), 280-336.

[27] C.T.C.Wall, 'Geometric properties of generic differentiable manifolds', in *Geometry and Topology, Rio de Janeiro 1976, Lecture Notes in Mathematics* 597,707-774, Springer-Verlag, New York.

[28] C.T.C.Wall, 'Finite determinacy of smooth map germs', *Bull. London Math. Soc.* 13 (1981), 481-539.

S. Bromberg & S. López de Medrano

C^r-sufficiency of quasihomogeneous functions

0 Introduction.

In the classification of singularities of smooth functions an important tool is provided by the results on sufficiency of jets, which give conditions for the Taylor polynomial of a function at a point to be equivalent (by a smooth change of coordinates) to the whole function in some neighbourhood of that point.

The first result of this kind follows from the Inverse Function Theorem and states that at a non-singular point a function is always equivalent to its linear part. The next is the Morse Lemma: at a singular point a function is equivalent to its quadratic part when the latter is non-degenerate. Powerful generalizations of this result for C^∞ functions have been given by Milnor, Mather, Tougeron and Arnold, and have been applied to classify a large number of singularities.

In the case of C^r functions the results are not as complete, and their connection with the C^∞ case are not clear. The main results obtained in this direction are the C^r Morse Lemma of Kuiper and the sufficiency results of Kuiper, Kuo, Takens and Paunescu. More recently, stronger results of this type have been obtained by Ibarra and Riestra. The present authors obtained in [B-LdM] a version of the C^r Morse Lemma which turns out to be the strongest possible of its kind. (See Theorem 2.1.3, below.) However, the (simple) method employed could not be generalized to give sufficiency results for higher order jets. Later on we discovered that the C^1 case of our Morse Lemma had been essentially obtained by Crouziex, Geymonat and Raugel ([C-G-R]), with a proof of the same nature as ours (their statement and proof being however less transparent), and also by Mercuri and Palmieri ([M-P]) by a different method.

In this article we will show that the method of Mercuri and Palmieri (which is in fact a more explicit version of the method already used by Kuiper [Kui]) can in fact be generalized and combined with methods of Paunescu ([P]) to obtain sufficiency results for quasihomogeneous jets, that are stronger than all previously known C^r results for such jets, and are again, in a certain sense, best possible. The proof is direct and simple —simpler in fact than that of Mercuri and Palmieri due to the fact that we work in finite-dimensional spaces— and gives sufficiency results not only for jets, but also for quasihomogeneous functions that are not necessarily polynomials, only requiring a certain degree of differentiability outside the singular point. This generalization includes some of the "blow-analytic" functions studied by Kuo, Bierstone and Milman and could have some implications for their theory. It should be said also that we obtain no new results for more general jets, and that any extension of our method would require new techniques which we have not yet developed.

We learned in Saõ Carlos that the same degree of sufficiency had been obtained for C^∞ quasihomogeneous jets by M. A. Ruas and M. J. Saia ([R-S]), their method being valid also for maps $\mathbb{R}^n \to \mathbb{R}^p$. It has still to be checked if our method works also for maps, or if theirs works with the weak differentiability conditions of our results.

We wish to thank the members of the Differential Analysis Seminar (P. Armendáriz, C. Ibarra, G. Izquierdo, L. Kushner and J. J. Rivaud) as well as professor T.-C. Kuo for many helpful suggestions and encouragements. Professor Marc Chaperon will recognize in the following pages echoes of conversations we have had over a number of years.

1 The main theorem.

To state our main result, we recall the following definitions:

Let d, d_i be real numbers, with $d_i > 0$. A function $f : \mathbb{R}^n \longrightarrow \mathbb{R}$ is called *quasihomogeneous of weights d_i and degree d* if for every $\lambda > 0$ and $(x_1, ..., x_n) \in \mathbb{R}^n$ $f(\lambda^{d_1}x_1, ..., \lambda^{d_n}x_n) = \lambda^d f(x_1, ..., x_n)$. (We can assume $f(0) = 0$, conventionally, even for the case $d < 0$.) In what follows the weights d_i will be fixed but we will consider functions of various degrees. A quasihomogeneous function is called *non-degenerate* if it has no critical points outside the origin. We define the quasihomogeneous functions of degree 1 (for every $a > 0$) $\rho_a(x_1, ..., x_n) = (|x_1|^{a/d_1} + ... + |x_n|^{a/d_n})^{1/a}$.

It is worth noting that any two such functions are equivalent in the sense that $\alpha_a, \beta_a > 0$ can be found such that $\alpha_a\rho_1(x_1, ..., x_n) \leq \rho_a(x_1, ..., x_n) \leq \beta_a\rho_1(x_1, ..., x_n)$. Observe that for a large enough ρ_a is of class C^r outside the origin.

We say that a germ $f : \mathbb{R}^n, 0 \longrightarrow \mathbb{R}, 0$ has *filtration* $> \delta$ if $\lim_{x\to 0} \frac{f(x)}{\rho_a(x)^\delta} = 0$.

The above condition does not depend on a.

Finally, two germs $f, g : \mathbb{R}^n, 0 \longrightarrow \mathbb{R}, 0$ are said to be C^r *equivalent* if there is a germ of a C^r diffeomorphism $\varphi : \mathbb{R}^n, 0 \longrightarrow \mathbb{R}^n, 0$ such that $f \circ \varphi = g$.

Main Theorem:

Let Q be a non-degenerate quasihomogeneous function of degree d and of class C^{r+1} $(r > 0)$ outside the origin. Let $f = Q + R$. Suppose that the residue R is a C^r function such that D^iR has a filtration $> \Delta_{r-i}$ for $i = 0, 1, \ldots, r$ where $\Delta_i = d + i\,M - m$ and $M = max\{d_i\}$, $m = min\{d_i\}$. Then f is C^r equivalent to Q by a diffeomorphism tangent of order r to the identity.

2 Remarks and examples.

To make clear the meaning of our main theorem we will show its relationship with other related results and give a few examples.

2.1 Special Cases.

In the case of a homogeneous function, where we can take all the $d_i = 1$, our result takes the following simple form:

2.1.1 Theorem.

Let Q be a non-degenerate homogeneous function of degree d and of class C^{r+1} outside the origin. Let $f = Q + R$ where the residue R is a C^r function which satisfies the following conditions:

$$\frac{D^i R(x)}{|x|^{d+r-i-1}} \to 0 \qquad i = 0, 1, ..., r.$$

Then f is C^r equivalent to Q.

Moreover, the conditions are necessary and sufficient for the existence of a C^r equivalence between f and Q which is r-tangent to the identity.

When $r = 1$ we obtain a more precise result:

2.1.2 Theorem.

Let Q be a non-degenerate homogeneous function of degree d and of class C^2 outside the origin. Let $f = Q + R$ where the residue R is a C^1 function which satisfies the following condition:

$$\frac{R(x)}{|x|^d} \to 0.$$

Then the following conditions are equivalent:

(i) f is C^1 equivalent to Q.

(ii) $DR(x)/|x|^{d-1} \to 0$.

This last result is a stronger version of a theorem of Kuiper [Kui] which asks for more differentiability conditions on Q and R to conclude the existence of the C^1 equivalence. For C^∞ functions this just means that any non-degenerate homogeneous jet of degree d is C^1 sufficient.

In the case $d = 2$ we obtain the strongest possible version of the classical Morse Lemma:

2.1.3 Theorem. (C^r Morse Lemma).

Let f be a C^r germ which satisfies $f(0) = 0$ and $Df(0) = 0$. Then the following conditions are equivalent:

(i) f is C^r equivalent to a non-degenerate quadratic polynomial.

(ii) $D^{r+1} f$ exists at $x = 0$ and $D^2 f(0)$ is non-degenerate.

Cf.[B-LdM].

2.2 Some examples.

In the application to C^∞ functions, our results are in a certain sense not best possible: Consider the case $Q(x,y) = x^4 + y^8$ and $f(x,y) = x^4 + y^8 + y^9$. Here the residue y^9 fails to satisfy the conditions of our main theorem, so we cannot conclude from it that f and Q are C^1 equivalent. But it is obvious that they are indeed C^∞ equivalent!

Nevertheless we apply our Main Theorem to obtain positive results such as the following:

Example. (Pham singularities). *Let $Q(x_1, \ldots, x_n) = x_1^{a_1} + \ldots + x_n^{a_n}$ with $a_1 \leq \ldots \leq a_n$. Then Q is C^1-sufficient as an a_n-jet for C^∞ functions.*

To obtain sharper results it would be necessary to combine our method with the C^∞ theory (specially that of Arnold, see [A-G-V]). Such a combination can actually give best possible results in cases like the C^r Morse Lemma stated above. In that case a direct application of our main theorem requires that the residue be of a rather high order to obtain a C^r equivalence. But then one can use the C^∞ Morse Lemma applied to the Taylor polynomial of f to obtain a C^∞ equivalent function whose residue does satisfy the required condition. (Cf.[B-LdM].)

As another example, our result implies that $x^4 + y^5 + \lambda x^2 y^3$ is C^1 equivalent to $x^4 + y^5$ for any λ, but it is known that these functions are not C^∞ equivalent for different values of λ ([A-G-V]). In this case a combination of both results shows that x^4+y^5 is indeed C^1-sufficient: Any function of the form $x^4+y^5+R(x,y)$ where $R(x,y)$ is a C^∞ function with greater filtration than the degree of x^4+y^5 is C^1 equivalent to x^4+y^5.

It is still not clear, however, how to put this combination into effect in order to obtain a general, best possible result.

Our main theorem is in fact best possible in the sense that the conditions cannot be weakened to give a more general statement of the same type. For example, it can be shown that $x^3 + y^7 + xy^5$ is not C^1 equivalent to $x^3 + y^7$. Many C^r examples can be given where the residue barely fails to satisfy the conditions and the function is not equivalent to Q. For instance take $Q(x,y) = xy$, considered as a quasihomogeneous function with weights $(s, 1)$ and degree $1 + s$. Then $f(x,y) = xy + y^{2r+1}\sin(1/y)$ is not C^r-equivalent to xy (since it does not have an $(r+1)$-th derivative at the origin), while the residue satisfies all the conditions of the Main Theorem, except the last one: $R^{(r)}(y)/y$ is actually bounded in a neighbourhood of 0, but does not tend to 0 with y. In addition, we have the fact that in the C^1 homogeneous case, and in the C^r Morse Lemma the conditions are strictly necessary.

2.3 Application to some non smooth functions.

The fact that our result does not require that the function or its quasihomogeneous principal part be smooth at the origin allows us to apply it not only to artifical ad-hoc examples but also to some functions that arise naturally, such as some kinds of real "meromorphic" functions. Consider for example the quasihomogeneous function

$$Q(x,y) = \frac{x^4 + y^4}{x^2 + y^2}.$$

This function is homogeneous of degree two, but is not a quadratic polynomial. In fact it is of class C^1 everywhere, but it has not a second derivative at the origin, so by the C^1 Morse Lemma above (or by a direct proof using homogeneity) it is not even C^1 equivalent to a quadratic polynomial.

Nevertheless it does have a typical property of a non degenerate quadratic polynomial: it satisfies a "Morse Lemma":

For any C^∞ function $R(x,y)$ of order ≥ 3 at the origin, the function $Q(x,y) + R(x,y)$ is C^1 equivalent to $Q(x,y)$.

(Actually it would be enough to have R of class C^2.)

Among these functions are the "blow-up analytic" functions studied by Bierstone, Kuo and Milman (cf.[Kuo]).

3 Proof of the main theorem.

We define a Riemannian metric on $\mathbb{R}^n \backslash 0$ associated with the weights d_i and the differentiability class r (cf. [P]),

$$\langle \partial/\partial x_i, \partial/\partial x_j \rangle_\rho = \rho^{-d_i - d_j} \delta_{ij}$$

where $\rho = \rho_a$ and a is chosen so that ρ_a is of class C^r outside the origin. Associated with this metric there is a gradient for real-valued functions:

$$Grad_\rho f(x) = \sum_{i=1}^{n} \rho^{2d_i} \frac{\partial f}{\partial x_i} \frac{\partial}{\partial x_i}.$$

We define on $\mathbb{R}^n \backslash 0$ a C^r vector field,

$$\Gamma(x) = \frac{Grad_\rho Q(x)}{\langle Grad_\rho Q(x), Grad_\rho Q(x) \rangle_\rho},$$

and we consider the differential equation

$$\dot{x} = \Gamma(x).$$

Let $\gamma(t;x)$ be the local flow associated with Γ, which is a C^r function defined in an open set of $\mathbb{R} \times (\mathbb{R}^n \backslash 0)$ whose derivative of order r is differentiable with respect to time. Moreover, γ also satisfies

$$Q(\gamma(t;x)) = t + Q(x).$$

We shall prove that $\gamma(t;x)$ is defined for $t = R(x)$ if $|x|$ is small and that

$$\varphi(x) = \begin{cases} \gamma(R(x);x) & \text{if } x \neq 0 \\ 0 & \text{if } x = 0 \end{cases}$$

is a local C^r diffeomorphism. This will prove the theorem, since

$$Q(\varphi(x)) = Q(\gamma(R(x),x)) = R(x) + Q(x) = f(x).$$

The i-th component of Γ is a quasihomogeneous function (with the same weights) of degree $-d + d_i$. And γ satisfies also a quasihomogeneous property:

$$\lambda^{d_i}\gamma_i(\frac{t}{\lambda^d}; \frac{x_1}{\lambda^{d_1}}, \cdots, \frac{x_n}{\lambda^{d_n}}) = \gamma_i(t; x_1, \cdots, x_n). \qquad (QH)$$

The ρ-sphere of radius 1, $S_\rho = \{x : \rho(x) = 1\}$, is a compact subset of $\mathbb{R}^n$. Therefore there exists $\epsilon > 0$ such that $\gamma(t;x)$ is defined for $|t| < \epsilon$ for every $x \in S_\rho$. We write x_ρ for the "projection" of x in S_ρ

$$x_\rho = (\frac{x_1}{\rho(x)^{d_1}}, \cdots, \frac{x_n}{\rho(x)^{d_n}}).$$

Now, $\gamma(t, x_\rho)$ is defined for $|t| < \epsilon$ and

$$\gamma_i(t,x) = \rho^{-d_i}\gamma_i(\frac{t}{\rho^d}, x_\rho)$$

is defined for $|\frac{t}{\rho^d}| < \epsilon$. By hypothesis,

$$\frac{|R(x)|}{\rho(x)^{d+rM-m}} \to 0.$$

We have, then, that

$$\frac{|R(x)|}{\rho(x)^d} < \epsilon$$

for $|x|$ small enough. Therefore φ is defined in a neighbourhood B of the origin and is of class C^r in $B\backslash 0$. Moreover,

$$\varphi(x) = \left(\rho^{d_1}\gamma_1(\frac{R(x)}{\rho^d}; x_\rho), \cdots, \rho^{d_n}\gamma_n(\frac{R(x)}{\rho^d}; x_\rho)\right).$$

Since

$$\gamma_i(\frac{R(x)}{\rho^d}; x_\rho) \qquad \text{is bounded}$$

$\varphi(x) \longrightarrow 0$ and, therefore, it is continuous.

For the differentiability, the partial derivatives of φ_i of order s in a deleted neighbourhood of the origin are linear combinations of terms of the form:

$$\frac{\partial^\sigma \gamma_i(R(x);x)}{\partial t^\tau \partial x_1^{\sigma_1}\cdots\partial x_n^{\sigma_n}}\left(\frac{\partial^{s_1}R(x)}{\partial x_1^{\lambda_{1,1}}\cdots\partial x_n^{\lambda_{1,n}}}\right)\cdots\left(\frac{\partial^{s_\tau}R(x)}{\partial x_1^{\lambda_{\tau,1}}\cdots\partial x_n^{\lambda_{\tau,n}}}\right)$$

where

$$\begin{aligned}\tau+\sigma_1+\cdots+\sigma_n &= \sigma\\ s_1+\cdots+s_\tau &= s+\tau-\sigma\\ \sum_{j=1}^{n}\lambda_{i,j} &= s_i.\end{aligned}$$

We are going to show that

$$\frac{\partial\varphi_i}{\partial x_j}\longrightarrow\delta_{ij}\qquad\text{when}\qquad x\longrightarrow 0$$

and for $\sigma>1$

$$\frac{\partial^\sigma\varphi_i}{\partial x_1^{\sigma_1}\cdots\partial x_n^{\sigma_n}}\longrightarrow 0\qquad\text{when}\qquad x\longrightarrow 0.$$

We are going to deal first with the case $s=1$. For $x\neq 0$

$$D\varphi_i(x)=\frac{\partial\gamma_i}{\partial t}(R(x);x)DR(x)+\frac{\partial\gamma_i}{\partial x}(R(x);x).$$

Taking derivatives in (QH), we obtain

$$\lambda^{d_i-d_j}\frac{\partial\gamma_i}{\partial x_j}(\frac{t}{\lambda^d};\frac{x_1}{\lambda^{d_1}},\cdots,\frac{x_n}{\lambda^{d_n}})=\frac{\partial\gamma_i}{\partial x_j}(t;x_1,\cdots,x_n).$$

We can then write the derivative,

$$\frac{\partial\varphi_i}{\partial x_j}(R(x);x)=\frac{1}{\rho^{d-d_i}}\frac{\partial\gamma_i}{\partial t}(\frac{R(x)}{\rho^d};x_\rho)\frac{\partial R(x)}{\partial x_j}+\frac{\partial\gamma_i}{\partial x_j}(\frac{R(x)}{\rho^d};x_\rho).$$

Finally,

$$\frac{\partial\gamma_i}{\partial t}(\frac{R(x)}{\rho^d};x_\rho)$$

is bounded because $(R(x)/\rho^d;x_\rho)$ belongs to a neighbourhood of the compact set $0\times S_\rho$. By hypothesis —for the $r=1$ case—,

$$\frac{DR(x)}{\rho^{d-m}}\longrightarrow 0$$

so that the first term of the derivative tends to 0 as x tends to 0.

To deal with the second term, we recall that $\gamma(0,x)=x$, therefore $\frac{\partial\gamma_i}{\partial x_j}(0,x)=\delta_{ij}$ and

$$\frac{\partial\gamma_i}{\partial x_j}(R(x);x)-\delta_{ij}=\frac{\partial\gamma_i}{\partial x_j}(R(x);x)-\frac{\partial\gamma_i}{\partial x_j}(0;x)=(\int_0^1\frac{\partial^2\gamma_i}{\partial t\partial x_j}(tR(x);x)\,dt)R(x).$$

Another differentiation in (QH) yields,

$$\lambda^{d_i-d_j-d}\frac{\partial^2\gamma_i}{\partial t\partial x_j}(\frac{t}{\lambda^d};\frac{x_1}{\lambda^{d_1}},\cdots,\frac{x_n}{\lambda^{d_n}})=\frac{\partial^2\gamma_i}{\partial t\partial x_j}(t;x_1,\cdots,x_n)$$

therefore,

$$\frac{\partial\gamma_i}{\partial x_j}(R(x);x)-\delta_{ij}=(\int_0^1\frac{\partial^2\gamma_i}{\partial t\partial x_j}(\frac{tR(x)}{\rho^d};x_\rho)\,dt)\frac{R(x)}{\rho^{d+d_j-d_i}}$$

Then,

$$|\frac{\partial\gamma_i}{\partial x_j}(R(x);x)-\delta_{ij}|\le\left(\int_0^1|\frac{\partial^2\gamma_i}{\partial t\partial x_j}(\frac{tR(x)}{\rho^d};x_\rho)|\,dt\right)\frac{|R(x)|}{\rho^{d+d_j-d_i}}$$

but

$$\frac{\partial^2\gamma_i}{\partial t\partial x_j}$$

is continous in a bounded neighbourhood of the compact set $0\times S_\rho$ and

$$\frac{|R(x)|}{\rho^{d+d_j-d_i}}\le\frac{|R(x)|}{\rho^{d+M-m}}\to 0.$$

Therefore

$$\frac{\partial\gamma_i}{\partial x_j}(R(x);x)-\delta_{ij}\longrightarrow 0.$$

We have just shown that

$$D\varphi(x)\longrightarrow Id\qquad x\longrightarrow 0$$

which implies that φ is a C^1 function in B and that

$$D\varphi(0)=Id$$

so that φ is a local C^1 diffeomorphism.

For $s>1$ the proof will also split into two cases, corresponding to $\tau>0$ and $\tau=0$.

Taking derivatives in (QH)

$$\lambda^{d_i-\tau d-\sigma_1 d_1-\cdots-\sigma_n d_n}\frac{\partial^\sigma\gamma_i}{\partial t^\tau\partial x_1^{\sigma_1}\cdots\partial x_n^{\sigma_n}}(\frac{t}{\lambda^d};\frac{x_1}{\lambda^{d_1}},\cdots,\frac{x_n}{\lambda^{d_n}})$$

$$=\frac{\partial^\sigma\gamma_i}{\partial t^\tau\partial x_1^{\sigma_1}\cdots\partial x_n^{\sigma_n}}(t;x_1,\cdots,x_n).$$

When $\tau > 0$:

$$\left|\frac{\partial^\sigma \gamma_i(R(x);x)}{\partial t^\tau \partial x_1^{\sigma_1}\cdots\partial x_n^{\sigma_n}}\left(\frac{\partial^{s_1}R(x)}{\partial x_1^{\lambda_{1,1}}\cdots\partial x_n^{\lambda_{1,n}}}\right)\cdots\left(\frac{\partial^{s_\tau}R(x)}{\partial x_1^{\lambda_{\tau,1}}\cdots\partial x_n^{\lambda_{\tau,n}}}\right)\right| =$$

$$\rho^{d_i-\tau d-\sigma_1 d_1-\cdots-\sigma_n d_n}\left|\frac{\partial^\sigma \gamma_i(\frac{R(x)}{\rho^d};x_\rho)}{\partial t^\tau \partial x_1^{\sigma_1}\cdots\partial x_n^{\sigma_n}}\right|\left|\frac{\partial^{s_1}R(x)}{\partial x_1^{\lambda_{1,1}}\cdots\partial x_n^{\lambda_{1,n}}}\right|\cdots\left|\frac{\partial^{s_\tau}R(x)}{\partial x_1^{\lambda_{\tau,1}}\cdots\partial x_n^{\lambda_{\tau,n}}}\right|.$$

By hypothesis,

$$\frac{\partial^{s_i}R(x)}{\partial x_1^{\lambda_{i,1}}\cdots\partial x_n^{\lambda_{i,n}}}$$

has a filtration $> \Delta_{r-s_i}$ therefore,

$$\frac{1}{\rho^{d+(r-s_i)M-m}}\left|\frac{\partial^{s_i}R(x)}{\partial x_1^{\lambda_{i,1}}\cdots\partial x_n^{\lambda_{i,n}}}\right| \to 0$$

whenever $x \to 0$. We shall show that

$$A = \sum_{i=1}^{\tau}(d+(r-s_i)M-m) + (d_i - \tau d - \sigma_1 d_1 - \cdots - \sigma_n d_n) \geq 0.$$

Remembering that

$$\begin{aligned} \tau + \sigma_1 + \cdots + \sigma_n &= \sigma \\ s_1 + \cdots + s_\tau &= s + \tau - \sigma, \end{aligned}$$

$\sum_{i=1}^{\tau}(d+(r-s_i)M-m) \geq \tau d - \tau m + \sum_{i=1}^{\tau}(s-s_i)M$.
Therefore

$$\begin{aligned} A &\geq d_i - \tau m - M(\sigma_1+\cdots+\sigma_n) + M\tau s - M(s+\tau-\sigma) \\ &\geq -(\tau-1)m - M(\sigma-\tau) + M\tau s - M(\tau-\sigma) - Ms \\ &\geq -(\tau-1)m + M\tau s - Ms = (\tau-1)(Ms-m) \end{aligned}$$

which is ≥ 0 since $\tau \geq 1$.

Finally,

$$\left|\frac{\partial^\sigma \gamma_i(R(x);x)}{\partial t^\tau \partial x_1^{\sigma_1}\cdots\partial x_n^{\sigma_n}}\left(\frac{\partial^{s_1}R(x)}{\partial x_1^{\lambda_{1,1}}\cdots\partial x_n^{\lambda_{1,n}}}\right)\cdots\left(\frac{\partial^{s_\tau}R(x)}{\partial x_1^{\lambda_{\tau,1}}\cdots\partial x_n^{\lambda_{\tau,n}}}\right)\right| \leq$$

$$\rho^A\left|\frac{\partial^\sigma \gamma_i(R(x)/\rho^d;x_\rho)}{\partial t^\tau \partial x_1^{\sigma_1}\cdots\partial x_n^{\sigma_n}}\right|\frac{1}{\rho^{\Delta_{r-s_1}}}\left|\frac{\partial^{s_1}R(x)}{\partial x_1^{\lambda_{1,1}}\cdots\partial x_n^{\lambda_{1,n}}}\right|\cdots\frac{1}{\rho^{\Delta_{r-s_\tau}}}\left|\frac{\partial^{s_\tau}R(x)}{\partial x_1^{\lambda_{\tau,1}}\cdots\partial x_n^{\lambda_{\tau,n}}}\right|$$

which tends to 0 when $x \to 0$.

When $\tau = 0$,

$$\frac{\partial^\sigma \gamma_i(R(x);x)}{\partial x_1^{\sigma_1} \cdots \partial x_n^{\sigma_n}} = \frac{\partial^\sigma \gamma_i(R(x);x)}{\partial x_1^{\sigma_1} \cdots \partial x_n^{\sigma_n}} - \frac{\partial^\sigma \gamma_i(0;x)}{\partial x_1^{\sigma_1} \cdots \partial x_n^{\sigma_n}} =$$

$$\left(\int_0^1 \frac{\partial^{\sigma+1}\gamma_i(tR(x);x)}{\partial t \partial x_1^{\sigma_1} \cdots \partial x_n^{\sigma_n}} dt\right)(R(x)).$$

Using again (QH),

$$\left|\frac{\partial^\sigma \gamma_i(R(x);x)}{\partial x_1^{\sigma_1} \cdots \partial x_n^{\sigma_n}}\right| \leq \frac{|R(x)|}{\rho^{d-d_i+\sigma_1 d_1+\cdots+\sigma_n d_n}} \left(\int_0^1 \left|\frac{\partial^{\sigma+1}\gamma_i(tR(x)/\rho^d;x_\rho)}{\partial t \partial x_1^{\sigma_1} \cdots \partial x_n^{\sigma_n}}\right| dt\right).$$

But $d - d_i + \sigma_1 d_1 + \cdots + \sigma_n d_n \leq \Delta_r = d + rM - m$ and $R(x)$ has a filtration $> \Delta_r$. Therefore $D^s\varphi_i(x) \to 0$ whenever $x \to 0$ for $1 < s \leq r$ and φ is a C^r function in a neighbourhood of the origin. And the theorem is proved.

References

[A-GZ-V] V. I. Arnold, S. Goussein-Zadé, A. Varchenko, *Singularités des applications différentiables.* Editions MIR, Moscou, 1986.

[B-LdM] S. Bromberg, S. López de Medrano, *Sur le Lemme de Morse et le Lemme de Séparation dans l'Espace de Hilbert.* C.R. Acad. Sci. Paris, t. 316, Série I, p. 909-912, 1993.

[C-G-R] M. Crouziex, G. Geymonat, G. Raugel , *Some remarks about the Morse Lemma in infinite dimensions.* Siam J. Math. Anal. Vol 19, No2, p. 358-371,March 1988 .

[Kui] N. Kuiper, *C^1-equivalence of functions.* Symposium on infinite dimensional topology.Baton Rouge (1967),Annals of Math. Studies, Princeton University Press (1972), p. 167-171.

[Kuo] T.-C. Kuo, *On classification of real singularities.* Invent. Math. 82, p. 257-262, 1985 .

[M-P] F. Mercuri, G. Palmieri, *Morse Theory with low differentiability.* Boll. Un. Mat. Ital. B(7) 1, p. 621-631, 1987.

[P] L. Paunescu, *A weighted version of the Kuiper-Kuo-Lojasiewicz Theorem.* Prepublication School of Mathematics and Statistics, The University of Sydney, March 1991.

[R-S] M. A. Ruas, M. J. Saia, *C^l-Determinacy of Weighted Homogeneous Germs.* Prepublication, U. de Saõ Paulo, 1993.

J.-P. Brasselet

Poincaré morphism and divisors of compact toric varieties

Introduction.

The aim of this lecture is to introduce the recent results of the author and Gottfried Barthel, Karl Fieseler and Ludger Kaup (from Konstanz University) concerning the relations between Weil and Cartier divisors classes and Poincaré homomorphism for toric varieties. We give also an interpretation of intersection homology of toric varieties in terms of divisors. In this lecture, we restrain ourself to the case of compact varieties, for simplification of demonstrations and results. These are given and proved in a general way (i.e. for any toric variety) in the paper of previously named authors [3]. The methods used in the present paper are also due to the same authors [2].

Firstly we recall the notions of Weil and Cartier divisors, and the construction of the Poincaré homomorphism for pseudovarieties. It is well known that, for a normal variety, there is an inclusion of the group of classes (modulo principal divisors) of Cartier divisors $\mathcal{C}(X)$ in the group of classes of Weil divisors $\mathcal{W}(X)$. On the other hand, for a compact normal m-dimensional variety X, there is a natural morphism $c^1 : \mathcal{C}(X) \longrightarrow H^2(X)$ and a natural morphism $\kappa : \mathcal{W}(X) \longrightarrow H_{m-2}(X)$. It is known that these two morphisms and the Poincaré morphism $H^2(X) \to H_{m-2}(X)$ provide a commutative diagram (see (3) ; Theorem 2). The main result proved in this paper can be formulated as follows : for compact toric varieties, the previous morphisms provide isomorphisms from classes of invariant (by the action of the torus $\mathbf{T}$) Cartier divisors in $H^2(X)$ and from classes of invariant Weil divisors in $H_{m-2}(X)$. We deduce a "geometrical" interpretation of the Poincaré homomorphism, for compact toric varieties (Theorem 3). Finally, we give the interpretation of intersection homology in terms of divisors [4].

I want to thank the organizers of the São Carlos Singularities Congress, specially Ton Marar. I want also to say how a big pleasure it was to discover the secrets of toric varieties with Gottfried Barthel, Karl Fieseler and Ludger Kaup.

1. Divisors.

In this section, we will denote by X a n-dimensional complex algebraic variety.

1.1. Weil and Cartier divisors.

A *Weil divisor* is an element of the free abelian group $W(X)$ whose generators are irreductible closed subvarieties of (complex) codimension 1 in X. Such a divisor can be written :

$$\sum n_i A_i - \sum m_j B_j \qquad \text{with } n_i, m_j > 0$$

where A_i and B_j are subvarieties of codimension 1 in X.

For example, in $\mathbf{C}^2$ with coordinates (z_1, z_2), let us consider the axis $z_1 = 0$ denoted by A, and the axis $z_2 = 0$ denoted by B. An example of Weil divisor is given by $2A - B$.

Let us denote by $\mathcal{R}(U)$ the ring of rational functions in the open set U in X. A *Cartier divisor* is given by a covering $X = \bigcup U_\alpha$ of X together with nonzero rational functions $f_\alpha \in \mathcal{R}(U_\alpha)$ such that for any α and β, we have $f_\alpha / f_\beta \in \mathcal{O}^*(U_\alpha \cap U_\beta)$ (nowhere zero holomorphic function). We denote by $D = (U_\alpha, f_\alpha)$ such a divisor. The set of Cartier divisors is a group denoted by $C(X)$.

Let us consider the example of $X = \mathbf{C}^2$ covered by only one open set $U = \mathbf{C}^2$. We define a Cartier divisor $D = (U, f)$ considering the rational function $f(z_1, z_2) = z_1^2 / z_2$ in U.

Proposition 1. ([8,§II.6],[10,§III,1]) *Let X be a normal variety, there is an inclusion*

$$C(X) \hookrightarrow W(X)$$

Let us explicit this inclusion in the previous example : If $A = \{f = 0\}$ is the set of zeroes of f with multiplicities and $B = \{1/f = 0\}$ is the set of poles of f with multiplicities, then the previous Weil divisor $2A - B$ corresponds to the previous Cartier divisor $D = (U, f)$. In fact, this example is an example of principal divisor : The subgroup of principal divisors, denoted by $P(X)$, is the subgroup of Cartier divisors corresponding to the nonzero rational functions. Let us consider the quotients :

$$\mathcal{C}(X) = C(X)/P(X) \qquad \text{and} \qquad \mathcal{W}(X) = W(X)/P(X)$$

Proposition 1 provides an inclusion $\mathcal{C}(X) \hookrightarrow \mathcal{W}(X)$, which is not an equality as shown by the following example : let X be the quotient variety of $\mathbf{C}^2$ by the subgroup Γ_m of mth roots of unity, the action of $\mu = \exp(2i\pi/m)$ on (z_1, z_2) is given by $\mu \cdot (z_1, z_2) = (\mu \cdot z_1, \mu \cdot z_2)$. We have :

$$\mathcal{C}(X) = 0 \qquad \text{and} \qquad \mathcal{W}(X) = \mathbf{Z}_m \, .$$

1.2. Divisors, homology and cohomology.

In this paragraph, we suppose that X is compact. Each Weil divisor can be considerated as a cycle in X. We define, by this way, a natural map $\kappa : W(X) \longrightarrow H_{2n-2}(X)$. The image of a principal divisor is zero, so we obtain an homomorphism, still denoted

$$\kappa : \mathcal{W}(X) \longrightarrow H_{2n-2}(X) \, . \tag{1}$$

On the other hand, for each normal variety, there is an isomorphism (cf. [7 : II, Prop. 6.15])

$$\alpha : \mathcal{C}(X) \xrightarrow{\cong} \mathrm{Pic}(X)$$

between the group of classes of Cartier divisors and the Picard group of X. This group is the group of classes of line bundles (or classes of inversible sheaves) modulo isomorphism. The isomorphism α is given by the map $D = (U_\alpha, f_\alpha) \mapsto \mathcal{O}(D)$ where $\mathcal{O}(D)$ is the line bundle whose transition functions $U_\alpha \to U_\beta$ are f_α / f_β. Converserly, for an inversible sheaf, we associate the class of the divisor of a global rational and non trivial section.

The long exact sequence associated to the "exponential" exact sequence

$$0 \to \mathbf{Z} \to \mathcal{O} \to \mathcal{O}^* \to 0$$

provides a natural map

$$\mathrm{Pic}(X) = H^1(X, \mathcal{O}^*) \longrightarrow H^2(X; \mathbf{Z}) \tag{2}$$

which associates to each line bundle ξ its Chern class $c^1(\xi)$. By composition with α, we obtain a map denoted

$$c^1 : \mathcal{C}(X) \longrightarrow H^2(X) . \tag{3}$$

1.3. Divisors and Poincaré homomorphism.

The toric varieties are examples of pseudovarieties with (real) even dimension. By definition, a $2n$-(real) dimensional pseudovariety X is a connected topological space with a closed subspace Σ such that :

(a) $X - \Sigma$ is an $2n$-dimensional oriented smooth variety, dense in X,

(b) $\dim \Sigma \leq 2n - 2$.

A compact pseudovariety admits a fundamental class $[X] \in H_{2n}(X)$. The Poincaré homomorphism

$$H^i(X) \longrightarrow H_{2n-i}(X)$$

is the cap-product by the fundamental class. If X is smooth, then it is an isomorphism.

An example of pseudovariety for which the Poincaré homomorphism is not an isomorphism is given by the (toric) variety $\mathbf{C}^2/\Gamma_m$ defined previously. We have $H^2(X) = 0$ and $H_2(X) = \mathbf{Z}_m$.

Theorem 2. [2] *Let X be a normal compact complex algebraic variety, there is a commutative diagram :*

$$\begin{array}{ccc} \mathcal{C}(X) & \hookrightarrow & \mathcal{W}(X) \\ \downarrow c^1 & & \downarrow \kappa \\ H^2(X) & \xrightarrow{\cap [X]} & H_{2n-2}(X) \end{array} \tag{4}$$

where the horizontal lower arrow is the Poincaré homomorphism.

2. Toric varieties.

Let us recall some basic definitions concerning toric varieties. For more details, see the references, namely [5], [6], [7] or [9].

Let $\mathbf{T} := (\mathbf{C}^*)^n$ denote the n-dimensional complex algebraic torus. A n-dimensional (complex) *toric variety* X is a normal complex algebraic variety (not necessarily compact) with an effective and almost transitive algebraic action of the torus $\mathbf{T}$. There is an open and dense orbit B_o isomorphic to $\mathbf{T}$. The toric varieties are representable by the way of elementary convex geometry, what is described as follows :

Let $N := \text{Hom}(\mathbf{C}^*, \mathbf{T}) \cong \mathbf{Z}^n$ denote the lattice of one-parameter subgroups of $\mathbf{T}$. The dual $M := \text{Hom}(\mathbf{T}, \mathbf{C}^*)$ is the character group. We denote by $\langle\ ,\ \rangle : M \times N \to \text{Hom}(\mathbf{C}^*, \mathbf{C}^*) \cong \mathbf{Z}$ the natural pairing, given by the composition, so we identify M with the dual lattice $\text{Hom}_{\mathbf{Z}}(N; \mathbf{Z}) \cong \mathbf{Z}^n$ of N.

A fan in the real vectorial space $N_{\mathbf{R}} := N \otimes_{\mathbf{Z}} \mathbf{R} \cong \mathbf{R}^n$ is a finite collection Δ of cones which are rational relatively to N, strictly convex in $N_{\mathbf{R}}$, such that the faces of every cone $\sigma \in \Delta$ belong also to Δ and that the intersection of two cones of Δ is a common face. The subset of cones of real dimension p will be denoted by $\Delta^{(p)} \subset \Delta$.

A fan Δ in $N_{\mathbf{R}}$ determines a toric variety, denoted $X = X_\Delta$. This is illustrated by an example, taken from [7], reference where the general construction will be find.

Let σ be the convex cone in $\mathbf{R}^2$ generated by the vectors $2e_1 - e_2$ et e_2. Consider the dual cone $\check{\sigma}$ of the“covectors” in $M_{\mathbf{R}}$ whose value on σ is nonnegative. The semigroup

$$S_\sigma = \{u \in M\ :\ \langle u, v\rangle \geq 0,\ \forall v \in \sigma\} \quad = \check{\sigma} \cap M\ ,$$

is generated by the covectors e_1^*, $e_1^* + e_2^*$, $e_1^* + 2e_2^*$. Consider the finitely generated commutative $\mathbf{C}$-algebra $\mathbf{C}[S_\sigma]$ whose generators correspond to the previous covectors, i.e. defined by :

$$\mathbf{C}[S_\sigma] = \mathbf{C}[X, XY, XY^2] = \mathbf{C}[x, y, z]/(y^2 - xz) = \mathbf{C}[x, y, z]/I$$

The toric variety X_σ associated to the cone σ is the affine variety $V(I) = \text{Spec}(\mathbf{C}[S_\sigma])$ of the common zeros of polynoms in I. In this example, it is the quadric cone defined by $y^2 - xz = 0$ in the affine space $\mathbf{C}^3$.

The affine toric variety X_τ associated to a face τ of a cone σ admits a natural open equivariant embedding in X_σ, this property provides an unique equivariant gluing of affine toric varieties corresponding to the cones of a fan $\Delta \subset N_{\mathbf{R}}$. The toric variety obtained is denoted $X = X_\Delta$, it admits a covering $(X_\sigma)_{\sigma \in \Delta}$ by the affine invariant open subsets associated to the cones, and each X_σ contains an unique orbit

B_σ of $\mathbf{T}$, closed in X_σ. The subfamily $(X_\sigma)_{\sigma\in\Delta^{(n)}}$ corresponding to the maximal cones of Δ covers X.

Examples : We give some examples of toric varieties associated to cones of $\mathbf{R}^2$. We will denote (e_1, e_2) an orthonormal basis of $\mathbf{R}^2$.

1) $\Delta = \{0\}$ (origin of $\mathbf{R}^2$) $\mapsto X = \mathbf{C}^* \times \mathbf{C}^*$

2) $\Delta =$ the ray (the half line) generated by e_1 $\mapsto X = \mathbf{C} \times \mathbf{C}^*$

3) $\Delta =$ the union of the rays generated by e_1 et e_2 $\mapsto X = \mathbf{C}^2 \setminus \{0\}$

4) $\Delta =$ the convex cone situated between the rays e_1 and e_2 $\mapsto X = \mathbf{C}^2$

5) $\Delta =$ the convex cone situated between the rays $me_1 - e_2$ and e_2 $\mapsto X = \mathbf{C}^2/\Gamma_m$

6) $\Delta =$ the union of the three convex cones situated between the rays e_1 and e_2, e_2 and $-e_1 - e_2$, $-e_1 - e_2$ and e_1 and their faces $\mapsto X = \mathbf{P}^2$

7) $\Delta =$ the union of the four cones situated between the rays e_1 and e_2, e_2 and $-e_1$, $-e_1$ and $(k+1)e_1 - e_2$, $(k+1)e_1 - e_2$ and e_1 and their faces $\mapsto X = \Sigma_k$ Hirzebruch Surface

8) $\Delta =$ the union of the four cones situated between the rays e_1 and e_2, e_2 and $-e_1$, $-e_1$ and $-e_2$, $-e_2$ and e_1 and their faces $\mapsto X = \mathbf{P}^1 \times \mathbf{P}^1$

In these simple examples, we verify the general following properties (for a fan in $\mathbf{R}^n$) :

Properties of Δ giving properties ...	of X_Δ
$\lvert\Delta\rvert = \mathbf{R}^n$ i.e. Δ is complete	Compact
For all σ in Δ, σ is generated by a part of a basis of $\mathbf{R}^n$ in N	Smooth
Δ is generated by a polytope, i.e. by the convex hull of a finite number of points	Projective

Let X be a toric variety, let us denote by $C^{\mathbf{T}}(X)$ and $W^{\mathbf{T}}(X)$ the Cartier and Weil divisor classes, equivariant by action of the torus $\mathbf{T}$. In the same way, let us denote by $P^{\mathbf{T}}(X)$ the subgroup of the equivariant principal divisors, we define $\mathcal{C}^{\mathbf{T}}(X) = C^{\mathbf{T}}(X)/P^{\mathbf{T}}(X)$ et $\mathcal{W}^{\mathbf{T}}(X) = W^{\mathbf{T}}(X)/P^{\mathbf{T}}(X)$. There is an inclusion

$$\mathcal{C}^{\mathbf{T}}(X) \hookrightarrow \mathcal{W}^{\mathbf{T}}(X)$$

We want to show the following result :

Theorem 3. [2] *Let X be a compact toric variety, there is a commutative diagram :*

$$\begin{array}{ccc} \mathcal{C}^{\mathbf{T}}(X) & \hookrightarrow & \mathcal{W}^{\mathbf{T}}(X) \\ \downarrow\cong & & \downarrow\cong \\ H^2(X) & \stackrel{\cap[X]}{\longrightarrow} & H_{2n-2}(X) \end{array}$$

where the vertical isomorphisms are induced by the morphisms c_1 and κ of Theorem 2.

We obtain, in an evident way, an interpretation of the Poincaré morphism in terms of divisors, for the compact toric varieties. In particular, the Poincaré morphism $H^2(X) \longrightarrow H_{2n-2}(X)$ is injective.

Theorem 3 can be deduced from the following Theorems 4 and 6, the commutativity of the diagram is obvious.

3. Interpretation of $H^2(X)$ in terms of divisors.

Let $X = X_\Delta$ be a compact toric variety. We recall the interpretation of the cohomology group $H^2(X)$ (with integer coefficients) in terms of invariant Cartier divisors and in terms of the fan Δ.

Theorem 4. ([2]) *Let $X = X_\Delta$ be a compact toric variety, the morphism c^1 induces an isomorphism :*

$$\mathcal{C}^{\mathbf{T}}(X) \stackrel{\cong}{\longrightarrow} H^2(X) \tag{5}$$

Proof. ([2]) Since the structural sheaf of a compact toric variety is acyclic (see [8 : Cor. 2.8]), the map (2) is an isomorphism. We deduce that c^1 is also an isomorphism.

The subvarieties X_σ corresponding to the maximal cones $\sigma \in \Delta^{(n)}$ are affine and contractible, we deduce from the exponential exact sequence that the restriction to X_σ of any Cartier divisor is (analytically) a principal divisor. By the "GAGA" theorem (cf. [8: §2.2]), every Cartier divisor on X is algebraic. On the other hand, the coordinate ring $\mathcal{O}_{\text{alg}}(B_o) \cong \mathbf{C}[z_1, z_1^{-1}, \ldots, z_n, z_n^{-1}]$ of the open dense orbit is factorial, this implies that every Cartier divisor on X is linearly equivalent to a Cartier divisor whose support is disjoint from B_o and is $\mathbf{T}$-equivariant. Therefore, the inclusion $C^{\mathbf{T}}(X) \hookrightarrow C(X)$ induces an isomorphism $\mathcal{C}^{\mathbf{T}}(X) \cong \mathcal{C}(X)$ and we obtain the theorem.

The group $\mathcal{C}^{\mathbf{T}}(X_\Delta)$ can be described in terms of characters of the fan Δ. We explicit this description in the following Theorem 5, proved in three steps.

1. Let us denote by $V(\Delta) = \{v_1, \ldots, v_k\} \subset N$ the unique minimal collection of generators of the fan Δ, they are *primitive* elements of the lattice N. Each one

generates a ray $\rho_i := \mathbf{R}_{\geq 0} v_i \in \Delta^{(1)}$ of Δ. For each $i = 1, \ldots, k$, let us denote by $B_i \subset X$ the orbit of (complex) codimension 1 corresponding to the ray ρ_i in the following way : since $v_i \in N$ is an one-parameter subgroup, the limit $x_i := \lim_{t \to 0} v_i(t)$ exists in X and its orbit is $B_i = \mathbf{T} \cdot x_i$.

Seen as a rational function in X, every character $\chi \in M = \mathrm{Hom}(\mathbf{T}, \mathbf{C}^*)$ is an inversible regular function in the torus. The order of this function along the orbit B_i corresponding to the generator $v_i \in V(\Delta)$ is $\langle \chi, v_i \rangle$. We define an injective homomorphism div $: M \to P^{\mathbf{T}}(X)$ by

$$\mathrm{div}(\chi) = \sum_{i=1}^{k} \langle \chi, v_i \rangle \bar{B}_i \,.$$

This homomorphism is in fact an isomorphism. It is sufficient to show that every equivariant principal divisor D is the image of a character χ. In fact, as the divisor D is equivariant, its support does not meet the open dense orbit $B_o \cong \mathbf{T}$. Consequently, the rational function which describes D is regular and inversible in B_o, it is a nonzero multiple of a character χ such that $\mathrm{div}(\chi) = D$.

2. We obtain Cartier equivariant divisors in X choosing, for each maximal cone σ, a character χ_σ such that the principal divisors $\mathrm{div}(\chi_\sigma)$ and $\mathrm{div}(\chi_{\sigma'})$ coincide in the affine open set $X_\sigma \cap X_{\sigma'} = X_{\sigma \cap \sigma'}$. In terms of fan, this compatibility condition is equivalent to write that the family (χ_σ) belongs to the finitely generated abelian group (of the "piecewise characters")

$$M^{[\Delta]} := \left\{ (\chi_\sigma)_{\sigma \in \Delta^{(n)}} \in M^{\Delta^{(n)}} \; ; \; \chi_\sigma|_{\sigma \cap \sigma' \cap N} = \chi_{\sigma'}|_{\sigma \cap \sigma' \cap N} \text{ for all } \sigma, \sigma' \in \Delta^{(n)} \right\} ,$$

which contains a direct factor canonically isomorphic to M, i.e. the subgroup of "constant" families. The map

$$\theta : M^{[\Delta]} \to \mathcal{C}^{\mathbf{T}}(X) \,, \qquad \theta(\chi_\sigma) = -\sum_{i=1}^{k} \langle \chi_\sigma, v_i \rangle \bar{B}_i$$

is an injective homomorphism (for the sign, cf. [8 : Prop. 2.1]). We want to show that θ induces an isomorphism

$$M^{[\Delta]}/M \;\cong\; \mathcal{C}^{\mathbf{T}}(X) \,. \tag{6}$$

It is sufficient to show that every equivariant Cartier divisor D defined in a maximal affine open X_σ (i.e., for $\dim \sigma = n$) is principal, i.e. :

$$\mathcal{C}^{\mathbf{T}}(X_\sigma) = P^{\mathbf{T}}(X_\sigma) \cong M \,.$$

We will show that the locally free sheaf $\mathcal{L} := \mathcal{O}(-D)$ is trivial on X_σ, i.e. it admits an everywhere nonzero section s. Let x_σ be the unique fixed point under the action of $\mathbf{T}$ in X_σ. Since X_σ is affine, there is a regular section s on X_σ such that $s(x_\sigma) \neq 0$. But the vectorial space $\mathcal{L}(X_\sigma)$ of such global sections admits a linear action of the torus and we can suppose that the section s belongs to one of the proper space of this action; then it is $\mathbf{T}$-equivariant and, since the point x_σ is in the closure of any orbit in X_σ, we have $s \neq 0$ everywhere. The isomorphism (6) follows.

3. The isomorphism

$$M^{[\Delta]}/M \;\cong\; \mathrm{Pic}(X_\Delta) \tag{7}$$

is obtained by composition of isomorphisms (6), (4) et (2). It admits the following direct interpretation : let us consider a family $(\chi_\sigma)_{\sigma\in\Delta^{(n)}}$, regarded as a family of rational functions and define an inversible sheaf $\mathcal{L}$ by

$$\mathcal{L}|_{X_\sigma} \;=\; (\mathcal{O}|_{X_\sigma})\cdot\chi_\sigma^{-1} \quad \text{for all } \sigma\in\Delta^{(n)} .$$

Note that $\mathcal{L}$ is the sheaf of rational functions $\mathcal{O}(-D)$, associated to the divisor $D = \theta\big((\chi_\sigma)\big)$. This defines a map $M^{[\Delta]} \to \mathrm{Pic}(X_\Delta)$ whose kernel is M.

Finally, we obtain the theorem (cf. [8 : Prop. 2.4]) :

Theorem 5. ([2]) *Let X_Δ denote a compact toric variety. There are isomorphisms*

$$M^{[\Delta]}/M \;\cong\; \mathcal{C}^{\mathbf{T}}(X_\Delta) \;\cong\; \mathrm{Pic}(X) \;\cong\; H^2(X_\Delta, \mathbf{Z})$$

4. Interpretation of $H_{2n-2}(X)$ in terms of divisors.

In this section, we proof the Theorem 6, giving an interpretation of the group $H_{2n-2}(X,\mathbf{Z})$ for a compact toric variety $X = X_\Delta$: it can be identified with the group of equivariant classes of Weil divisors.

If X is a toric variety, the homomorphism $\kappa : W(X) \longrightarrow H_{2n-2}(X,\mathbf{Z})$ (see (1)) induces an homomorphism still denoted $\kappa : W^{\mathbf{T}}(X) \longrightarrow H_{2n-2}(X,\mathbf{Z})$. A Weil divisor, *equivariant* by the action of $\mathbf{T}$ can be written a sum of closures $\bar{B}_i$ of codimension one orbits. Therefore, the free abelian group $W^{\mathbf{T}}(X)$ of these divisors is the direct sum $\bigoplus_{i=1}^{k} \mathbf{Z}\bar{B}_i$ where $k := k(\Delta)$ is the number of cones of dimension 1, i.e. the number of rays of Δ. We can write κ as

$$\kappa : W^{\mathbf{T}}(X) \;=\; \bigoplus_{i=1}^{k(\Delta)} \mathbf{Z}\bar{B}_i \longrightarrow H_{2n-2}(X,\mathbf{Z}) .$$

The principal equivariant divisors group $P^{\mathbf{T}}(X)$ is a subgroup of $W^{\mathbf{T}}(X)$ contained in $\operatorname{Ker} \kappa$. We obtain an induced homomorphism

$$\bar{\kappa} : \mathcal{W}^{\mathbf{T}}(X) := W^{\mathbf{T}}(X)/P^{\mathbf{T}}(X) \longrightarrow H_{2n-2}(X, \mathbf{Z}) .$$

On the other hand, we have already seen that there is an isomorphism (div) between the characters group M and $P^{\mathbf{T}}(X)$. We want to show the following result :

Theorem 6. ([2]) *Let X_Δ denote a compact toric variety, the homomorphism $\bar{\kappa}$ is an isomorphism*

$$\bar{\kappa} : \mathcal{W}^{\mathbf{T}}(X_\Delta) = \Big(\bigoplus_{i=1}^{k(\Delta)} \mathbf{Z}\bar{B}_i\Big) \Big/ (M) \xrightarrow{\cong} H_{2n-2}(X_\Delta; \mathbf{Z}) .$$

The demonstration, due to [2], will be performed in the Propositions 7 (κ is onto) and 10 ($\operatorname{Ker} \kappa = M$).

Proposition 7. *The natural homomorphism $\kappa : W^{\mathbf{T}}(X) \longrightarrow H_{2n-2}(X, \mathbf{Z})$ is onto.*

Proof : If X is smooth, the Poincaré homomorphism $P_{2n-2}(X)$ is an isomorphism between $H^2(X, \mathbf{Z}) \cong \operatorname{Pic}(X) = \sum_{i=1}^k \mathbf{Z}[\mathcal{O}(\bar{B}_i)]$ ([8 : Cor. 2.5]) and $H_{2n-2}(X, \mathbf{Z}) = \sum_{i=1}^k \mathbf{Z}[\bar{B}_i]$. The result follows.

If X is a singular variety, we use the smooth case by resolving singularities. Let us consider an equivariant proper modification

$$\pi : \hat{X} = X_{\hat{\Delta}} \longrightarrow X = X_\Delta$$

where the fan $\hat{\Delta}$ is finer than Δ. The closed subsets

$$S := \{x \in X;\ \dim \pi^{-1}(x) \geq 1\} \subset X \quad \text{and} \quad E := \pi^{-1}(S) \subset \hat{X}$$

are equivariant, and $\pi|_{\hat{X} \setminus E} : \hat{X} \setminus E \to X \setminus S$ is an isomorphism. Let us denote by $\hat{B}_1, \ldots, \hat{B}_{k(\Delta)}$ those of codimension one orbits $\hat{B}_1, \ldots, \hat{B}_{\hat{k}}$ (with $\hat{k} := k(\hat{\Delta})$) in $\hat{X}$ which correspond to the rays in $\Delta^{(1)} \subset \hat{\Delta}^{(1)}$. Then the closures $\overline{\hat{B}_j}$ for $j = k+1, \ldots, \hat{k}$ are components of (complex) dimension $n-1$ in E. Let us consider the following commutative diagram where the upper line is an exact sequence :

$$\begin{array}{ccccccccc}
0 & \longrightarrow & \bigoplus_{j=k+1}^{\hat{k}} \mathbf{Z}\bar{\hat{B}}_j & \longrightarrow & \bigoplus_{j=1}^{\hat{k}} \mathbf{Z}\bar{\hat{B}}_j & \longrightarrow & \bigoplus_{j=1}^{k} \mathbf{Z}\bar{B}_j & \longrightarrow & 0 \\
 & & \| & & \downarrow{\scriptstyle \hat{\kappa}} & & \downarrow{\scriptstyle \kappa} & & \\
0 & \longrightarrow & H_{2n-2}(E, \mathbf{Z}) & \longrightarrow & H_{2n-2}(\hat{X}, \mathbf{Z}) & \longrightarrow & H_{2n-2}(X, \mathbf{Z}) & \longrightarrow & 0 ,
\end{array}$$

Since $\hat{X}$ is smooth, the arrow $\hat{\kappa}$ is an epimorphism, and κ is also. Proposition 7 is an immediate consequence of this diagram as soon as we know that the lower line is exact. We will show this fact in Lemma 11, which is a consequence of the previous lemmas :

Lemma 8. *Using the previous notations, let* $\pi : (\hat{X}, E) \longrightarrow (X, S)$ *a proper equivariant modification of compact toric varieties of (complex) dimension* n*. There is a long exact sequence*

$$(8) \quad \ldots \to H_{i+1}(X) \to H_i(E) \to H_i(\hat{X}) \oplus H_i(S) \to H_i(X) \xrightarrow{\partial} H_{i-1}(E) \to \ldots \quad ,$$

and, if the fans Δ *and* $\hat{\Delta}$ *have the same number of rays, there are isomorphisms*

$$(9) \qquad H_{2n-j}(\hat{X}) \cong H_{2n-j}(X) \quad \text{for } j \leq 2 \; .$$

Proof : The map $(\hat{X}, E) \longrightarrow (X, S)$ is a relative isomorphism, it induces isomorphisms in relative homology, and one obtains easily the long exact sequence (8) by comparison of the long exact homology sequences of $(\hat{X}, E)$ and (X, S) (cf. [1 : 3.A.7]). For the proof of (9), let us remark that a toric variety is normal, that implies $\dim S \leq n-2$ and $H_{2n-j}(S) = 0$ for $j \leq 3$. If the fan $\hat{\Delta}$ has the same number of rays than Δ, the exceptional subset E does not contain any codimension one orbit. Therefore $\dim E \leq n-2$ and consequently $H_{2n-j}(E) = 0$ for $j \leq 3$, which proves (9).

We prove now the following result, which is an refined version of the "Caratheodory Theorem" (cf. [9 : Thm. A.3]) :

Lemma 9. *Every fan* Δ *admits a subdivision as a simplicial fan* $\hat{\Delta}$ *with the same rays than* Δ.

Proof : We obtain such a simplicial fan $\hat{\Delta}$ as a growing union $\bigcup_{q=0}^{n} \hat{\Delta}_q$ of simplicial subfans, where each $\hat{\Delta}_q$ is a refinement of the q-skeleton of Δ :

$$(10) \qquad \Delta^{(\leq q)} := \{\sigma \in \Delta \, , \, \dim \sigma \leq q\} = \bigcup_{i \leq q} \Delta^{(i)}$$

These fans $\hat{\Delta}_q$ are built by induction, beginning by simplicial fans $\hat{\Delta}_q := \Delta^{(\leq q)}$ for $q \leq 2$. For $q \geq 2$, suppose that the simplicial subdivision $\hat{\Delta}_q$ of $\Delta^{(\leq q)}$ is already defined. Then, it suffices to choose in each cone $\sigma \in \Delta^{(q+1)}$ a ray $\rho_\sigma \in \Delta^{(1)}$ and to write

$$\hat{\Delta}_{q+1} \quad := \quad \hat{\Delta}_q \cup \bigcup_{\sigma \in \Delta^{(q+1)}} \{\rho_\sigma + \tau \; ; \; \tau \in \hat{\Delta}_q, v_\sigma \notin \tau, \tau \subset \sigma\}$$

which proves the lemma.

Lemma 10. *For a compact toric variety $X = X_\Delta$ of (complex) dimension n, we have*

$$H_{2n-1}(X, \mathbf{Z}) = 0 \qquad et \qquad b_{2n-2}(X) = k(\Delta) - n \quad .$$

Proof: Firstly, for a complex variety X of dimension n, the abelian group $H_{2n-1}(X, \mathbf{Z})$ has no torsion. It is sufficient to show

$$b_{2n-1}(X) = 0 \quad \text{and} \quad b_{2n-2}(X) = k(\Delta) - n \; . \tag{11}$$

Lemma 9 shows that there is a *simplicial* fan $\hat{\Delta}$ finer than Δ, with the same number of rays. Applying this result to the proper equivariant modification $\hat{X} \to X$, and using (9) we have $H_{2n-j}(\hat{X}) \cong H_{2n-j}(X)$ for $j \leq 2$; in particular, the corresponding Betti numbers are equal. Then it is sufficient to prove the formula (11) in the simplicial case, and that has been made in [5 : Prop. 12.11].

Lemma 11. *Let, as in lemma 8, $\pi : (\hat{X}, E) \longrightarrow (X, S)$ a proper equivariant modification of compact toric varieties with $\dim_{\mathbf{C}} X = n$. There is a short exact sequence*

$$0 \to H_{2n-2}(E) \to H_{2n-2}(\hat{X}) \to H_{2n-2}(X) \to 0 \; , \tag{12}$$

Proof : The left exactitude of the sequence (12) comes from the sequence (8), lemma 10 and $H_{2n-2}(S) = 0$ (proof of lemma 8). The equality $H_{2n-2}(E) = \bigoplus_{k+1}^{\hat{k}} \mathbf{Z}[\bar{B}_j]$ implies $b_{2n-2}(E) = k(\hat{\Delta}) - k(\Delta)$, and

$$b_{2n-2}(E) - b_{2n-2}(\hat{X}) + b_{2n-2}(X) = 0$$

by the second result of lemma 10. Therefore the homomorphism $\partial : H_{2n-2}(X) \to H_{2n-3}(E)$ has rank 0, and since the group $H_{2n-3}(E)$ is free, we have $\partial = 0$ and the right exactitude of sequence (12).

Proposition 7 is then a consequence of Lemma 11. For the proof of Theorem 6, we have to show :

Proposition 12. *The kernel of the epimorphism $\kappa : W^{\mathbf{T}}(X) \to H_{2n-2}(X)$ is* $\operatorname{Ker} \kappa = P^{\mathbf{T}}(X)$.

Proof : Lemma 10 shows that the two groups

$$\mathcal{W}^{\mathbf{T}}(X) = \bigoplus_{i=1}^{k} \mathbf{Z}\bar{B}_i \Big/ (M) \quad \text{and} \quad H_{2n-2}(X) \cong W^{\mathbf{T}}(X)/\operatorname{Ker} \kappa = \bigoplus_{i=1}^{k} \mathbf{Z}\bar{B}_i/\operatorname{Ker} \kappa$$

have same rank $b_{2n-2}(X) = k - n$; then it is sufficient to show that they have the same torsion.

The singular set of the normal variety X is contained in the equivariant subvariety Y, union of orbits of dimension $\leq n-2$. From the homology exact sequence and the relative Poincaré duality theorem, we deduce the isomorphisms

$$H_{2n-j}(X) \cong H_{2n-j}(X,Y) \cong H^j(X \setminus Y) \quad \text{for} \quad j \leq 2 .$$

On the other hand, since the torsion subgroups of $H^2(X \setminus Y)$ and of $H_1(X \setminus Y)$ are isomorphic, it is sufficient to compute the second one. We remark that the invariant smooth open set $X \setminus Y$ is the toric variety associated to the subfan $\Delta^{(\leq 1)}$ (the 1-skeleton of Δ, see (10)). We know that the fundamental group of a toric variety X_Γ, determinated by a fan Γ non necessarily complete, is the abelian group

$$\pi_1(X_\Gamma) \cong H_1(X_\Gamma) = N/N_0 \qquad \text{with} \quad N_0 := \langle \bigcup_{\sigma \in \Gamma} N \cap \sigma \rangle = \sum_{\sigma \in \Gamma} \langle N \cap \sigma \rangle$$

(where we denote by $\langle \ldots \rangle$ the subgroup of N generated by a subset) see [8 : Prop.1.9].

In our case $\Gamma = \Delta^{(\leq 1)}$ and the group N_0 is the subgroup $\langle v_1, \ldots, v_k \rangle := \sum \mathbf{Z} v_i \subset N$ generated by the primitive vectors $v_i \in V(\Delta)$. This group has finite index because $\Delta^{(n)} \neq \emptyset$. Therefore $H_1(X \setminus Y)$ is a torsion group and there are isomorphisms

$$\operatorname{Tor} H_{2n-2}(X, \mathbf{Z}) \cong H_1(X \setminus Y) \cong N/\langle v_1, \ldots, v_k \rangle \cong \mathbf{Z}^n/\langle C \rangle,$$

where $\langle C \rangle \subset \mathbf{Z}^n$ is the subgroup generated by the columns of the coefficient matrix

$$C := \begin{pmatrix} \langle \varepsilon_1, v_1 \rangle & \ldots & \langle \varepsilon_1, v_k \rangle \\ \vdots & \ddots & \vdots \\ \langle \varepsilon_n, v_1 \rangle & \ldots & \langle \varepsilon_n, v_k \rangle \end{pmatrix}$$

where the character $\varepsilon_j : \mathbf{T} = (\mathbf{C}^*)^n \to \mathbf{C}^*$ is defined by $\varepsilon_j(t_1, \ldots, t_n) = t_j$. The subgroup $P^{\mathbf{T}}(X) \subset W^{\mathbf{T}}(X)$ is generated by the elements $(\varepsilon_j) = \sum_{i=1}^{k} \langle \varepsilon_j, v_i \rangle \bar{B}_i$, i.e. there is an isomorphism $M \cong P^{\mathbf{T}}(X) = \langle (\varepsilon_1), \ldots, (\varepsilon_n) \rangle$, which implies

$$\mathcal{W}^{\mathbf{T}}(X) = \bigoplus_{i=1}^{k} \mathbf{Z} \bar{B}_i \Big/ (M) \cong \mathbf{Z}^k/\langle C^{\mathrm{tr}} \rangle ,$$

where C^{tr} is the transposed matrix of C. With these notations, it suffices to show the isomorphism

$$\operatorname{Tor} \mathbf{Z}^k/\langle C^{\mathrm{tr}} \rangle \cong \mathbf{Z}^n/\langle C \rangle .$$

Let us consider the short exact sequence

$$0 \longrightarrow \mathbf{Z}^n \xrightarrow{C^{\mathrm{tr}}} \mathbf{Z}^k \longrightarrow L \longrightarrow 0$$

and the induced exact sequence

$$0 \longrightarrow \mathrm{Hom}(L, \mathbf{Z}) \longrightarrow \mathrm{Hom}(\mathbf{Z}^k, \mathbf{Z}) \xrightarrow{C} \mathrm{Hom}(\mathbf{Z}^n, \mathbf{Z}) \longrightarrow \mathrm{Ext}^1(L, \mathbf{Z}) \longrightarrow 0\,.$$

The group $\mathrm{Tor}(L) \cong \mathrm{Ext}^1(L, \mathbf{Z})$ is isomorphic to the torsion group $\mathbf{Z}^n/\langle C \rangle \cong \mathrm{Tor} H_{2n-2}(X)$ and the result follows.

This proves the Theorem 6 and then the Theorem 3.

In [4], we prove a general result giving the interpretation of the intersection homology in terms of divisors. The following Theorem gives a particular case of this result.

Let us denote by $X_4 \subset X$ the invariant open subset, which is union of orbits of dimension at least $2n-4$ and by $\mathcal{C}^{\mathbf{T}}(X_4) \subset \mathcal{W}^{\mathbf{T}}(X)$ the subgroup of equivariant Weil divisors whose restricyion to X_2 is a cartier divisor. The intersection homology group $IH_{2n-2}(X)$ relative to the middle perversity is interpreted as follows :

Theorem 13. ([4]) *Let $X = X_\Delta$ be a compact toric variety, there is a commutative diagram :*

$$\begin{array}{ccccc} \mathcal{C}^{\mathbf{T}}(X) & \hookrightarrow & \mathcal{C}^{\mathbf{T}}(X_4) & \hookrightarrow & \mathcal{W}^{\mathbf{T}}(X) \\ \downarrow \cong & & \downarrow \cong & & \downarrow \cong \\ H^2(X) & \longrightarrow & IH_{2n-2}(X) & \longrightarrow & H_{2n-2}(X) \end{array}$$

where the vertical isomorphisms are induced by the morphisms c_1 and κ of Theorem 2.

References

[1] G. Barthel, L. Kaup : *Topologie des surfaces complexes compactes singulières.* in S. Kilambi, G. Barthel, L. Kaup : *Sur la topologie des surfaces complexes compactes.* Sém. Math. Sup. **80**, Les Presses de l'Université de Montréal, 1982, pp. 61–297.

[2] G. Barthel, J.-P. Brasselet, K.-H. Fieseler et L. Kaup : *Variétés toriques et homomorphisme de Poincaré.* Preprint Konstanz University, 1992.

[3] G. Barthel, J.-P. Brasselet, K.-H. Fieseler et L. Kaup : *Diviseurs invariants et homomorphisme de Poincaré des variétés toriques complexes.* Preprint Konstanz University, 1994.

[4] G. Barthel, J.-P. Brasselet, K.-H. Fieseler et L. Kaup : *Invariante Divisoren und Schnitthomologie von torischen Varietäten.* To appear in "Banach Center Publications", Warsaw, 1994.

[5] J.L. Brylinski : *Éventails et variétés toriques.* in *Séminaire sur les Singularités des Surfaces*, Palaiseau 1976–1977, éd. par M. Demazure, H. Pinkham et B.

Teissier. Lecture Notes in Math. **777**, Springer-Verlag, Berlin etc., 1980, pp. 247–288.

[6] V.I. Danilov : *The Geometry of Toric Varieties.* Russian Math. Surveys **33:2** (1978), 97–154.

[7] W. Fulton : *Introduction to Toric Varieties.* Annals of Math. Studies, Princeton Univ. Press 1993.

[8] R. Hartshorne : *Algebraic Geometry.* Graduate Texts in Math. **52**, Springer Verlag, New York etc., 1977.

[9] T. Oda : *Convex Bodies and Algebraic Geometry.* Ergebn. Math.Grenzgeb. (3. Folge), Bd. **15**, Springer-Verlag, Berlin etc., 1988.

[10] I.R. Shafarevich : *Basic Algebraic Geometry* Springer-Verlag, Berlin etc. (Grundlehren der mathematischen Wissenschaften ; 213), 1977.

J. Basto Gonçalves[1]

Singularities of Euler equations and implicit Hamilton equations

Abstract

Singularities of the Euler equations in one variable, with generic non regular lagrangean, are studied and classified. To these second order equations correspond first order differential equations not solved for the derivative, implicit Hamilton equations; their typical phase portraits around singularities are studied.

Keywords: Singularities, lagrangean submanifolds, normal forms, generic hamiltonians.

1 Basic results and definitions

A symplectic manifold is an even dimensional manifold M with a non degenerated closed 2-form ω. As in every case here only the local situation is relevant, M will always be $T^*\mathbf{R}^n$ or $T\mathbf{R}^n$, identified with $\mathbf{R}^{2n}$ with coordinates (x,p) or $(x,\dot{x})$ respectively, and $\omega = dp \wedge dx = \sum_{i=1}^n dp_i \wedge dx_i$ on $T^*\mathbf{R}^n$; in the situations to be considered, there is a standard way of defining a symplectic form on $T\mathbf{R}^n$ from the one on the cotangent space.

Assume $M = T^*\mathbf{R}^n$, with coordinates (x,y); denoting the corresponding coordinates on $T^*M = \mathbf{R}^{4n}$ by $((x,y),(p,q))$, the standard symplectic form on T^*M is given by $\omega = dp \wedge dx + dq \wedge dy$. The map Φ from TM to T^*M defined by $((x,y),(\dot{x},\dot{y})) \mapsto ((x,y),(\dot{y},-\dot{x}))$ is a diffeomorphism, and the standard form on TM will be the pull-back by Φ of the standard form on T^*M, $\omega = d\dot{y} \wedge dx - d\dot{x} \wedge dy$. In general, the tangent space of a symplectic manifold M is a new symplectic manifold with a standard 2-form induced by the original symplectic form on M.

A lagrangian submanifold l of $M = T^*\mathbf{R}^n$ is a n-dimensional submanifold of M such that the restriction $\omega|l \equiv 0$.

A hamiltonian vector field with hamiltonian H on M is given by

$$\dot{x} = \frac{\partial H}{\partial p}, \quad \dot{p} = -\frac{\partial H}{\partial x}$$

Its graph in TM is a lagrangean submanifold, as the graph of dH is a lagrangean submanifold of T^*M. Thus lagrangian submanifolds of the tangent space generalize hamiltonian vector fields as they generalize differentials of functions in the cotangent space, or functions in the base space.

[1]Financial support from JNICT and INVOTAN

2 Euler equations

It is usually assumed when dealing with the Euler equations on a manifold N that the lagrangean $L : TN \longrightarrow \mathbf{R}$ is regular; using the usual coordinates $(x, \dot{x})$ on TN this can be expressed by the condition:

$$\frac{\partial^2 L}{\partial \dot{x}^2}(x, \dot{x}) > 0$$

meaning positive definite in the matrix case, where the dimension of N is bigger than one. The lagrangean is hyper-regular if it is regular and the fibre derivative induces a diffeomophism $\mathcal{L}$ of the tangent and cotangent spaces of N:

$$(x, \dot{x}) \mapsto (x, p) = \left(x, \frac{\partial L}{\partial \dot{x}}(x, \dot{x})\right)$$

using the usual coordinates (x, p) on T^*N.

Under these assumptions, to the Euler second order differental equations on N

$$\frac{\partial^2 L}{\partial \dot{x}^2}(x, \dot{x})\ddot{x} + \frac{\partial^2 L}{\partial x \partial \dot{x}}(x, \dot{x})\dot{x} = \frac{\partial L}{\partial x}$$

correspond the Hamilton first order equations on the cotangent space:

$$\dot{x} = \frac{\partial H}{\partial p}, \quad \dot{p} = -\frac{\partial H}{\partial x}$$

where H is the Legendre transformation of L, defined by:

$$H(x, p) = p\dot{x} - L(x, \dot{x}), \quad (x, \dot{x}) = \mathcal{L}^{-1}(x, p)$$

A point $(x, \dot{x}) \in TN$ is a singular point of the Euler equations if at that point

$$\det \frac{\partial^2 L}{\partial \dot{x}^2}(x, \dot{x}) = 0$$

and consequently the equation cannot be solved for the higher order derivative. In particular, the usual existence and uniqueness theorems for ordinary differential equations do not apply if the initial condition is a singular point; also $\mathcal{L}$ is not a diffeomorphism around that point.

As L does not depend on t, there exists a first integral for the Euler equations:

$$h(x, \dot{x}) = \dot{x}\frac{\partial L}{\partial \dot{x}}(x, \dot{x}) - L(x, \dot{x})$$

From now on, n is assumed to be 1. If $x(t)$ is a solution of those equations, then the image of $t \mapsto (x(t), \dot{x}(t))$ is contained in a level set of h; these level sets are curves around the regular points, determining the integral curves of the Euler equation outside singular points.

On the set of singular points we can have three different behaviours, as dh is non zero and defines a direction transverse to that set, or tangent to that set, or $dh = 0$ and h has a critical point; in the first case the points are said to be regular singular points.

At all points where both are well defined and non zero, the vector field

$$v(x,\dot{x}) = \left(\dot{x}, \frac{\partial^2 L}{\partial \dot{x}^2}(x,\dot{x})^{-1}\left(\dot{x}\frac{\partial^2 L}{\partial x \partial \dot{x}}(x,\dot{x}) - \frac{\partial L}{\partial x}(x,\dot{x})\right)\right)$$

corresponding to the Euler equation and the hamiltonian vector field X^h corresponding to the function h define the same direction and therefore the same integral curves, disregarding orientation.

Proposition 1 *For an open and dense set in $C^\infty(M,N)$ with the Whitney topology, L is such that the set Σ of singular points is smooth, the set of non regular singular points is discrete, and the critical points of h are non degenerate.*

Proof:

$$S = \left\{(x,\dot{x}) : \frac{\partial^2 L}{\partial \dot{x}^2}(x,\dot{x}) = 0\right\}$$

Non smoothness implies two extra conditions:

$$\frac{\partial^3 L}{\partial x \partial \dot{x}^2}(x,\dot{x}) = 0, \quad \frac{\partial^3 L}{\partial \dot{x}^3}(x,\dot{x}) = 0$$

Critical points of h on Σ are defined by just one extra condition:

$$\frac{\partial h}{\partial x}(x,\dot{x}) = \dot{x}\frac{\partial^2 L}{\partial x \partial \dot{x}}(x,\dot{x}) - \frac{\partial L}{\partial x}(x,\dot{x}) = 0$$

as on Σ:

$$\frac{\partial h}{\partial \dot{x}}(x,\dot{x}) = \dot{x}\frac{\partial^2 L}{\partial \dot{x}^2}(x,\dot{x}) \equiv 0$$

The critical points are degenerate if an extra condition is verified:

$$\dot{x}\left(\dot{x}\frac{\partial^3 L}{\partial x \partial \dot{x}^2}(x,\dot{x})^2 - \frac{\partial^3 L}{\partial \dot{x}^3}(x,\dot{x})\left(\dot{x}\frac{\partial^3 L}{\partial \dot{x} \partial x^2}(x,\dot{x}) - \frac{\partial^2 L}{\partial \dot{x}^2}(x,\dot{x})\right)\right) = 0$$

Note that if $\dot{x} = 0$, the critical point is always degenerate.

Tangency points are also characterized by just one extra condition: let X^h be the hamiltonian vector field corresponding to the function h; then the tangency points on Σ are defined by:

$$d\frac{\partial^2 L}{\partial \dot{x}^2}(x,\dot{x}).X^h(x,\dot{x}) = 0$$

that can be written as

$$\frac{\partial^3 L}{\partial \dot{x}^3}(x, \dot{x}) = 0$$

taking in account that they are not critical points of h.

When discussing existence of solutions of the Euler equation, a certain degree of differentiabillity is assumed, at least C^2.

At a singular point, if the second derivative of the solution is defined and continuous, in particular finite, a first order differential equation must be satisfied, namely:

$$\dot{x}\frac{\partial^2 L}{\partial x \partial \dot{x}}(x, \dot{x}) = \frac{\partial L}{\partial x}(x, \dot{x})$$

therefore there are no solutions through the regular singular points and the tangency points considered above, as this condition is exactly the condition for a critical point of h on Σ.

At a critical point, assumed to be non degenerate, the characteristic polynomial for the linear part of X^h is:

$$\lambda^2 + \dot{x}^2\left(\frac{\partial^3 L}{\partial x \partial \dot{x}^2}(x, \dot{x})\frac{\partial^3 L}{\partial \dot{x}^3}(x, \dot{x}) - \frac{\partial^3 L}{\partial x \partial \dot{x}^2}(x, \dot{x})^2\right) - \dot{x}\frac{\partial^3 L}{\partial \dot{x}^3}(x, \dot{x})\frac{\partial^2 L}{\partial x^2}(x, \dot{x})$$

and therefore the only possibilities are saddles and centres (imaginary eigenvalues).

If the critical point $(x_0, \dot{x}_0)$ is a saddle, then the condition for one of the eigenvectors of the linear part of X^h to be vertical is:

$$\frac{\partial^2 h}{\partial \dot{x}^2}(x_0, \dot{x}_0) = \dot{x}_0\frac{\partial^3 L}{\partial \dot{x}^3}(x_0, \dot{x}_0) = 0$$

Assume this is not verified; then both separatrices of the saddle can be parametrized so that they become solutions of the Euler equation with initial condition $(x_0, \dot{x}_0)$:

Let $(x(\tau), \dot{x}(\tau))$ be an arbitrary regular parametrization of one of the separatrices, with $(x(0), \dot{x}(0)) = (x_0, \dot{x}_0)$; there exists $\tau(t)$ with $\tau(0) = 0$, defined on a neighbourhood of 0, such that:

$$\frac{d}{dt}(x(\tau(t)) \equiv \dot{x}(\tau(t))$$

$\tau(t)$ is the solution of the first order differential equation:

$$\frac{d}{dt}\tau = \dot{x}(\tau)\left(\frac{d}{d\tau}x(\tau)\right)^{-1}, \quad \tau(0) = 0$$

So now $(x(t), \dot{x}(t)) = (x(\tau(t)), \dot{x}(\tau(t)))$, with a slight abuse of notation, is a new parametrization of the separatrix, $(x(0), \dot{x}(0)) = (x_0, \dot{x}_0)$, and for $t \neq 0$, is a solution of the Euler equation; for $t = 0$, and since $(x_0, \dot{x}_0)$ is a critical point:

$$\begin{aligned} \frac{\partial^2 L}{\partial x^2}(x(0), \dot{x}(0)) &= 0 \\ \dot{x}(0)\frac{\partial^2 L}{\partial x \partial \dot{x}}(x(0), \dot{x}(0)) - \frac{\partial L}{\partial x}(x(0), \dot{x}(0)) &= 0 \end{aligned}$$

therefore the Euler equation is also verified. Since the separatrices are at least C^2, $(x(t), \dot{x}(t))$ is a solution of the Euler equation. This proves:

Theorem 1 *For an open and dense set in $C^k(M, N)$, $k \geq 4$, L is such that:*

- *there are no solutions of the Euler equation if the initial condition is a regular, a tangency or a centre singular point.*
- *there are two solutions of the Euler equation if the initial condition is a saddle singular point.*

3 Implicit Hamilton equations

As seen before, the map

$$(x, \dot{x}) \mapsto (x, p) = (x, \frac{\partial L}{\partial \dot{x}}(x, \dot{x}))$$

induced by the fibre derivative is not a diffeomophism of the tangent and cotangent spaces of $N = \mathbf{R}$ at a singular point $(x, \dot{x}) \in \Sigma \subset TN$ of the Euler equations, where:

$$\frac{\partial^2 L}{\partial \dot{x}^2}(x, \dot{x}) = 0$$

and consequently the corresponding hamiltonian is not well defined.

Consider the map $i : TN \longrightarrow TN \oplus T^*N$, where $TN \oplus T^*N$ is the Whitney sum of the tangent and cotangent spaces, with coordinates $(x, \dot{x}, p)$, defined by:

$$(x, \dot{x}) \mapsto \left(x, \dot{x}, \frac{\partial L}{\partial \dot{x}}(x, \dot{x})\right)$$

and on $TN \oplus T^*N$ the function $\mathcal{H}(x, \dot{x}, p) = p\dot{x} - L(x, \dot{x})$. Then $\mathcal{H} \circ i = h$, and, if π is the restriction of the projection $(x, \dot{x}, p) \mapsto (x, p)$ to the image W, the hamiltonian is defined by $H = \mathcal{H} \circ \pi^{-1}$ whenever π is a diffeomorphism.

The image by i of the set Σ of singular points of the Euler equation defines a curve $\mathcal{C}$ on W, the criminant curve, and is exactly the set of critical points of π; its projection on T^*N is the discriminant curve $\mathcal{D}$ (not necessarily smooth), the set of critical values of π.

The form

$$\theta = \frac{\partial L}{\partial x}(x, \dot{x})dx - \dot{x}dp$$

on $TN \oplus T^*N$ defines a direction field on an open set of W containing the criminant curve, with discrete singularities located on that curve, for L in an open dense subset of $C^k(TN \oplus T^*N)$:

$$\theta = 0 \quad \Leftrightarrow \quad \frac{\partial L}{\partial x}(x(0), \dot{x}(0)) = 0 \text{ and } \dot{x} = 0$$

$$\theta \perp TW \quad \Leftrightarrow \quad \frac{\partial^2 L}{\partial \dot{x}^2}(x, \dot{x}) = 0 \text{ and } \dot{x}\frac{\partial^2 L}{\partial x \partial \dot{x}}(x, \dot{x}) = \frac{\partial L}{\partial x}(x, \dot{x})$$

Those singularities are just the image by i of the saddle and centre singular points of the Euler equation. The field direction defined by θ on W is spanned by the image $i_* v$ of the vector field corresponding to the Euler equation, when non zero, or similarly by $i_* X^h$.

If π_C is the restriction of π to the criminant curve $\mathcal{C}$, the image by i of the tangency singular points are exactly the critical points of π_C, where the tangent to $\mathcal{C}$ is parallel to the $\dot{x}$ axis: $\mathcal{C}$ is defined by the equations:

$$p - \frac{\partial L}{\partial \dot{x}}(x, \dot{x}) = 0, \quad \frac{\partial^2 L}{\partial \dot{x}^2}(x, \dot{x}) = 0$$

and its tangent is parallel to the $\dot{x}$ axis if:

$$\frac{\partial}{\partial \dot{x}}\left(p - \frac{\partial L}{\partial \dot{x}}(x, \dot{x})\right) = \frac{\partial^2 L}{\partial \dot{x}^2}(x, \dot{x}) = 0, \quad \frac{\partial^3 L}{\partial \dot{x}^3}(x, \dot{x}) = 0$$

which are the equations defining the tangency singular points.

The exponent of a nondegenerate critical point of a direction field is defined to be the ratio of the eigenvalue of largest modulus of the linearization of a vector field spanning the direction field to the smallest, for saddles and nodes, and as the modulus of the ratio of the imaginary part to the real, for a focus.

In the framework of [4] for studying the normal forms of differential equations not solved for the derivative, this means that the generic (in the context of all implicit differential equations) assumption made there, namely that the exponent of the critical points is well defined and different from ± 1, is violated: for hamiltonian vector fields the exponent is one for saddles, and is not defined for a centre. To obtain normal forms and typical phase portraits for the implicit differential equations corresponding to the Euler equation a different framework is needed.

Consider the map $i : TN \longrightarrow T(T^*N)$ defined by:

$$(x, \dot{x}) \mapsto \left(x, \frac{\partial L}{\partial \dot{x}}(x, \dot{x}), \dot{x}, \frac{\partial L}{\partial x}(x, \dot{x})\right)$$

Its image is a lagrangean submanifold l of $T(T^*N)$, but if the lagrangean L is not regular it does not correspond to a vector field: around the image of the singular points, the submanifold l does not project diffeomorphically onto the base $M = T^*N$.

An implicit Hamilton equation on a symplectic manifold M is a lagrangean submanifold of the tangent space TM.

When the dimension of l is two, the criminant curve $\mathcal{C}$ of l, or of the implicit Hamilton equation, is the set of critical points for its projection on the base space, and the discriminant curve $\mathcal{D}$ is the set of the critical values, the projection of $\mathcal{C}$.

The form $\theta = \dot{p}dx - \dot{x}dp$ on $T(T^*N)$ is closed, therefore locally is the differental of a function; if l is constructed from a lagrangean L, then:

$$\theta|l = -dh, \quad h(x,\dot{x}) = \dot{x}\frac{\partial L}{\partial \dot{x}} - L(x,\dot{x})$$

It defines a direction field on l, with singularities at the points where the tangent space of l is contained in $\theta = 0$. Proceeding as before, the following theorem is proved:

Theorem 2 *Let l be the implicit Hamilton equation corresponding to the Euler equation in N with lagrangean L; the criminant curve $\mathcal{C}$ of l corresponds to the curve $\Sigma \subset TN$, the singularities of the direction field induced by $\theta = \dot{p}dx - \dot{x}dp$ on l correspond to the saddle and centre singular points in Σ, and the critical points of the projection of $\mathcal{C}$ to the tangency singular points. The integral curves of the direction field are defined by the level curves of the restriction of $\mathcal{H}(x,\dot{x},p) = p\dot{x} - L(x,\dot{x})$ to l.*

Note that the level sets of h and $\mathcal{H}$ on l are identical (using the variables $(x,\dot{x})$ to parametrize l).

4 Normal forms

A map $F : \mathbf{R}^k \times \mathbf{R}^n \times \mathbf{R}^n \longrightarrow \mathbf{R}$ is the generating family of a lagrangean submanifold l of TM, where $M = T^*N$ and $N = \mathbf{R}^n$, if:

$$l = \left\{(x,p,\dot{x},\dot{p}) : \dot{x} = \frac{\partial F}{\partial p}(\lambda,x,p), \dot{p} = -\frac{\partial F}{\partial x}(\lambda,x,p), \frac{\partial F}{\partial \lambda}(\lambda,x,p) = 0\right\}$$

Two generating families F_1 and F_2 are R^+-equivalent if:

$$F_1(\lambda,x,p) = F_2(\varphi(\lambda,x,p),\psi_1(x,p),\psi_1(x,p)) + S(x,p)$$

and R-equivalent if in addition $S \equiv 0$. They are stably equivalent if they become (R,R^+)-equivalent after joining quadratic forms Q_i on new variables λ:

$$F_1(\lambda_1,\ldots,\lambda_k,x,p) + Q_1(\lambda_{k+1},\ldots,\lambda_s) \sim F_2(\lambda_1,\ldots,\lambda_l,x,p) + Q_2(\lambda_{l+1},\ldots,\lambda_s)$$

A lagrangean map $l \longrightarrow M$ is the composition of the embedding of the lagrangean submanifold l in TM with the projection $\pi : TM \longrightarrow M$; a caustic is the set of critical values of a lagrangean map.

Two lagrangean maps (germs) are Lagrange equivalent if there exist a map (germ) $S : T^*N \longrightarrow \mathbf{R}$ and (the germ of) a diffeomorphism $g : T^*N \longrightarrow T^*N$ such that the (germ of a) diffeomophism $g_* + \pi^*dS : T(T^*N) \longrightarrow T(T^*N)$ transforms one lagrangean submanifold into the other. They are strictly equivalent if $S \equiv 0$. Unless otherwise stated, all germs are taken at the origin.

It is proved in [1] that:

Theorem 3 *All lagrangean submanifolds l can be constructed from a generating family, and two lagrangean germs are (lagrangean, strictly) equivalent if the germs of their generating families are stably (R^+,R)-equivalent. If the dimension of l is two, the germs of the generic lagrangean maps are Lagrange equivalent to the germs of the projection on M of the lagrangean submanifolds defined by the generating families:*

- $F(\lambda, x, p) = \lambda^3 + p\lambda.$
- $F(\lambda, x, p) = \pm\lambda^4 + x\lambda^2 + p\lambda.$

with $\lambda \in \mathbf{R}$.

Dealing with implicit Hamilton equations, the first objective is to obtain a list of the generic germs up to strict equivalence. From theorem 3 it follows that then the possible generating families are:

- $F(\lambda, x, p) = \lambda^3 + p\lambda + S(x, p).$
- $F(\lambda, x, p) = \pm\lambda^4 + x\lambda^2 + p\lambda + S(x, p).$

for some smooth function S on $M = T^*N$.

The origin is said to be a fold point of an implicit Hamilton equation if:

$$F(\lambda, x, p) = \lambda^3 + p\lambda + S(x, p), \quad \frac{\partial S}{\partial x}(0,0) \neq 0$$

Then F can be written as $F(\lambda, x, p) = \lambda^3 + p\lambda + ax + s(x, p)$, $a \neq 0$, for some function s such that $s(0,0) = 0$ and $s_x(0,0) = 0$, where s_x denotes the derivative with respect to x. Since $\lambda^3 + p\lambda + x$ is a R-miniversal unfolding of λ^3 [1], it is possible to find a change of coordinates $(\lambda, x, p) \mapsto (\varphi(\lambda, x, p), \psi_1(x, p), \psi_2(x, p))$, preserving the origin, such that:

$$F(\varphi(\lambda, x, p), \psi_1(x, p), \psi_2(x, p)) + a\psi_1(x, p) + s(\psi_1(x, p), \psi_2(x, p)) = \lambda^3 + p\lambda + x$$

In general the condition on the derivative of S fails at discrete points on the caustic: the origin is said to be a folded critical point of an implicit Hamilton equation if:

$$F(\lambda, x, p) = \lambda^3 + p\lambda + S(x, p), \quad \frac{\partial S}{\partial x}(0,0) = 0, \quad \frac{\partial^2 S}{\partial x^2}(0,0) \neq 0, \quad \frac{\partial S}{\partial p}(0,0) \neq 0$$

and it is a centre or a saddle as S_p and S_{xx} have, or have not, the same sign at the origin.

Then F can be written as $F(\lambda, x, p) = \lambda^3 + p\lambda + ax^2 + bp + s(x, p)$, $a \neq 0$, for some function s such that $s(0,0) = s_x(0,0) = s_{xx}(0,0) = s_p(0,0) = 0$, where s_x denotes the derivative with respect to x. It is possible, using the parametric Morse lemma [3], to find a change of coordinates such that:

$$F(\lambda, \psi(x, p), p) + a\psi(x, p) + bp + s(\psi(x, p), p) = \lambda^3 + p\lambda + bp \pm x^2 + \varphi(p)$$

where $\varphi(0) = \varphi_p(0) = 0$.

The origin is said to be a pleat point of an implicit Hamilton equation if:

$$F(\lambda, x, p) = \pm\lambda^4 + x\lambda^2 + p\lambda + S(x, p), \quad \frac{\partial S}{\partial x}(0,0) \neq 0$$

and it is elliptic or hyperbolic as $S_x > 0$ or $S_x < 0$ respectively.

The phase portraits of two implicit Hamilton equations on $M = T^*N$ are said to be equivalent if there exists a diffeomophism of M taking one into the other; a similar definition can be made for germs.

If the two generating families F_1 and F_2 are R-equivalent the corresponding phase portraits are equivalent, but that is not necessary, F_1 can be just R-equivalent to a multiple of F_2, for instance: only a difference in the parametrization of the integral curves is involved. From the above it follows:

Theorem 4 *The phase portrait around a point in the caustic of the germ of a generic implicit Hamilton equation on $M = T^*N$ is equivalent to the phase portrait of the germ of the implicit Hamilton equation defined by one of the generating families:*

- $F(\lambda, x, p) = \lambda^3 + p\lambda + x$ *at a fold point.*
- $F(\lambda, x, p) = \lambda^3 + p\lambda + p + x^2 + \varphi(p)$ *at a folded centre point.*
- $F(\lambda, x, p) = \lambda^3 + p\lambda + p - x^2 + \varphi(p)$ *at a folded saddle point.*
- $F(\lambda, x, p) = \lambda^4 + x\lambda^2 + p\lambda + x + \varphi(x, p)$ *at an elliptic pleat point.*
- $F(\lambda, x, p) = \lambda^4 + x\lambda^2 + p\lambda - x + \varphi(x, p)$ *at a hyperbolic pleat point.*

Remarks:

- It appears that there is a functional modulus for folded critical points in the hamiltonian case, when, for the general implicit equations, the exponent is the only differentiable modulus [4].
- The existence of a functional modulus in the pleat case has already been proved [4].

To eliminate the moduli, a different notion of equivalence can be used [2]: instead of allowing changes of coordinates depending only on (x, p), special changes of coordinates (x, p, z), preserving the surface

$$z = F(\lambda, x, p), \quad \frac{\partial F}{\partial \lambda}(\lambda, x, y) = 0$$

are allowed. Then the above list becomes:

Theorem 5 *The phase portrait around a point in the caustic of the germ of a generic implicit Hamilton equation on $M = T^*N$ is equivalent to the phase portrait of the germ of the implicit Hamilton equation defined by one of the generating families:*

- $F(\lambda, x, p) = \lambda^3 + p\lambda + x$ *at a fold point.*
- $F(\lambda, x, p) = \lambda^3 + p\lambda + p + x^2$ *at a folded centre point.*
- $F(\lambda, x, p) = \lambda^3 + p\lambda + p - x^2$ *at a folded saddle point.*
- $F(\lambda, x, p) = \lambda^4 + x\lambda^2 + p\lambda + x$ *at an elliptic pleat point.*
- $F(\lambda, x, p) = \lambda^4 + x\lambda^2 + p\lambda - x$ *at a hyperbolic pleat point.*

Assuming $N = \mathbf{R}$, a straightforward computation shows that the list of situations considered in Theorem 5 can be obtained from Euler equations with suitable lagrangeans $L(x, \dot{x})$, the corresponding generating families being $\mathcal{H}(x, \dot{x}, p) = p\dot{x} - L(x, \dot{x})$ (taking $\lambda = \dot{x}$):

- Fold point: $L(x, \dot{x}) = -\dot{x}^3 - x$, $\mathcal{H}(x, \dot{x}, p) = \dot{x}^3 + p\dot{x} + x$.
- Folded centre: $L(x, \dot{x}) = -(\dot{x} - 1)^3 - x^2$, $\mathcal{H}(x, \dot{x}, p) = (\dot{x} - 1)^3 + p\dot{x} + x^2$.
- Folded saddle: $L(x, \dot{x}) = -(\dot{x} - 1)^3 + x^2$, $\mathcal{H}(x, \dot{x}, p) = (\dot{x} - 1)^3 + p\dot{x} - x^2$.
- Elliptic pleat: $L(x, \dot{x}) = -\dot{x}^4 - x\dot{x}^2 - x$, $\mathcal{H}(x, \dot{x}, p) = \dot{x}^4 + x\dot{x}^2 + p\dot{x} + x$.
- Hyperbolic pleat:$L(x, \dot{x}) = -\dot{x}^4 - x\dot{x}^2 + x$, $\mathcal{H}(x, \dot{x}, p) = \dot{x}^4 + x\dot{x}^2 + p\dot{x} - x$.

Remark: for the folded centre and saddle, the relevant point is $x = 0$, $\dot{x} = 1$, and the origin in the remaining ones; as noted before when discussing the critical points of h, at $\dot{x} = 0$ the critical points are always degenerate.

Aknowledgements

Part of this work has been done when on leave at the Mathematics Institute, University of Warwick; its hospitality is very thankfully aknowledged.

References

[1] V. Arnold, A. Varchenk, S. Goussein-Zade - *Singularités des applications différentiables* vol. 1, Éditions MIR, 1986

[2] V. Arnold - *Wave front evolution and equivariant Morse lemma*, Comm. Pure App. Math., 39 (1976), pp557-581

[3] M. Chaperon - *Géometrie Différentielle et Singularités de Sistèmes Dynamiques*, Astérisque 138-139 (1986)

[4] A. Davydov - *Normal form for a differential equation, not solvable for the derivative, in a neighbourhood of a singular point*, Funct. Anal. Appl., 19 (1985), pp 81-89